# ENERGY REFERENCE HANDBOOK

Edited by
Thomas F.P. Sullivan, Esq.
and
Martin L. Heavner

June 1981


966 Hungerford Drive, #24, Rockville, Maryland 20850

Library of Congress Catalog No. 80-84728

International Standard Book Number: 0-86587-082-9

Printed and bound in the United States of America

**Library of Congress Cataloging in Publication Data**
Main entry under title:

Energy reference handbook.

1. Power resources--Dictionaries. 2. Power (Mechanics)--Dictionaries. 3. Force and energy--Dictionaries. I. Sullivan, Thomas F. P. II. Heavner, Martin L. 1955- .
TJ163.16.E53 1981 621.042'03'21 80-84728
ISBN 0-86587-082-9 AACR2

## ACKNOWLEDGEMENTS

A great deal of research went into the compilation of this glossary, and a list of the publications most widely referenced as source material is given on page viii. Those who have assisted us are too numerous to acknowledge; however, we are very grateful to Carl Hall, Dean of Engineering at Washington State University, and Elliot Boardman, Vice President of Government Institutes for their technical assistance on this Third Edition. Special thanks to the Government Institutes' research staff: Roland Schumann, for his dedicated attention for gathering new source material and to Charlene Ikonomou for her editorial production of the publication.

We also would like to thank the readers of our First and Second Editions of the ENERGY REFERENCE HANDBOOK for their constructive comments and suggestions, and we hope that they and others will offer the same for this Edition.

## PREFACE

The field of energy technology has literally exploded in the 1970's. We find ourselves now using terms that did not even exist several years ago. In addition, many words have developed uses somewhat different from their commonly accepted meanings.

In June of 1974, Government Institutes conducted and sponsored the 1st Energy Technology Conference in Washington, D.C. One result of that conference was the recognition that the need existed for a reference source incorporating the key words and terms frequently used in the various segments of the energy field. So, the First Edition of the ENERGY REFERENCE HANDBOOK was created to fill that need. The First Edition in 1974 was so well received that it sold out, necessitating publication of a Second Edition in 1977. The success of the Second Edition resulted in our preparing this expanded and even more comprehensive Third Edition.

Many readers have provided constructive guidance and suggestions on terms and materials to be added to the book that would increase its value as a reference—to these readers we are indebted. Out-of-date terms have been culled and replaced with the most current terms. Special attention has been paid to those terms peculiar to each area within the energy technology field, namely: coal, nuclear, oil, gas, solar, wind, ocean power, geothermal, shale, and the environment.

In addition to the definitions, we have added many tables and charts on the various aspects of energy which forecast the reserves of various fuel resources, plus other information which we believe is relevant to a handy reference volume; the conversion tables are especially important as the United States enters the metric age.

We have endeavored to compile and define the most common terms to result in a better understanding of the terminology used in each energy application. In many cases it is sometimes difficult to settle upon a single definition acceptable to all concerned, who may represent broad and varied areas of pursuits and interests. We have in many instances, therefore, chosen to give several definitions with the anticipation that the final effort will be acceptable to all.

There is a need to recognize the interdependence of the various forms of energy and the interactions of the countries and regions of the world. We ascribe to the concept that all forms of energy will be needed in the future to satisfy the world's energy supply needs and that we all must view this supply/demand scenario as a world energy system.

It is our desire that the Third Edition of the Energy Reference Handbook will be of use to a broad audience in the new fields of energy and technology and, with its use, will provide a greater understanding of various words, procedures and theories. Most important, it is hoped that persons engaged in one area of energy will be made aware of the terminology of other areas of energy. Since the expanding need of technology in solving energy problems increases almost daily, proper communication within these areas is essential. We sincerely hope that our efforts will make some small contribution to solving the world's energy problems.

## TABLE OF CONTENTS

## SOURCE MATERIAL FOR GLOSSARY


A — Arbuckle, et. al., Environmental Law Handbook, 6th ed., Government Institutes, Inc., Washington, DC, 1979.

B — Lewis, R.S. and Spinrad, B. I., eds., The Energy Crisis, Educational Foundation for Nuclear Science, Chicago, IL, 1972.

C — Lovins, Armory B., Soft Energy Paths, Friends of the Earth International, 1977. Reprinted with permission from Ballinger Publishing Company.

D — Thrush, Paul W. and Bureau of Mines Staff, Eds., Dictionary of Mining, Mineral and Related Terms, revised ed., U.S. Department of Interior, U.S. Government Printing Office, Washington, DC, 1968.

E — Common Environmental Terms - A Glossary, U.S. Environmental Protection Agency, Washington, DC, 1977.

F — Conversion Factors Table, Engineering Standards, Phillips Petroleum Company, Bartlesville, OK, 1966.

G — Energy Facts, Science Policy Research Division, Congressional Research Service, U.S. Government Printing Office, Washington, DC, 1973.

H — A Floridian's Guide to Solar Energy, State of Florida Energy Office, Tallahassee, FL, 1975.

I — Gas Facts - A Glossary, American Gas Association, Arlington, VA.

J    Geothermal Energy: A National Proposal for Geothermal Resources Research, University of Alaska, U.S. Government Printing Office, Washington, DC, 1972.

K    Glossary of Electric Utility Terms, Edison Electric Institute, New York, New York.

L    Glossary of Important Power and Rate Terms, Abbreviations and Units of Measurement, Federal Power Commission, U.S. Government Printing Office, Washington, DC, 1965.

M    Glossary of Terms Used in Petroleum and Refining, American Petroleum Institute, Washington, DC, 1962.

N    Listing of Solar Radiation Measuring Equipment and Glossary, ERDA (under NASA contract), 1978.

O    Nuclear Terms: A Glossary, U.S. Atomic Energy Commission, U.S. Government Printing Office, Washington, DC, 1974.

P    The Shallow Land Burial of Low-Level Radioactively Contaminated Solid Waste, National Academy of Sciences, Washington, DC, 1976.

Q    Solar Energy, A United Kingdom Assessment, UK Section of International Solar Energy Society, London, 1976.

R    Solar Energy Utilization for Heating & Cooling, Chapter 59, 1974 Applications Handbook, National Science Foundation, U.S. Government Printing Office, Washington, DC, 1974.

S    Solar Heating and Cooling Demonstration Program: A Descriptive Summary of HUD Cycle 2 Solar Residential Projects, HUD, (in co-operation with ERDA), U.S. Government Printing Office, Washington, DC, 1976.

T    Wind Energy Systems: A Non-Pollutive, Non-Depletable Energy, Public Information Office, NASA Lewis Research Center, December 1973.

U Wind Machines, National Science Foundation, U.S. Government Printing Office, Washington, DC, 1975.

V "Factsheet #18," National Science Teachers Association, Alternate Energy Sources, A Glossary of Terms.

W Nuclear Terms: A Brief Glossary, U.S. Atomic Energy Commission booklet, 1974.

X Nuclear Power Issues and Choices, (Report of the Nuclear Policy Study Group).

Y Hall, Carl W., Dictionary of Drying, Marcel Dekker Publishers, New York, NY, 1979.

Z Science Council of Canada, Report #30, June, 1979.

AA Environmental Glossary, Government Institutes, Inc., 1980.

BB Field, Edwin M., Oil Burners, Theodore Audel Company, 1977.

CC Harrison, George R., The Conquest of Energy, Morrow Publishing Company, 1968.

DD Gas Facts - A Glossary, American Gas Association, Arlington, VA.

EE Wind Energy Systems: A Non-Pollutive, Non-Depletable Energy, Public Information Service, NASA Lewis Research Center, December 1973.

FF Hunt, V. Daniel, Energy Dictionary, Van Nostrand Reinhold Co, New York, NY, 1979.

# GLOSSARY OF ENERGY TERMS

**AAAS**
American Association for the Advancement of Science.

**AAPG**
American Association of Petroleum Geologists.

**AAQS**
Ambient air quality standards.

**AAR**
Association of American Railroads, the trade association for the railroad industry.

**abatement**
The method of reducing the degree of intensity of pollution. E

**abrasion**
The wearing away by friction; the act of wearing by rubbing or friction, the chief agents being currents of water laden with sand and other rock debris and by glaciers. D

**abrasion drilling**
An oil-drilling technique in which the use of abrasive material under pressure "cuts" the substrata, instead of using the conventional drill steam and bit. M

**absolute**
(1) In chemistry, free from impurity or admixture. (2) In physics, not dependent on any arbitrary standard. D

**absolute pressure**
Pressure measured with respect to zero pressure, as distinct from pressure measured with respect to some standard pressure. An example of such a standard pressure is atmospheric pressure (30 lb pressure psi is equivalent to 44.7 lb pressure psia). M

**absolute temperature**
Temperature at which zero is a condition absolutely free of heat and equivalent to -459.72°F or -273.18°C. To convert temperatures on Fahrenheit or Centigrade scales to degrees absolute, add 459.72 or 273.18, respectively. M

**absolute viscosity**
The force which will move 1 sq cm of plane surface with a speed of 1 cm per sec relative to another parallel plane surface from which it is separated by a layer of liquid 1 cm thick. This viscosity is expressed in dynes per sq cm, its unit being the poise, which is equal to 1 dyne-sec per sq cm. A unit of one-hundredth of a poise, designated as a centipoise, is more convenient and is commonly used. M

**absolute zero**
A point which has been determined on the thermometer scale (by theoretical considerations), beyond this a further decrease in temperature is inconceivable. It is that temperature at which the volume of a gas would have become zero or it would have lost all the molecular vibration which manifests itself as heat. The temperature of absolute zero is -459.72°F or -273.18°C. BB

**absorbed dose**
When ionizing radiation passes through matter, some of its energy is imparted to the matter. The amount absorbed per unit mass of irradiated material is called the absorbed dose, and is measured in rems and rads. W

**absorber**
Any material that absorbs or diminishes the intensity of ionizing radiation. Neutron absorbers, like boron, hafnium and cadmium are used in control rods for reactors. Concrete and steel absorb gamma rays and neutrons in reactor shields. A thin sheet of paper or metal will absorb or attenuate alpha particles and all except the most energetic beta particles. W

**absorbite**
The trade name for activated charcoal. D

**absorptiometer**
A device for measuring the solubility of a gas in a liquid. D

**absorption**
(1) The process by which one substance draws into itself another substance. Example: a sponge picking up water. (2) In solar energy, the process in which incident radiant energy is retained by a substance. N

**absorption gasoline**
Gasoline extracted from natural gas or refinery gas, e.g., by contacting the absorbed gas with an oil and subsequently distilling the gasoline from the heavier oil. M

**absorption loss**
(1) The loss of water occurring during initial filling of a reservoir in wetting rocks and soil. (2) That part of the transmission loss which is due to the dissipation or the conversion of sound energy into some other form of energy, usually heat. D

**absorption oil**
An oil used to separate the heavier components from a vapor mixture by absorption of the heavier components during intimate contacting of the oil and vapor. It is used in recovering gasoline from wet gas. M

**absorptivity**
The ratio of the radiant energy absorbed by a body to that falling upon it. It is equal to the emissivity for radiation of the same wavelength. D

**ac**
Alternating current.

**accelerated weathering test**
A test to indicate the effect of weather on coal, in which the coal is alternately exposed to freezing, wetting, warming, and light. D

**accelerator**
A device for increasing the velocity and energy of charged elementary particles; for example, electrons or protons, through application of electrical and/or magnetic forces. Accelerators have made particles move at velocities approaching the speed of light. Types of accelerators include betatrons, Cockcroft-Walton accelerators, cyclotrons, linear accelerators, synchrocyclotrons, synchrotrons, Van de Graff generators and others. D

**accelerometer (acceleration pickup)**
An electroacoustic transducer that responds to the acceleration of the

surface to which the transducer is attached, and delivers essentially equivalent electric waves. E

**acceptor levels**
Energy levels formed within the energy gap by a deficiency of electrons. D

**acclimation**
The physiological and behavioral adjustments of an organism to changes in its immediate environment. E

**ACES**
Annual cycle energy systems.

**Acheson furnace**
A resistance-type furnace for the production of silicon carbide and synthetic graphite. D

**acid**
(1) A compound that dissociates in a water solution to furnish hydrogen ions. (2) Having acid-forming constituents present in excess of the proportion required to form a neutral or normal compound. D

**acid clay**
A naturally occurring clay which, after evaluation, usually with acid, is used mainly as a decolorant or refining agent, and sometimes as a desulfurizer, coagulant, or catalyst. D

**acid fracture**
Treatment combining physical and chemical agencies to improve permeability of sand-carbonate reservoirs.

**acidizing**
A process of pumping acid directly into the producing oil well. When it reaches the producing formation, the acid reacts with certain materials in the reservoir rock, etching out channels through which the oil and gas can flow toward the well bore. Hydrochloric acid in a modified form is most frequently used. The process allows trapped oil and gas to move toward the well bore. M

**acid mine drainage**
Acidic drainage from bituminous coal mines containing a high concentration of acidic sulfates, especially ferrous sulfate. D

**ACIL**
American Council of Independent Laboratories.

**acoustical absorptivity**
The ratio of the sound energy absorbed by a surface to that incident upon it. D

**accoustical privacy**
(1) The science of sound, including the generation, transmission and effects of sound waves, both audible and inaudible. (2) The acoustics of an auditorium or of a room, the totality of those physical qualities (such as size, shape, amount of sound absorption, and amount of noise) which determine the audibility and perception of speech and music. E

**acre**
A measure of surficial area, usually of land. The statute acre of the U.S. and England contains 43,560 square feet (4,840 square yards of 160 square rods); abbreviation, a. D

**acre-foot**
A quantity of water that would cover 1 acre, 1 foot deep. Contains 43,560 cubic feet, 1,233 cubic meters, 32,580 gallons (U.S.). One acre-foot of water can satisfy the municipal and industrial energy demands of four people for 1 year. G

**acrometer**
An instrument for determining the density of gases. D

**ACRS**
Advisory Committee on Reactor Safeguards.

**ACS**
American Chemical Society.

**actinides**
A group name for the radioactive series of heavy elements starting with the element actinium of atomic number 89, and continuing to element 103, lawrencium. The name is taken from actinium, the first member of the series. Z

**activated sludge**
Sediment waste that has been aerated and subjected to bacterial action to remove organic matter. E

**activation**
The process of making a material radioactive by bombardment with neutrons, protons, or other nuclear particles. Also called radioactivation. O

**activation energy**
The energy that molecules must acquire in order to react. M

**active solar system**
A solar system in which an energy resource, in addition to solar, is used for the transfer of thermal energy. This additional energy, generated on or off the site, is required for pumps, blowers, or other heat transfer medium moving devices necessary for system operation. S

**activity**
A measure of the rate at which a material is emitting nuclear radiations; usually given in terms of the number of nuclear disintegrations occurring in a given quantity of material over a unit of time; the standard unit of activity is the curie, which is equal to $3.7 \times 10^{10}$ disintegrations per second. P

**acute radiation sickness syndrome**
An acute organic disorder that follows exposure to relatively severe doses of ionizing radiation. It is characterized by nausea, vomiting, diarrhea, blood cell changes, and in later stages by hemorrhage and loss of hair. W

**adaptation**
A change in structure or habit of an organism that produces better adjustment to the environment. E

**additive**
An agent used for imparting new, or for improving existing, characteristics of oils. M

**adiabatic**
Referring to a temperature change that does not involve an exchange of heat with the surroundings, warming by compression, cooling by expansion. E

**adiabatic compression**
Compression in which no heat is added to or subtracted from the air and the internal energy of the air is increased by an amount equivalent to the external work done on the air. The increase in temperature of the air during adiabatic compression tends to increase the pressure on account of the decrease in volume alone; therefore, the pressure during adiabatic compression rises faster than the volume diminishes. D

**ADP**
Atmospheric dew point or automatic data processing.

**adsorption**
The adhesion of the molecules of gases or dissolved substances to the surface of solid bodies, resulting in relatively high concentration of the gas or solution at the place of contact. In the oil industry: a separation process, similar to solvent extraction, in oil refining. A solid solvent is used which must be porous in order to adsorb, or hold, the undesired petroleum components on its surface. M

**adsorption gasoline**
Natural gasoline obtained by the adsorption process, adsorbed from wet gas by activated carbon or charcoal. M

**adsorption water**
Water held on the surface of solid particles by molecular forces with emission of heat (heat of wetting). E

**AEC**
Atomic Energy Commission.

**AEE**
Association of Energy Engineers (no longer exists; it was absorbed into ERDA which was then absorbed into the U.S. DOE).

**aeolian**
See eolian.

**aeration**
The process of being supplied or impregnated with air. Aeration is used in waste water treatment to foster biological and chemical purification. E

**aeration cell**
An electrolytic cell, the electromotive force which is due to a difference in air (oxygen) concentration at one electrode as compared with that at another electrode of the same material. Also called oxygen cell. D

**aerobic**
Refers to life or processes that can occur only in the presence of oxygen. A

**aerosol**
A dispersion of solid or liquid particles of microscopic size in gaseous media, such as smoke, fog, or mist. E

**aerosphere**
The atmosphere considered as a spherical shell of gases surrounding the earth. E

**AFB**
Atmospheric fluidized bed.

**AFDC**
Allowable funds used during construction.

**afterburner**
(1) An air pollution abatement device that removes undesirable organic gases through incineration. (2) A ramjet coupled to a jet engine to provide additional power. E

**afterheat**
The heat produced by the continuing decay of radioactive atoms in a nuclear reactor after the fission chain reaction has ceased. Most of the afterheat is due to the decay of fission products. D

**AF/Y**
Acre foot/year.

**AGA**
American Gas Association.

**aggregate**
(1) To bring together; to collect or to gather into a mass. It can be sand, gravel, or any clastic material in a bedded iron ore, sometimes so abundant as to make it resemble a puddingstone. (2) Uncrushed gravel, crushed stone or rock, sand, or artificially produced inorganic materials, which form the major part of concrete. D

**agricultural geology; agrogeology**
The application of geology to agricultural problems and to soil improvement. D

**AHAM**
Association of Home Appliance Manufacturers.

**AIChE**
American Institute of Chemical Engineers.

**AIF**
Atomic Industrial Forum.

**AIME**
American Institute of Mining Engineers. American Institute of Mining, Metallurgical and Petroleum Engineers.

**air**
The mixture of gases that surrounds the earth and forms its atmosphere; composed by volume of 21 percent oxygen and 78 percent nitrogen; by weight about 23 percent oxygen and 77 percent nitrogen. It also contains about 0.03 percent carbon dioxide, some aqueous vapor, and some argon. D

**air-blown asphalt**
Asphalt produced by blowing air through residual oils or similar mineral oil products at moderately elevated temperatures. M

**airborne pollutants**
Pollutants may be classified as: (1) airborne particulates, also known as aerosols, and (2) gases or vapors, including the permanent gases and those compounds having a boiling point below about 200°C. The aerosols are air suspensions, including dusts, smoke mists, and fumes. Vapors are gaseous materials derived from materials usually solid or liquid, such as gasoline. Gases and vapors will diffuse throughout the atmosphere. The more important gases are sulfur dioxide, hydrogen fluoride, chlorine, oxide of nitrogen, aldehydes, carbon monoxide, and organic vapors. E

**air cleaning**
A coal cleaning method that utilizes air tables to remove the dust and waste from coal. Air cleaning requires that the coal contain less than 5% of surface moisture as a rule. It is effective only in the coarsest sizes (plus 10 to 28 mesh) and is best suited to coals having a sharply defined line between coal and refuse material. It is a less expensive and also a less accurate method of cleaning coal than the wet cleaning method. D

**air conditioning**
The simultaneous control, within prescribed limits, of the quality, quantity, and temperature-humidity of the air in a designated area. It is essentially atmospheric environmental control. D

**air curtain**
(1) A method for mechanical containment of oil spills. Air is bubbled through a perforated pipe causing an upward water flow that retards the spreading of oil. (2) Used as a barrier to prevent fish from entering a polluted body of water. E

**airdox**
A system for breaking down coal by which compressed air, generated locally by a portable compressor at 10,000 pounds per square inch, is used in a releasing cyclinder, which is placed in a hole drilled in the coal. Thus, slow breaking results, with no flame, in producing a larger amount of lump coal than by explosives. D

**air gas**
A combustible gas made by charging air with the vapor of some volatile hydrocarbon mixture (as gasoline) and used for lighting and heating. D

**air horsepower**
The rate at which energy is used in horsepower units, in moving air between points. D

**air mass**
A widespread body of air with properties that were established while the air was situated over a particular region of the earth's surface and that undergoes specific modifications while in transit away from that region. E

**air pollution**
The presence of contaminants in the air in concentrations that prevent the

normal dispersive ability of the air and that interfere directly or indirectly with man's health, safety or comfort or with the full use and enjoyment of his property. A

**air quality control region**
An area designated by the Federal government where two or more communities—either in the same or different states—share a common air pollution problem. A

**air quality criteria**
A compilation of the scientific knowledge of the relationship between various concentrations of pollutants in the air and their adverse effects. E

**air quality standards**
The prescribed level of pollutants in the outside air that cannot be exceeded legally during a specified time in a specified geographical area. A

**air sampling**
Collection and analysis of samples of air to measure its radioactivity or to detect the presence of substances. CC

**air wave**
The acoustic energy pulse transmitted through the air as a result of the detonation of a seismic shot. D

**albedo**
The ratio of the light reflected by a surface to the light falling on it. H

**albedometer**
The general name for the instrument used to measure the ratio of the radiation reflected by a surface to that incident on it. N

**alcohol**
An alkyl compound containing a hydroxyl group; classified as primary, C(OH); secondary, $C_2(OH)$; tertiary, $C_3(OH)$. Methyl alcohol (methanol) and ethyl alcohol (ethanol) are the principal alcohols being considered for engine fuels. Y

**alcohol fuel**
Lower alcohols used as blending agents in gasoline, particularly where petroleum is scarce or gasoline is low in octane number. M

**algal coal**
Coal composed mainly of algal remains, marine, brackish and freshwater plants. Also called boghead coal. D

**aliphatics**
A basic type of organic petrochemical of the straight-chain type—their carbon atoms are lined up in a row. The most important aliphatic building blocks are ethylene, butylene, acetylene and propylene. Hydrocarbons from these gases are chemically synthesized into a myriad of end products—from plastics and synthetic rubber to drugs and detergents. AA

**alkali**
Certain soluble salts, usually of sodium, potassium, magnesium, or calcium, that can combine with acids to form neutral salts. They are often used in water and wastewater treatment plants to control acidity. E

**alkylation**
A refining method which alters the structure of hydrocarbons, rearranging petroleum molecules to form high-octane products, usually the union of an olefin with isobutene to yield branched-chain paraffinic hydrocarbons. M

**allotrophic**
Applied by Berzelius to those substances which exist in two or more forms, such as carbon existing diamond and graphite. D

**alluvium**
Sand, silt or similar detrital material deposited in flowing water, or the

permanent unconsolidated deposits thus formed. E

**alpha particle [ Symbol α ]**
A positively charged particle emitted by certain radioactive materials. It is made up of two neutrons and two protons bound together, hence is identical with the nucleus of a helium atom. O

**alternating current**
Current, the direction of which is reversed at regular intervals, usually 120 reversals per second or 60 cycle current. D

**alternating-current ampere**
That current which will produce heat at the same rate as a direct-current ampere, when flowing through a given ohmic resistance. D

**alternating-current transformer**
A device used to raise or lower the voltage of an alternating circuit. It consists of an induction coil having a primary and secondary winding and a closed iron core. D

**alternator**
A generator producing alternating current by the rotation of its rotor, which is driven by a steam or water turbine. A gas turbine or a diesel engine can be used as a prime mover in certain cases. D

**alum**
A common name for aluminum sulfate. A coagulant. E

**aluminum oleate**
A soap used in grease making. M

**AMA**
American Medical Association or the Automobile Manufacturers Association.

**ambient**
Surrounding on all sides.

**ambient air**
Any unconfined portion of the atmosphere; the outside air. A

**ambient noise level**
The ambient noise level is that level which exists at any instant, regardless of the source surrounding a body. E

**ambient temperature**
Temperature of the surrounding cooling medium, such as gas or liquid, which comes into contact with the heated parts of the apparatus. L

**American melting point**
An arbitrary temperature 3° F above the ASTM Method D 87 paraffin wax melting point. The latter is also known as the English melting point. M

**amp**
Shortened form of ampere.

**AMP**
American melting point.

**ampere**
(1) A unit of measure for an electric current. (2) The amount of current which flows in a circuit in which the electromotive force is one volt and the resistance is one ohm. G

**amperemeter; ammeter**
An instrument for measuring the strength of an electric current in amperes. D

**amperometric titration**
The electrometric detection of the equivalence point in a titration by observation of the change in diffusion current at a suitable applied voltage as a function of the volume of titrating solution. E

**amplitude**
The maximum displacement from the mean position in connection with vibration. D

**amu**
Atomic mass unit.

**anabolism**
A process by which energy is stored in chemicals as the simpler elements are combined in molecules which are more complex. Y

**anaerobic**
Refers to life or processes that occur in the absence of oxygen. A

**anaerobic digestion**
The degradation of organic matter by micro-organisms in the absence of air (oxygen), usually in a water environment, under appropriate chemical and thermal conditions, producing methane, hydrogen sulfide, and carbon dioxide. Y

**anemometer**
An instrument for measuring the force or velocity of wind; a wind gauge. E

**angle of incidence**
The angle between the direction of the sun and the perpendicular to the surface on which sunlight is falling. It affects (1) the intensity of the direct radiation component normal to the receiving surface (the smaller the angle the greater the intensity); and (2) the ability of the surface to reflect, transmit, or absorb the sun's rays. H

**angle of inclination**
The angle at which a solar collector is elevated from the horizontal. Also called angle of tilt. H

**angle of polarization**
That angle, the tangent of which is the index of refraction of a reflection substance. D

**angle of reflection**
That angle at which a reflected ray of energy leaves a reflecting surface, measured between the direction of the outgoing ray and a perpendicular to the surface at the point of reflection. N

**angstrom [ Symbol Å or A ]**
A unit of length, used in measuring electromagnetic radiation, equal to $10^{-8}$ centimeter. Named for A.J. Angstrom, Swedish spectroscopist. O

**animal waste conversion**
The process for obtaining oil from animal wastes. A Bureau of Mines experiment has obtained 80 gallons of oil per ton from cow manure. In comparison, average oil shale yields 25 gallons of oil per ton of ore. G

**animate energy**
Energy produced by living creatures, especially that resulting from operation of the muscles of animals. EE

**annealing**
Heating and slowly cooling to increase the ductility or remove internal stresses, as of metal or glass. M

**annual maximum demand**
The greatest of all demands of the electrical load under consideration which occurred during a prescribed demand interval in a calendar year. K

**anode**
The positive pole or electrode of an electrolytic cell, vacuum tube, etc. M

**ANS**
American Nuclear Society.

**ANSI**
American National Standards Institute.

**anthracite**
A hard, black, lustrous coal that burns efficiently, containing a high percentage of fixed carbon and a low percentage of volatile matter differing from bituminous. Commonly referred to as hard coal, it is mined in

the U.S., mainly in eastern Pennsylvania, although in small quantities in other states. D

**anthracite silt**
Minute particles of anthracite too fine to be used in ordinary combustion. D

**anthracite stove**
A closed-in type of domestic stove specially designed to burn anthracite. It is used mainly for heating purposes, and is very economical in fuel consumption. The stove can be kept burning for long periods with only the occasional removal of ash and refueling. D

**anthracitization**
The process of transformation of bituminous coal into anthracite. D

**anthracology**
The science of coal. D

**anthracosis**
A disease of the lungs caused by inhalation and accumulation of carbon particles (coal dust). E

**antichlors**
Reagents, such as sulfur dioxide, sodium bisulfite, and sodium thiosulfate, which can be used to remove excess chlorine residuals from water or watery wastes by conversion to an inert salt. E

**anticline**
An arch-like upward fold in the earth's strata. M

**antifoam agent**
An additive used for controlling foam. Antifoam agents are used in some lubricating oils. M

**antifreeze solution**
A fluid, such as methanol or ethylene glycol, added to or used to replace the water in the cooling systems of engines in order to prevent freezing. M

**anti-icing compounds**
Gasoline additives which prevent or protect the engine from ice which can form during engine warm-up in cold weather. M

**antiknock**
Resistance to detonation or pinging in spark-ignition engines. M

**antiknock agent**
A chemical compound such as tetraethyllead which, when added in small amounts to the fuel charge of an internal-combustion engine, tend to lessen knocking. M

**antimatter**
Matter in which the ordinary nuclear particles (neutrons, protons, electrons, etc.) are conceived of as being replaced by their corresponding antiparticles (antineutrons, antiprotons, positrons, etc.). An antihydrogen atom, for example, would consist of a negatively charged antiproton with an orbital positron. Normal matter and antimatter would mutually annihilate each other upon contact, being converted totally into energy. O

**antioxidants and inhibitors**
Gasoline additives which prevent the formation of gum deposits in the engine. M

**antirust compound**
A gasoline additive which protects the fuel system against corrosion from the minute quantities of water which condense from the air in fuel tanks. M

**APCA**
Air Pollution Control Association.

**APHA**
American Public Health Association.

**API**
American Paper Institute or American Petroleum Institute.

**API engine service classification system**
Classifications and designations for lubricating oils for automotive engines, approved by the API Lubrication Committee and adopted by the API Division of Marketing. Service conditions are segregated into broad classifications for the purpose of guiding the choice of types of motor oils desirable for each service. M

**API gravity**
An arbitrary scale expressing the gravity or density of liquid petroleum products. The measuring scale is calibrated in terms of degrees API. M

**apogean tides**
Tides with decreased range which occur at the time when the moon is in apogee. E

**apogee**
The highest or most distant point.

**apparent day**
Solar day; interval between successive transits of sun's center across observer's meridian. The time thus measured is not uniform or clock time. D

**aquifer**
An underground bed or stratum of earth, gravel or porous stone that contains water. A

**area mining**
A term applied to strip mining in flat or rolling regions. Land reclamation has been most successful in areas of this characteristic. B

**arenaceous**
Applied to rocks that have been derived from sand or that contain sand. D

**argillaceous**
Clay-bearing substances or rock. D

**aromatics**
A type of organic petrochemicals of the ring type--their carbon atoms are in a circular arrangement. Most of the organic chemicals thus far synthesized are derivatives of aromatics, so named because of their somewhat pleasant odor. The principal aromatics are commonly referred to as the BTX group—benzene, toluene, and xylene. M

**artificial liquid fuels**
Fuels created by the hydrogenation of coal, the destructive distillation of coal, lignite, or shale at low temperature, and by a recombination of the constituents of water gas in the presence of a suitable catalyst. D

**ASA**
American Standards Association.

**ASAE**
American Society of Agricultural Engineers or American Society of Association Executives.

**A-Scale sound level**
The measurement of sound approximating the auditory sensitivity of the human ear. The A-scale sound level is used to measure the relative noisiness or annoyance of common sounds. A

**ASCE**
American Society of Civil Engineers.

**ASCS**
Agricultural Stabilization and Conservation Service.

**ash**
All mineral matter left after the complete combustion of fuel. Principal constituents of ashes usually are silica, alumina, lime, oxide and bisulphide or iron. BB

**ash content**
The percent by weight of residue left after combustion of a sample of a fuel

oil or other petroleum oil; it is usually determined in the United States by ASTM Method D 482. M

**ASHRAE**
American Society of Heating, Refrigerating, and Air Conditioning Engineers.

**ASM**
American Society for Metals.

**ASME**
American Society of Mechanical Engineers.

**ASNT**
American Society for Non-Destructive Testing.

**asphalt**
A petroleum product, extracted as a refining residue or by solvent precipitation from residual fractions. Solid or semi-solid at normal temperatures, asphalt liquefies when heated. It is a powerful binding agent, a sticky adhesive, and a highly waterproof, durable material in which the predominating constituents are bitumens. M

**asphalt base crude oil**
Crude oil containing large proportions of asphaltic matter. M

**asphalt cement**
A fluxed or unfluxed asphalt especially prepared as to quality and consistency for direct use in the manufacture of bituminous pavements. M

**aspirator**
An apparatus which serves to create a partial vacuum through pumping a jet of water, steam, or some other fluid or gas past an orifice opening out of the chamber in which the vacuum is to be produced. M

**associated gas**
Free natural gas in immediate contact, but not in solution, with crude oil in the reservoir. G

**ASTM**
American Society for Testing and Materials.

**ASTM coal classification**
A system, based on proximate analysis, in which coals containing less than 31 percent volatile matter on the mineral matter free basis (Parr formula), are classified only on the basis of fixed carbon, that is, 100 percent volatile matter. They are divided into five groups; above 98 percent fixed carbon; 98 to 92 percent fixed carbon; 92 to 86 percent fixed carbon; 86 to 78 percent fixed carbon; and 78 to 69 percent fixed carbon. The first three of these groups are called anthracites and the last two are called bituminous coals. The remaining bituminous coals, the subbituminous coals and the lignites, are then classified into groups as determined by the calorific value of the coals containing their natural bed moisture, that is, the coals as mined but free from any moisture on the surface of the lumps. D

**ATGAS**
A process for coal gasification. The primary feature of the process is the dissolving of coal in a bath of molten iron. G

**atm**
Abbreviation of atmosphere.

**atmosphere**
The layer of air surrounding the earth. A

**atmospheric pressure**
The pressure of the air at sea level. As a standard, the pressure at which the mercury barometer stands at 760mm, or 30 in. (equivalent to approximately 14.7 psi). M

**atmospheric water**
Water in the atmosphere in gaseous, liquid, or solid state. E

**atom**
A particle of matter indivisible by chemical means. It is the fundamental building block of the chemical elements. The elements, such as iron, lead, and sulfur, differ from each other because they contain different kinds of atoms. There are about six sextillion (6 followed by 21 zeros, or 6 x $10^{21}$) atoms in an ordinary drop of water. According to present-day theory, an atom contains a dense inner core (the nucleus) and a much less dense outer domain consisting of electrons in motion around the nucleus. O

**atomic bomb**
A bomb whose energy comes from the fission of heavy elements such as uranium-235 and plutonium-239. W

**atomic charge**
Electrical charge density due to gain or loss of one or more electrons. D

**atomic energy**
Nuclear energy; the energy released by a nuclear reaction or by radioactive decay. O

**atomic mass unit (amu)**
One-twelfth the mass of a neutral atom of the most abundant isotope of carbon, carbon-12. W

**atomic number**
(1) The number of protons in the nucleus of an atom, (2) its positive charge. Each chemical element has its characteristic atomic number, and the atomic numbers of the known elements form a complete series from 1 (hydrogen) to 91 (protactinium). To date there are 106 elements (the last 15 elements are synthesized). O

**atomic reactor**
See nuclear reactor.

**atomic weight**
The mass of an atom relative to other atoms. The present-day basis of the scale of atomic weights is carbon; the most common isotope of this element has arbitrarily been assigned an atomic weight of 12. The unit of the scale is 1/12 the weight of the carbon-12 atom, or roughly the mass of one proton or one neutron. The atomic weight of any element is approximately equal to the total number of protons and neutrons in its nucleus. O

**atomize**
To divide a liquid into extremely minute particles, either by impact with a jet of steam or compressed air, or by passage through some mechanical device. M

**attenuation of solar radiation**
Loss of energy suffered by a beam of radiant energy which traverses the earth's atmosphere. Losses are caused by scattering and selective absorption when the energy hits certain air molecules and aerosols. N

**audiometer**
Instrument for measuring hearing sensitivity. A

**auger mining**
A method of coal extraction using instruments to bore horizontally into coal seams; for example, into the side of a hill.

**autofining**
A fixed-bed catalytic process for desulfurizing distillates. G

**autogenous**
Self-generated; produced independently of external aid. M

**automatic heating**
A central heating system operated without manual attention. It usually means oil-, gas-, or stoker-fired furnaces and boilers. D

**autoradiograph**
A photographic record of radiation from radioactive material in an

object, made by placing the object very close to a photographic film or emulsion. The process is called autoradiography. It is used, for instance, to locate radioactive atoms or tracers in metallic or biological samples. W

**autotrophic**
Self-nourishing; denoting those organisms capable of constructing organic matter from inorganic substances. A

**available energy**
That part of the total energy which can be usefully employed. In a perfect engine, that part which is converted to work. D

**available power loss**
The available power loss of a transducer connecting an energy source and an energy load is the transmission loss measured by the ratio of the source power to the output power of the transducer. D

**average boiling point**
Unless otherwise indicated the sum of the ASTM distillation temperatures from the 10-percent point to the 90-percent point, inclusive, divided by 9. Sometimes half the initial and half the maximum distillation temperatures are also added, and the sum then divided by 10. M

**average demand**
The demand on, or the power output of, an electric system or any of its parts over any interval of time, as determined by dividing the total number of kilowatt hours by the number of units of time in the interval. K

**avg**
Abbreviation for average (in formulas and tables only).

**avgas**
Common expression for aviation gasoline. M

**aviation gasoline**
Any of the special grades of gasoline suitable for use in certain airplane engines, as given in ASTM Method D 910. M

**avoirdupois**
The system of weights used in the United States and England for the ordinary purposes of trade. The fundamental unit is the pound of 16 ounces or 7,000 grains. D

**awg**
American wire gauge which has been adopted as a standard for gauging the size of wires for electrical purposes.

# B

**back**
(1) The roof or upper part in any underground mining cavity. (2) The ore body between a level and the surface, or between two levels. D

**back arch**
A concealed arch carrying the backing or inner part of a wall in a mine where the exterior facing material is carried by a lintel. D

**back balance**
A kind of self-acting incline in a mine. D

**backcast stripping**
A strip mining method using two draglines, one of which strips and casts the overburden while the other recasts a portion of the overburden. D

**backfill**
The material used to refill a ditch or other excavation, or the process of doing so. E

**background level**
With respect to air pollution, amounts of pollutants present in the ambient air due to natural sources. A

**background radiation**
The radiation in man's natural environment, including cosmic rays and radiation from the naturally radioactive elements, both outside and inside the bodies of humans and animals. Also called natural radiation. O

**back pressure**
Resistance transferred from rock into drill stem when the bit is being fed at a faster rate than the bit can cut. D

**backscatter**
When radiation of any kind strikes matter (gas, solid or liquid), some of it may be reflected or scattered back in the general direction of the source. An understanding or exact measurement of the amount of backscatter is important when beta particles are being counted in an ionization chamber, in medical treatment with radiation, or in the use of industrial radioisotopic thickness gauges. W

**backup**
A reserve generating capacity of a power system. G

**BACT**
Best available control technology.

**bad ground**
Soft, highly fractured or cavernous rock formations in which drilling a borehole is a slow procedure involving time-consuming cementing or casing operations. D

**baffle**
A device such as a steel plate, used to check, retard, or divert the flow of materials. D

**bagasse**
The fibrous material remaining after the extraction of the juice from sugarcane. Used as a fuel and as a mix in making lightweight refractories. The dried and pulverized or shredded sugarcane fibers sometimes added to a drilling fluid to plug crevices in and prevent loss of circulation liquid from a borehole. D

**bag filter**
An apparatus for removing dust from dust-laden air, employing cylinders of closely woven material which permit passage of air but retain solid particles. D

**baghouse**
Chamber in which exit gases from roasting, smelting, calcining or other emitters are filtered through membranes (bags) which arrest solids. D

**balanced winding**
The conventional method of winding in a mine shaft. As the cage containing the loaded cars ascends, the other cage containing the empties descends and thus, the cages and cars are balanced. D

**band**
(1) Slate or other rock interstratified with coal. Commonly called middle band in Arkansas. (2) Dirt band, sulfur band, or other band, as the case may be. D

**banded coal types**
Banded bituminous coal consists of bands made from various types of coal, formerly known as bright coal, dull coal, and mother of coal. D

**bank**
The top of the shaft (mine shaft), or out of the shaft. D

**Banka method**
A manual method of boring used for sampling alluvial deposits. Also called empire method (undesirable usage). D

**bare**
(1) To cut coal by hand. (2) To hole by hand. D

**barn [ Symbol b ]**
A unit area used in expressing the cross sections of atoms, nuclei, electrons, and other particles. One barn is equal to $10^{-24}$ square centimeter. O

**baro**
Short form of barometer.

**barometer**
An instrument employed to determine atmospheric pressure. M

**barometer holiday**
In mining, any day on which no work is carried on underground, owing to the very low state of the barometer (for instance, when it drops below 29 inches), as much firedamp may be expected to be given off in the mine. D

**barometric condenser**
A device for condensing steam by direct contact with water. It produces a partial vacuum in refinery equipment such as vacuum pipe still. M

**barrel ( bbl )**
Volumetric unit of measurement equivalent to 42 U.S. gallons. One barrel equals 5.6 cubic feet of 0.159 cubic meters. For crude oil, 1 bbl is about 0.136 metric tons, 0.134 long tons, and 0.150 short tons. The energy values of petroleum products per barrel are: crude petroleum, 5.6 million Btu/bbl; residual fuel oil, 6.29; distillate fuel oil, 5.83; gasoline, 5.25; jet fuel (kerosine type), 5.67; jet fuel (naptha type), 5.26; kerosine, 5.67; petroleum coke, 6.02; and asphalt, 6.64. G

**barrier**
Blocks of coal left between the workings of different mine owners and within those of a particular mine for safety and the reduction of operational costs. It helps to prevent disasters of inundation by water, of explosions, or fire involving an adjacent mine or another part of a mine and to prevent water running from one mine to another or from one section to another of the same mine. D

**barrier shield**
A wall or enclosure shielding the operator from an area where radioactive material is being used or processed by remote control equipment. W

**barring**
The end and side timber bars used for supporting a rectangular shaft. D

**basal metabolism**
The amount of heat liberated by a person at rest in a comfortable environment (about 40 BTU per hour). D

**base load**
The minimum load of a utility (electric or gas) over a given period of time. G

**base load plant**
Power plant which is normally operated to carry base load and which, consequently, operates essentially at a constant load. L

**base load unit**
An electric generating facility designed to operate at constant output with little hourly or daily fluctuation. X

**base metal**
Any of the nonprecious metals. D

**basicity**
Of an acid, the number of hydrogen atoms per molecule of it which can be replaced by a metal. D

**basin**
The lowest part of a mine or area of coal lands; a general region with an overall history of subsidence and thick sedimentary section. D

**batch oil**
A pale, lemon-colored, low-viscosity mineral oil used particularly in the manufacture of cordage. M

**batch process**
Any process in which a quantity of material is handled or considered as a unit. Such processes involve intermittent, as contrasted to continuous, operation. Examples include operations in batch cracking units or cokers. M

**BATEA**
Best available technology economically achievable.

**bathyclinograph**
In oceanography, an instrument for measuring vertical currents in the deep sea. D

**bathymeter**
This instrument measures temperature, pressure, and sound velocity to depths up to 7 miles. Modern devices are completely transistorized and use frequency modulation for telemetering. D

**battery**
In mining terminology, a bulkhead or structure of timber for keeping coal in place. D In electrical industry, see electrochemical cell.

**Baumé scale**
A device for determining the specific gravity of liquids, particularly petroleum products. It has been superseded to a considerable extent by the American Petroleum Institute scale. D

**B-B fraction**
Butane-butene fraction.

**BBL, Bbl, bbl**
Barrel(s).

**b/cd, bpcd**
Barrels per calendar day.

**BCR**
Bituminous Coal Research Institute.

**BCURA**
British Coal Utility Research Association.

**b/d, B/D**
Barrels per day.

**BDT**
Bone dry tons.

**Be**
Baumé (degree).

**beam**
A bar or straight girder used to support a span of roof between two support props or walls in a mine. D

**bearing**
Undercutting the coal face by holing. D

**Beaufort scale**
A scale, graded from 0 to 12, devised by Admiral Beaufort in the 19th century to indicate wind strength. Thus, zero on this scale represents a calm, 12 represents a hurricane, in which the wind velocity exceeds 75 mph. This scale has been adopted internationally. D

**Becke method; Becke test**
(1) In optical mineralogy, a method or test for determining relative indices of refraction. (2) A method for determining microscopically the index of refraction of a mineral compared with that of an oil or another substance, such as Canada balsam, in which it is immersed, or of two adjacent minerals in a microscopic thin section. D

**bedding**
The arrangement of rock in layers, strata, or beds. D

**bedrock**
The solid rock beneath the loose material (soil and subsoil) with which most of the land surface of the earth is covered. It is sometimes several hundred feet beneath the surface, but is usually found at a much smaller depth; in places, especially on steep slopes, it has no soil cover at all. E

**BEIR**
Biological effects of ionizing radiation.

**Belknap process**
Old method of coal cleaning in a bath of heavy liquid, produced by dissolving calcium chloride in water. The shale sinks and the coal floats. D

**belt conveyor**
A moving endless belt that rides on rollers and on which coal or other materials can be carried for various distances. D

**bench**
One of two or more divisions of a coal seam, separated by slate, etc., or simply separated by the process of cutting the coal, one bench or layer being cut before the adjacent one. D

**bending moment**
The algebraic sum of the couples or the moments of the external forces, or both, to the left or to the right of any section on a member subjected to bending by couples or transverse forces, or both. D

**benzene ($C_6H_6$)**
A colorless liquid hydrocarbon, made from coal tar and by catalytic reforming of napthenes. It is used in the manufacture of phenol, styrene, nylon, detergents, aniline, phthalic anhydride and other compounds; as a solvent; and as a component of high-octane gasoline. G

**Bergius process**
Manufacture of liquid fuels from coal by mixing powdered coal with oil and hydrogen under moderately high temperatures (about 800°F) and high pressures (about 10,000 psi). D

**berm**
A horizontal shelf or ledge built into an embankment or sloping wall of an open pit or quarry to break the continuity of an otherwise long slope for the purpose of strengthening and increasing the stability of the slope or to catch or arrest slope slough material. D

**beta particle [ Symbol β (beta) ]**
An elementary particle emitted from a nucleus during radioactive decay, with a single electrical charge and a mass equal to 1/1837 that of a proton. A negatively charged beta particle is identical to an electron. A positively charged beta particle is called a positron. W

**betatron**
A doughnut-shaped accelerator in which electrons, traveling in an orbit of constant radius, are accelerated by a changing magnetic field. Energies as high as 340 Mev have been attained. W

**Bev**
Symbol for billion (or $10^9$) electron volts. Also written as BeV. W

**BHT**
Bottom-hole temperature.

**BI-GAS**
(1) A process for coal gasification. (2) A two-stage, high-pressure oxygen-blown system reacts with pulverized coal and steam in an entrained flow. FF

**binding energy**
(1) The binding energy of a nucleus is the minimum energy required to dissociate it into its component neutrons and protons. (2) Neutron or proton binding energies are those required to remove a neutron or a proton, respectively, from a nucleus. (3) Electron binding energy is that required to remove an electron from an atom or a molecule. W

**biochemical oxygen demand (BOD)**
A measure of the amount of oxygen consumed in the biological processes that break down organic matter in water. Large amounts of organic waste use up large amounts of dissolved oxygen; thus the greater the degree of pollution, the greater the BOD. E

**bioconversion**
See biomass conversion.

**biodegradable**
Used in sewage disposal and water pollution to describe those substances that can be quickly broken down by the bacteria used for this purpose at sewage disposal plants. D

**biogas**
The product of digestion of organic materials from digester, usually anaerobic, generally with a methane content of 60 - 70%. Y

**biological dose**
The radiation dose absorbed in biological material, which is measured in rems. W

**biological half-life**
The time required for a biological system, such as a human or animal, to eliminate by natural processes half the amount of a substance (such as a radioactive material) that has entered it. W

**biological oxidation**
The process by which bacterial and other microorganisms feed on complex organic materials and decompose them. Self-purification of waterways and activated sludge and trickling

filter waste water treatment processes depend on this principle. The process is also called biochemical oxidation. E

**biological shield**
A mass of absorbing material placed around a reactor or radioactive source to reduce the radiation to a level safe for humans. W

**biomass conversion**
The production of fuel or energy from organic waste, whether it be plant material, animal manure, municipal sewage sludge or solid waste. Various techniques exist but can be classified into two categories: thermal and biological systems. Thermal systems include direct or mass burning, pyrolysis and hydrogenation. Biological systems use digestion by anaerobic bacteria to produce methane gas. N

**biomonitoring**
The use of living organisms to test the suitability of effluent for discharge into receiving waters and to test the quality of such waters downstream from a discharge. E

**biophotolysis**
The production of hydrogen and oxygen from water and sunlight by biological catalysts. Y

**biosphere**
The oceans, upper land mass, and lower atmosphere in which life exists. Y

**bio-synfuel**
Synthetic fuels produced artificially from biological materials. Y

**bio-synthoil**
Synthetic oil produced artificially from biological materials. Y

**biota**
All the species of plants and animals occurring within a certain area. E

**bit**
Any device that may be attached to, or is, an integral part of a drill string and is used as a cutting tool to bore into or penetrate rock or other materials by utilizing power applied to the bit percussively or by rotation. D

**bitumen**
(1) A general name for various solid and semisolid hydrocarbons. (2) A native substance of dark color, comparatively hard and nonvolatile, composed principally of hydrocarbons. G

**bituminous**
Containing bitumen or constituting the source of bitumen. M

**bituminous coal**
(1) A coal which is high in carbonaceous matter, having between 15 and 50 percent volatile matter. (2) Soft coal. (3) A coal other than anthracite and low-volatile coal and lignite, dark brown to black in color and burning with a smoky, luminous flame. D

**bituminous lignite**
Term used in coal literature to apply to a certain kind of lignite. D

**bituminous sand**
Naturally occurring mixtures of asphalt and loose sand grains. The bituminous cementing material associated with the sand may run as high as 12 percent. M

**bituminous shale**
A shale containing hydrocarbons or bituminous material; when rich in such substances, it yields oil or gas on distillation (for example, oil shale). D

**black body**
As applied to heat radiation, this term signifies that the surface in question emits radiant energy at each wave length at the maximum rate possible for the temperature of the surface, and at the same time absorbs all

incident radiation. Only when a surface is a black body can its temperature be measured accurately by means of an optical pyrometer. D

**black box**
A separate and self-contained electronic unit or element of an electronic device which can be treated as a single package. The name comes from the fact that the housings of such units are often black. D

**black light**
Electromagnetic radiation not visible to the human eye; used by miners and prospectors for ultraviolet light. D

**black oil**
Any of the black-colored oils used for the lubrication of heavy, slow-moving, rough machinery where it would be impractical or uneconomical to use higher grade lubricants. M

**black oil shale**
Oil shale in the Eastern United States particularly common around the bituminous coal and petroleum regions. D

**black shale**
Usually a very thin bedded shale, rich in sulfides (especially pyrite which may have replaced fossils) and rich in organic material, deposited under barred basin conditions causing anaerobic accumulation. D

**blasting**
The operation of breaking coal, ore, or rock by boring a hole in it, inserting an explosive charge, and detonating or firing it. Also called shot firing. D

**blend**
Any mixture prepared for a special purpose; e.g., the products of a refinery are blended to suit market requirements. M

**blended fuel oil**
A mixture of residual and distillate fuel oils. M

**blending stock**
Any of the stocks used to make commercial gasoline. These include natural gasoline, straight-run gasoline, cracked gasoline, polymer gasoline, alkylate, and aromatics. M

**blinding**
(1) In uranium leaching, reduced permeability or ion-exchange resins due to adherent slimes. (2) In sieving, blocking of screen apertures by particles. D

**BLM**
Bureau of Land Managment.

**block rate schedule**
Rate schedule which provides different unit charges for various blocks of demand or energy. L

**bloom**
(1) Fluorescence. (2) The color of an oil by reflected light when this differs from its color by transmitted light. For certain purposes, the oil trade has preferred oils of yellowish-green rather than bluish-green bloom. This demand can be met by special processing. M

**blow**
A general term to describe a strong wind, especially associated with squalls and storms.

**blow-by**
In internal-combustion engines, the escape of combustion gases or unburned fuel from the combustion chamber past the pistons and rings into the crankcase during the power stroke or the compression stroke. M

**blown oil**
Oil oxidized by blowing air through it while hot. This procedure is used for increasing the viscosity of fatty oils such as rapeseed, castor, china wood, and similar oils. M

**blowout preventer stack**
A series of large valves used to seal off the well bore so that gases and

liquids under pressure in the hole can be controlled. There are at least two blowout preventers on each well and a maximum number per well will range from four to seven. M

**BLS**
Bureau of Labor Statistics in the Department of Labor.

**blue smoke**
A blue exhaust smoke from a diesel engine, indicating that only a part of the fuel is being burned; also called cold smoke. M

**BLWR**
Boiling light-water reactor.

**BM**
Bureau of Mines in the Department of the Interior.

**BNFL**
British Nuclear Fuels, Limited.

**BNL**
Brookhaven National Laboratory (NY).

**BOD**
Biochemical oxygen demand.

**body burden**
The amount of radioactive material present in the body of a human or an animal. W

**body of coal**
A term frequently used to indicate the "fatty" flammable property in coal, which is the basis of the phenomenon called combustion. D

**BOE**
Barrels of oil equivalent.

**boiler feedwater**
Water forced into a boiler to take the place of that evaporated in the generation of steam. E

**boiler horsepower**
A unit of rate of water evaporation. One boiler horsepower equals the evaporation, per hour, of 34-1/2 lb of water at a temperature of 212°F into steam at 212°F. M

**boiling point**
When the vapor pressure of a liquid becomes equal to the pressure of the atmosphere above it, the liquid is said to have reached its boiling point; the temperature at which a substance boils. E

**boiling water reactor**
A reactor in which water, used as both coolant and moderator, is allowed to boil in the core. The resulting steam can be used directly to drive a turbine. O

**bolometer**
Instrument for measuring the intensity of radiant energy. Its principle is based on the variation of electrical resistance with the incoming radiation, of one or both of the metallic strips which the instrument comprises. N

**BOM**
Bureau of Mines.

**bond**
(1) The cohesion of adhesion that develops between particles of ceramic materials in the unfired or fired state. (2) The overlapping of brick in various ways in a structure so as to provide strength. D

**bone**
A hard coal-like substance high in noncombustible mineral matter; often found above or below, or in partings between layers of relatively pure coal. D

**bone seeker**
A radioisotope that tends to accumulate in human bones when it is introduced into the body. An example is strontium-90, which behaves chemically like calcium. W

**boom**
A floating device that is used to contain oil on a body of water. E

**borehole**
A hole with a drill, auger, or other tools for exploring strata in search of minerals, for water supply, for blasting purposes, for proving the position of old workings, faults, and for releasing accumulations of gas or water. D

**bottled gas**
Ordinarily, butane or propane, or butane-propane mixtures, liquefied and bottled under pressure for domestic use. M

**bottom-hole pressure**
The load, expressed in pounds or tons, applied to a bit or other cutting tool while drilling. D

**bottoming cycle**
A means to increase the thermal efficiency of a steam electric generating system by converting some waste heat from the condenser into electricity rather than discharging all of it to the environment. G

**bottom water**
In oil wells, the water that lies below the productive sand and is separated from it. D

**Boyle's Law**
States: The volume of a gas varies inversely with its pressure at constant temperature. E

**BPA**
Bonneville Power Administration.

**BPD, BPCD, BPSD**
Barrels per day, per calendar day, per stream day.

**Bradford breaker**
A machine which combines coal crushing and screening. D

**brake horsepower**
That horsepower delivered by an engine to a brake or dynamometer. It is less than the indicated horsepower by the amount lost in transmission bearings, gear teeth, belts, etc. M

**branded gasoline**
Well-known, or brand name gasoline of the major oil companies.

**brattice**
A board of plank lining, or other partition, in any mine passage to confine the air and force it into the working places. Its object is to keep the intake air from finding its way by a short route into the return airway. Temporary brattices are often made of cloth. Also spelled brettice; brettis; brattish. D

**breaker**
In anthracite mining, the structure in which the coal is broken, sized, and cleaned for market. Also known as coalbreaker. D

**break-in oil**
Low-viscosity oil used to lubricate newly made engines during initial operation. M

**breeder reactor**
A reactor that produces fissionable fuel as well as consuming it, especially one that creates more than it consumes. The new fissionable material is created by capture in fertile materials of neutrons from fission. The process by which this occurs is known as breeding. O

**breeding ratio**
The ratio of the number of fissionable atoms produced in a breeder reactor to the number of fissionable atoms consumed in the reactor. O

**breeze**
A light, gentle wind, moving at speeds from 4 to 30 miles per hour.

**bridge**
A device to measure the resistance of wire or other conductors forming a part of an electric circuit. D

**bright coal**
The constituent of banded coal which is of a jet black, pitchy appearance, more compact than dull coal, and breaking with a conchoidal fracture when viewed macroscopically, and which in thin section always shows preserved cell structure of woody plant tissue, either of stem branch, or root. Same as anthraxylon. D

**brine**
(1) Water saturated or strongly impregnated with salt. (2) Concentrated brackish saline or sea waters. E

**briquet**
A small brick made by compaction of fine materials and a binder. M

**British thermal unit (Btu)**
The amount of heat required to raise the temperature of one pound of water one degree Fahrenheit under stated conditions of pressure and temperature (equal to 252 calories).

**brown coal**
A low-rank coal which is brown, brownish-black, but rarely black. It commonly retains the structures of the original wood. It is high in moisture, low in heat value, and checks badly upon drying. D

**BS, BSI**
British Standards and British Standards Institution, respectively.

**b/sd**
Barrels per stream day.

**Btu, BTU**
British thermal unit(s).

**BTX group**
Benzene, toluene, and xylene, principle aromatics. M

**bulk density**
The weight of solid particles which can be held by a container of known dimensions compared to the weight of water which can be held by the same container. M

**bulkhead**
A tight partition of wood, cork, and mud or concrete in mines for protection against gas, fire and water. D

**bulk mining**
A method of mining in which large quantities of low-grade ore are mined without attempt to segregate the high-grade portions. D

**bulk plant**
A wholesale distributing unit for petroleum products, often having facilities on railroad sidings. Generally it has tank storage for light oils and a warehouse with storage for products sold in barrels and packages. G

**bunker "C" fuel oil**
(1) A heavy residual fuel oil used by ships, industry, and for large-scale heating installations. The United States Navy calls it "Navy heavy". (2) In industry it is often referred to as No. 6 fuel. M

**burden**
All types of rock or earthy materials overlying bedrock. D

**burner reactor**
A converter reactor. O

**burning oil**
An illuminating oil, such as kerosine, mineral seal oil, etc., suitable for burning in a wick lamp. M

**burnup**
A measure of reactor fuel consumption. It can be expressed as (a) the percentage of fuel atoms that have undergone fission, or (b) the amount of energy produced per unit weight of fuel in the reactor. O

**burning point**
The temperature at which a material ignites. BB

**Burton process**
A thermal cracking process formerly used, in which oil was cracked in a pressure still and any condensation of the products of cracking which also took place under pressure. M

**bus**
An electrical conductor which serves as a common connection for two or more electrical circuits. A bus may be in the form of rigid bars, either circular or rectangular in cross section, or in the form of stranded-conductor overhead cables held under tension. K

**busbar**
An electrical conductor in the form of rigid bars located in a switchyard or power plant, serving as a common connection for two or more electrical circuits. G

**busbar cost**
Cost of producing electricity at the output terminals of a generator or power station as opposed to cost of electricity delivered to consumers. C

**butt**
Opposite of face, coal exposed at right angles to the face, and in contrast to the face, generally having a rough surface. D

**BWR**
Boiling-water reactor.

**byproduct power**
Electric power produced as a byproduct incidental to some other operation. L

**byproducts**
(1) Secondary products which have commercial value and are obtained from the processing of a raw material. They may be the residues of the gas production process, such as coke, tar, and ammonia, or they may be the result of further processing of such residues, such as ammonium sulphate. (2) As a nuclear term, any radioactive material (except source material or fissionable material) obtained during the production or use of source material or fissionable material. It includes fission products and many other radioisotopes produced in nuclear reactors. DD

# C

**°C**
Degree Celsius, formerly, and still more commonly, known as degree centigrade.

**CAA**
Clean Air Act.

**cable system**
One of the well-known drilling systems, sometimes designated as the American or rope system. The drilling is performed by a heavy string of tools suspended from a flexible manila or steel cable to which a reciprocating motion is imparted by an oscillating "walking beam" through the suspension rope or cable. D

**cable tool**
One of the two basic types of drilling equipment used in drilling for oil, it is the oldest and least used today. A cable tool rig is made up of machinery and gear that raise and drop a "string" of tools consisting of a "bit" and steam on the end of the cable. The heavy bit pounds its way into the earth, pulverizing soil and rock. At intervals, the string of tools is removed. Then the hole is flushed, and the resulting "slurry" of drilling cuttings is removed by bailing. At appropriate intervals, the hole is lined with steel casing to prevent caving in, and to protect underground fresh water strata encountered during drilling. M

**cage**
Mining term for elevator. D

**caisson**
A cylindrical steel section of shaft, used for sinking through running or waterlogged ground. D

**caking coal**
Coal which softens and agglomerates on heating and, after volatile matter has been driven off at high temperatures, produces a hard gray cellular mass of coke. All caking coals are not good coking coals. G

**cal**
Shortened form of calorie (small).

**calamine**
A commercial, mining, and metallurgical term comprising the oxidized ores of zinc, as distinguished from the sulfide ores or blends. D

**calendar day**
(1) The time from midnight of one day to midnight of the next day. (2) The 1/365 part of a normal year. (3) Relates to total lapsed time of a period measured in days. M

**California polymerization**
A process developed by the Standard Oil Company of California for converting $C_3$-$C_4$ olefins to motor fuel. The catalyst is phosphoric acid on quartz chips. M

**calorie**
(1) The amount of heat required to raise the temperature of 1 g of water 1°C, at or near the temperature of maximum density. This unit is called a "small calorie," or "gram-calorie." (2) The amount of heat required to raise the temperature of 1 kg of water 1°C, this unit is called a "large calorie" or "kilogram-calorie." (3) Capitalization of the word calorie indicates a kilogram-calorie. M

**calorific intensity**
The temperature of a fuel attained by its combustion. D

**calorific power**
The quantity of heat liberated when a unit weight or a unit volume of a fuel is completely burned. D

**calorifics**
The science of heating. D

**calorific value**
The heat liberated by the combustion of a unit quantity of a fuel.

**calorimeter**
Any apparatus for measuring the quantity of heat generated in a body or emitted by it, as by friction, combustion, etc. D

**cam**
A rotating piece, either noncircular or eccentric, used to convert rotary into reciprocating motion, often of irregular outline and giving motion that is irregular in direction, rate, or time. D

**candlepower**
The illuminating power of a standard candle employed as a unit for determining the illuminating quality of kerosene and other illuminants. One international candle or one American candle equals 1.11 Hefner candles. M

**CANDU**
Canadian deuterium-uranium reactor. A nuclear reactor of Canadian design, which uses natural uranium as a fuel and heavy water as a moderator and coolant. X

**cap**
(1) A piece of plank or timber placed on top of a prop, stull, or post. (2) A flat piece of wood inserted between the top of the prop and the roof. In addition, the horizontal member of a set of timber used as a roadway support. D

**capability**
The maximum load which a generating unit, generating station, or other electrical apparatus can carry under specified conditions for a given period of time, without exceeding approved limits of temperature and stress. K

**capability margin**
(1) The difference between net electrical system capability and system maximum load requirements (peak load). (2) The margin of capability available to provide for scheduled maintenance, emergency outages, system operating requirements, and unforeseen loads. (3) On a regional or national basis, it is the difference between aggregate net system capability of the various systems in the region or nation and the sum of system maximum (peak) loads without allowance for time diversity between the loads of the several systems. However, within a region, account is taken of diversity between peak loads of systems that are operated as closely coordinated groups. K

**capacitance**
A measure of the electrical charge of a capacitor consisting of two plates separated by an insulating material called the dielectric. Capacitance

measurement is widely used in liquid level measurement, in interface detection in pipelines, and in continuous process analysis. M

**capacitor**
Electrical appliance working on the condenser principle. Two conducting plates are separated by an insulating layer. When alternating current is applied the capacitor is adjusted so that its leading current balances the lag of the circuit giving a high-power factor. D

**capacity**
The load for which a generating unit, generating station, or other electrical apparatus is rated either by the user or by the manufacturer. K

**capacity factor**
The ratio of energy actually produced to that which would have been produced in the same period (usually a year) had the unit operated continuously at rated capacity. X

**capacity of the wind**
The total amount of detrital material of a given kind that can be sustained (per unit volume of air) by a wind of a given velocity. In the aggregate, wind transports more material than water, although water at the same speed of flow is capable of transporting much larger particles. D

**cap rock**
Hard formation lying above certain reservoirs that prevents upward migration of oil.

**capture**
A process in which an atomic or nuclear system acquires an additional particle; for example, the capture of electrons by positive ions, or capture of electrons or neutrons by nuclei. O

**carbide miner**
A pushbutton mining machine with a potential range of 1,000 feet into the seam from the highwall, production of some 600 tons per shift, and a recovery of 65 to 75 percent of the coal within the reach of the machine. This unit is a continuous miner working from a control stand outside the seam of coal. D

**carbohydrate**
Various organic compounds made up of the elements of carbon, hydrogen and oxygen including primarily starches, sugar, and celluloses, most of which are formed from green plants. Y

**carbon**
A nonmetallic element existing in diamonds, graphite, and numerous amorphous forms, combined as carbon dioxide, carbonates, and in all living things. Carbon is unique in forming an almost infinite number of compounds since it is present in all organic compounds. M

**carbonaceous**
Containing carbon or carbon compounds. The term is applied especially to shale or rock containing some particles of carbon distributed evenly throughout the whole mass.

**carbon black**
(1) An inorganic member of the petroleum chemicals produced by the burning of oil or natural gas in the presence of just enough oxygen to prevent all the carbon from being consumed. The end result is the blackest substance known to man. Its uses include the strengthening of rubber and cement, manufacture of paint, ink, batteries, radio and television tubes, explosives, and steel alloys. (2) Used in the polishing of lenses, prisms, and reflectors used in precision optical equipment. M

**carbon cycle**
(1) The flow of carbon in living beings in which carbon dioxide is fixed by photosynthesis in green plants from organic nutrients and is ultimately returned to the inorganic state by

respiration and decay. Also known as the Calvin cycle. (2) The carbon cycle also refers to a cycle of thermonuclear reactions in which four hydrogen atoms synthesize into a helium atom with the release of nuclear energy and which is considered to be the source of most of the energy given up by the sun and stars. Y

**carbon dioxide ($CO_2$)**
A colorless, odorless, nonpoisonous gas that is a normal part of the ambient air. $CO_2$ is a product of fossil fuel combustion, and some researchers have theorized that excess $CO_2$ raises atmospheric temperatures. E

**carbonization**
The conversion of an organic compound, such as oil, into char or coke by heat in the substantial absence of air. M

**carbon monoxide (CO)**
A colorless, odorless, highly toxic gas that is a normal byproduct of incomplete fossil fuel combustion. CO, one of the major air pollutants, can be harmful in small amounts if breathed over a period of time. E

**carcinogenic**
Cancer producing. E

**carman**
In anthracite and bituminous coal mining, a worker who handles mine or railroad cars underground or at the surface of a mine. D

**case**
(1) A small fissure, admitting water into the mine workings. (2) To line a borehole with steel tubing, such as casing or pipe. D

**casing**
(1) A zone of material altered by vein action and lying between the unaltered country rock and the vein. (2) Special steel tubing welded or screwed together and lowered into a borehole to prevent entry of loose rock, gas, or liquid into the borehole or to prevent loss of circulation liquid into porous, cavernous, or crevassed ground. D

**casinghead gas**
The natural gas which issues from the casinghead (the mouth or opening) of an oil well. M

**cask**
A heavily shielded container used to store and/or ship radioactive materials. W

**catabolism**
Those chemical and physical processes continuously taking place in living organisms and cells by which protoplasm is used and broken down into simpler substances with the release of energy. E

**cataclysmic**
Accompanied with violent disruption. D

**catalysis**
(1) A process in which the rate of chemical reaction is hastened or retarded by contact with an unrelated substance called the catalyst. (2) A substance capable of changing the rate of a reaction without itself undergoing any net change. M

**catalyst**
A chemical which when present in the chemical reaction is not consumed as a part of the molecular structure of interaction, but is capable of changing the rate of that reaction. Metal catalysts are often used in gasification processes. Some of the metal ions in solid waste serve as a catalyst in the formation of gases during gasification process. Y

**catalytic converter**
An air pollution abatement device that removes organic contaminants by oxidizing them into carbon dioxide and water through chemical reaction. Can be used to reduce nitrogen oxide emissions from motor vehicles. G

**catalytic cracking**
The breakdown of chemicals into high boiling point liquids accompanied by the production of large quantities of gaseous hydrocarbons in the presence of a catalyst at high temperatures. Y

**catalytic hydrogenation of coal tar**
A process being developed to convert sulfur-bearing coal into non-polluting fuel.

**catalytic reforming**
A refining method which alters the structure of hydrocarbons, rearranging petroleum molecules to form high-octane products. M

**cathode**
(1) The electrode where electrons enter (current leaves) an operating system, such as a battery, an electrolyte cell, an X-ray tube, or a vacuum tube. (2) In a battery or electrolytic cell, it is the electrode where reduction occurs. (3) Opposite of anode. D

**cathode rays**
A stream of electrons emitted by the cathode, or negative electrode, of a gas-discharge tube or by a hot filament in a vacuum tube, such as a television tube. W

**caustic**
(1) The alkali, caustic soda, or sodium hydroxide used in many refining processes. (2) Corrosive. M

**CBO**
Congressional Budgeting Office.

**CBR**
Commercial breeder reactor.

**CCPS**
Combined-cycle power system.

**CDBR**
Commercial demonstrator fast breeder reactor.

**CEA**
Council of Economic Advisors.

**CECA**
Consumers Energy Council of America.

**cell**
(1) The smallest cooling tower subdivision which can function as an independent unit with regard to air and water flows bounded by exterior walls or partition walls. Each cell may have one or more fans or stacks and one or more distribution systems. (2) A location providing isolation of an engine stand to reduce noise, hazard, etc. M

**cellulose**
(1) A hydrocarbon consisting of $C_6H_{10}O_5$ with wood being a principal source of cellulose. (2) Along with xylan, a principal constituent in the cell walls of green plants. Wood is 43% cellulose. Y

**center of gravity**
(1) The point in a body through which the resultant of the parallel forces of gravity acting on all particles of the body passes no matter in what position the body is held. (2) The center of mass. E

**center of pressure**
The point of application of the resultant of all normal pressures acting on a surface. E

**centigrade (Celsius) scale**
A thermometer scale on which the interval between the freezing point and boiling point of water is divided into 100 parts or degrees centigrade, so that 0°C corresponds to 32°F and 100°C to 212°F. Also called Celsius after Anders Celsius who first described it. M

**centipoise**
A unit of absolute viscosity. M

**centistoke**
A unit of kinematic viscosity. M

**central station service**
Electric service supplied from an electric system rather than by self-generation. L

**centrifugal force**
The force exerted as a material particle moving along a curve reacts to the body that constrains the motion and is impelled by inertia to move away from the center of curvature, the force being directed outwardly along the radius of curvature. D

**centrifuge**
A rotating device for separating liquids of different specific gravities or for separating suspended colloidal particles, such as clay particles in an aqueous suspension, according to particle-size fractions, by centrifugal force. D

**centrifuge isotope separation**
An isotope enrichment process that separates lighter molecules containing uranium-235 from heavier molecules containing uranium-238 by means of ultra high speed centrifuges. X

**CEQ**
Council on Environmental Quality.

**CFFC**
Clean fuel from coal.

**cfm, cfs**
Cubic feet per minute; cubic feet per second.

**CGCC**
Coal gasification combined cycle.

**cgs system**
A metric system of physical measurements in which the fundamental units of length, mass, and time are the centimeter, the gram, and the mean solar second, respectively. E

**chain reaction**
(1) A reaction that stimulates its own repetition. (2) In a fission chain reaction a fissionable nucleus absorbs a neutron and fissions, releasing additional neutrons. These in turn can be absorbed by other fissionable nuclei, releasing still more neutrons. A fission chain reaction is self-sustaining when the number of neutrons released in a given time equals or exceeds the number of neutrons lost by absorption in non-fissioning material or by escape from the system. O

**Chance process**
A method of cleaning coal by using a fluid mixture of sand and water which floats off a clean coal product but allows slate and other impurities to sink. Named after Thomas M. Chance, American mining engineer. D

**char**
The porous and highly reactive residue from carbonization of coal at low temperatures (450° to 700°C).

**characteristic speed**
A speed or velocity of revolution, expressed in revolutions per minute, at which the runner of a given type of turbine would operate if it were so reduced in size and proportion that it would develop one horsepower under a one-foot head. E

**charcoal**
A dark-colored or black porous form of carbon made from vegetable or animal substances or from wood by charring in a kiln or retort from which air is excluded and used for fuel and in various mechanical, artistic, and chemical processes. D

**charged particle**
(1) An ion. (2) An elementary particle that carries a positive or negative electric charge. O

**charging stock**
Oil that is to be treated in a particular unit. M

**Charles' Law**
The volume of a gas at constant pressure varies in direct proportion to the absolute temperature. E

**chemical energy**
Energy released or absorbed when atoms form compounds. Generally becomes available when atoms have lost or gained electrons, and often appears in the form of heat. D

**chemical symbol**
A single capital letter, or a combination of a capital letter and a lowercase letter, which is used to represent either an atom or a gram atom of an element; for example, the symbol for sodium is NA (from the Latin natrium). D

**chemisorption**
Adsorption, especially when irreversible, by means of chemical forces in contrast with physical forces. E

**cherry coal**
(1) A soft noncaking coal which burns readily. (2) A deep black, dull or lustrous bituminous coal, with a somewhat conchoidal fracture, readily breaking up into cuboidal fragments. D

**chilling effect**
The lowering of the earth's temperature due to the increase of atmospheric particulates that inhibit penetration of the sun's energy. E

**chinook**
A warm, dry wind blowing eastward from a mountain chain, most often the Cascades in Washington and Oregon, but also applied to winds blowing from the Rocky Mountains into Montana, Wyoming, and Colorado.

**chokedamp**
A mine atmosphere that causes choking, or suffocation, due to insufficient oxygen. As applied to "air" that causes choking, does not mean any single gas or combination of gases. D

**Christmas tree**
Surface installation on a production well consisting of several pipe connections and valves. Z

**chute**
A channel or shaft underground, or an inclined trough above ground, through which ore falls or is shot by gravity from a higher level to a lower level. Also spelled shoot. D

**Ci**
Curie.

**CIP**
Cascade Improvement Program or cast iron pipe.

**circuit breaker**
An overload protective device installed in the positive circuit to interrupt the flow of electric current when it becomes excessive or merely exceeds a predetermined value. D

**circumsolar radiation**
Radiation scattered by the atmosphere into the area of the sky immediately adjacent to the sun. It causes the solar aureole, and its areal extent is directly related to the atmospheric turbidity, being greater with higher turbidity. N

**cladding**
The outer jacket of nuclear fuel elements. It prevents corrosion of the fuel and the release of fission products into the coolant. Aluminum or its alloys, stainless steel and zirconium alloys are common cladding materials. O

**cladding waste**
Fuel rods in most nuclear reactors today are made up of fissionable materials clad in a protective alloy sheathing that is relatively resistant to radiation and the physical and

chemical conditions that prevail in a reactor core. The spent fuel rods are chopped up, and the residues of the fissionable materials are leached out chemically. The remaining residues, principally the now-radioactivated cladding material and insoluble residues of nuclear fuel, fission products, and transuranium nuclides, are left behind as cladding waste, which is a special category of transuranium radioactive waste. P

**clarified oil**
The heavy oil which has been taken from the bottom of a fractionator in a catalytic cracking process, and from which residual catalyst has been removed. M

**class rate schedule**
Rate schedule which is applicable to one or more specified classes of business or customer use. L

**clay refining**
A treating process in which vaporized gasoline or other light petroleum product is passed through a bed of granular clay such as fuller's earth. Certain olefinic compounds are polymerized to gums and are adsorbed by the clay. M

**cleaned coal**
Coal produced by a cleaning process (wet or dry). D

**cleaning coal, dry**
The mechanical separation of impurities from coal by methods which avoid the use of a liquid. D

**cleaning coal, wet**
The mechanical separation of impurities from coal by methods involving the use of a liquid. D

**clear gasoline**
A gasoline which is free from antiknock additives such as tetraethyllead. In making comparative engine tests between leaded and unleaded fuels, the clear, unleaded gasoline is sometimes referred to as straight gasoline base, base fuel, or as gasoline "neat." M

**clearness number**
A solar energy concept introduced to express the ratio between the actual clear-day direct radiation intensity at a specified location and the intensity calculated for the standard atmosphere for the same location and date. R

**cleat**
Main joint in a coal seam along which it breaks most easily. Runs in two directions, along and across the seam. D

**closed-cycle reactor system**
A reactor design in which the primary heat of fission is transferred outside the reactor core to do useful work by means of a coolant circulating in a completely closed system that includes a heat exchanger. O

**closed steam**
Steam used in a heating coil so that no direct contact takes place between the steam and the material to be heated. Contrast this to open steam, where intimate contact between the media is possible. M

**CNG**
Compressed natural gas.

**coal**
A solid, brittle, more or less distinctly stratified combustible carbonaceous rock, formed by partial to complete decomposition of vegetation; varies in color from dark brown to black; not fusible without decomposition and very insoluble. Campbell divides coals into the following ranks or classes: (1) lignite, (2) sub-bituminous, (3) bituminous, (4) semibituminous, (5) semianthracite, and (6) anthracite. D

**coal alkylation**
A process being developed to convert sulfur-bearing coal into a non-polluting fuel. G

**coal analysis**
The determination by chemical methods, of the proportionate amounts of various constituents of coal. Two kinds of coal analyses are ordinarily made: (1) proximate analysis and (2) ultimate analysis. D

**coal ash**
Noncombustible matter in coal. D

**coal auger**
A special type of continuous miner. It consists essentially of a large diameter screw drill which cuts, transports, and loads the coal onto vehicles or conveyors. D

**coal bank**
An exposed seam of coal. D

**coal basin**
Depressions in older rock formations in which coal-bearing strata have been deposited. D

**coalbed**
A bed or stratum of coal. The term coal seam is more commonly used in the United States and Canada. D

**coal blasting**
There are two methods of breaking coal with explosives, namely, (1) blasting cut coal, which is the method most commonly used, and (2) blasting off the solid, or grunching. D

**coalbreaker**
A building containing the machinery for breaking coal with toothed rolls, sizing it with sieves, and cleaning it for marketing. D

**coal breccia**
Coal broken into angular fragments by natural processes occurring within the coal bed. Polished and slickensided surfaces may be common. D

**coal briquettes**
Coal made more suitable for burning by a process which forms it into regular square- or oval-shaped pieces. D

**coal, chemical constituents of**
Carbon, hydrogen, oxygen, nitrogen, and inorganic matter that constitutes the ash. Sulphur in the free state is sometimes present in coal. BB

**coal classification**
The grouping of coals according to certain qualities or properties, such as coal type, rank, carbon-hydrogen ratio, volatile matter, etc. See also ASTM coal classification system and coal fuel ratio. D

**coal cleaning plant**
A plant where raw or run-off mine coal is washed, graded, treated to remove impurities and reduce ash content. D

**coal cutter machine**
A machine powered by compressed air or electricity which drives a cutting chain to undercut or overcut the seam, or to remove a layer of shale. Precussive cutters bore holes or make vertical cuts.

**coal deposits**
Usually termed beds, coal deposits range from a fraction of an inch to several hundred feet in thickness. Colloquially, they are called seams and veins. The differences in coals are due to age, pressure (folding and/or depth of burial), and heat, which may have been supplied by transecting dikes or by movement in the rocks.

**coal drill**
Usually an electric rotary drill of a light, compact design. Aluminum and

its alloys usually are used to reduce the weight. Where dust is a hazard, wet drilling is employed. D

**coal dryer**
A plant or vessel in which water or moisture is removed from fine coal. D

**coal equivalent of fuels burned**
The quantity of coal (tons) of stated kind and heat value which would be required to supply the Btu equivalent of all fuels burned. In determining this coal equivalent, the Btu content of other fuels is generally divided by the representative heat value per ton of coal burned. K

**coal fuel ratio**
The content of fixed carbon divided by the content of volatile matter is called the fuel ratio. According to their fuel ratios coals have been classed: Anthracite, not less than 10, Semianthracite, 6 to 10, Semi-bituminous, 3 to 6, and Bituminous, 3 or less. D

**coal gas**
Manufactured gas made by distillation or carbonization of coal in a closed coal gas retort, coke oven, or other vessel. G

**coal gasification**
The conversion of coal to a gas which is suitable for use as a fuel. The gas produced may be either a high-Btu or a low-Btu fuel. High-Btu gas is similar to natural gas and will range in energy content from 900 to 1,000 Btu per cubic foot. Low-Btu gas may range as low as 200 Btu per cubic foot. G

**coalification**
(1) The processes involved in the genetic and metamorphic history of coalbeds. The plant materials that form coal may be present in vitrinized or fusinized form. Materials contributing to coal differ in their response to diagenetic and metamorphic agencies. The three essential processes of coalification are called incorporation, vitrinization, and fusinization. In essence, coalification is the natural process in which plant materials are made into coal. Peat is the initial stage, anthracite is the final stage. Y

**coalite; semicoke**
A trade name for a smokeless fuel produced by carbonizing coal at a temperature of about 600°C. It has a calorific value per pound of about 13,000 Btu's and is used for domestic purposes. D

**coal liquefaction; coal hydrogenation**
The conversion of coal into liquid hydrocarbons and related compounds by hydrogenation at elevated temperatures and pressures. In essence, this involves putting pulverized bituminous coal into an oily paste, which is treated with hydrogen gas under appropriate conditions of temperature and pressure to form the liquid molecules of carbon and hydrogen which constitute oil. D

**coal measures**
Strata containing coalbeds, particularly those of the Pennsylvanian period. D

**coal measure unit**
The coal measure strata disclose a rough repetition or cycle of different kinds of rock in the same regular manner. Broadly, the cycle of strata upwards in coal, shale, sandstone and coal. This sequence is sometimes referred to as a unit. D

**coal mine**
Any and all parts of the property of a mining plant, on the surface or underground, which contributes, directly or indirectly, under one management to the mining or handling of coal. D

**coal mining methods**
The methods of working coal seams have been gradually evolved and

progressively improved or modified as knowledge and experience were gained and power machines became available. Over the years, a very large number of methods of mining coal have been developed to suit the seam and local conditions, and they may be divided, broadly, into long-wall and pillar methods of working. D

**coal oil**
(1) Oil obtained by the destructive distillation of bituminous coal. (2) Archaic term of kerosene made from petroleum. M

**coal rash**
Very impure coal containing much argillaceous material, fusain, etc. D

**coal rate**
The weight in pounds of coal, including the coal equivalent of other fuels burned for electric generation divided by the resulting net generation. K

**coal saw**
A coal cutter employing a very thick chain and bits, on a saw which cuts a kerf 2 inches wide. The coal saw is for use where hydraulic devices could be employed to break down the coal and thus eliminate most or all of the shooting ordinarily required. D

**coal seam**
A bed or stratum of coal. D

**coal slurry**
Finely crushed coal mixed with sufficient water to form a fluid. D

**coal tar**
A gummy, black substance produced as a byproduct when bituminous coal is distilled. G

**coal type**
A variety of coal, such as common banded coal, cannel coal, algal coal, and splint coal. The distinguishing characteristics of each type of coal arise from the differences in the kind of plant material that produced it. D

**coal washing**
The process of separating particles of various sizes, densities, and shapes by allowing them to settle in a fluid. D

**co-conversion or co-combustion**
Biomass blended with fossil fuel such as pellated or briquetted sawdust with coal, which may have advantages of hotter flame, will dilute pollutants (sulfur) in coal and provide a more complete combustion of biomass. Y

**cod**
Chemical oxygen demand.

**COE**
Corps of Engineers.

**coefficient of expansion**
The ratio of the increase of length, area, or volume of a body for a given rise in temperature to the original length, area, or volume of the body. M

**coefficient of friction**
The quotient which is obtained when the force just necessary to move two solid bodies past each other (static friction) or the force necessary to keep them moving at a steady rate (sliding friction) is divided by the force pressing the solids together. M

**coefficient of haze (COH)**
A measurement of visibility interference in the atmosphere. E

**coffin**
A heavily shielded shipping cask for spent (used) nuclear fuel elements. Some coffins weigh as much as 75 tons. O

**cogeneration**
The process in which fuel is used to produce heat for boiler-turbine or gas for turbine, to drive a generator to produce electricity, with the rejected heat used for process heat. Y

**COH**
Coefficient of haze.

**cohesion**
That force or interattraction by which molecules of the same kind or of the same body are held together so that the body resists being pulled apart. M

**coh unit**
A measure of light absorption by particles, defined as that quantity of light scattering solids producing an optical density of 0.1. E

**coke**
(1) The residue of coal left after destructive distillation and used as fuel. (2) A similar residue left by other materials (as petroleum) distilled to dryness.

**coking**
(1) Any cracking process in which the time of cracking is so long that coke is produced as the bottom product. (2) Thermal cracking for conversion of heavy, low-grade oil into lighter products and a residue of coke. (3) The undesirable building up of coke or carbon deposits on refinery equipment. M

**coking coal**
The most important of bituminous coals, which burns with a long yellow flame, giving off more or less smoke, and creates an intense heat when properly attended. It is usually quite soft, and does not bear handling well. In the fire, it swells, fuses, and finally runs together in large masses, which are rendered more or less porous by the evolution of the contained gaseous hydrocarbons. D

**cold point**
The temperature at which any given grade of oil will freeze or become cloudy. BB

**cold reserve**
Thermal generating capacity available for service but not maintained at operating temperature. L

**collector efficiency**
The output (energy collected) divided by the input (the solar energy falling on the collector surface) within a specified period of time. H

**collector plates**
The primary function of the solar collector plates is to absorb as much as possible of the radiation reaching it through the glazing, to lose as little heat as possible upward to the atmosphere and downward through the back of the container, and to transfer the retained heat to the transport fluid. R

**collimator**
A device for focusing or confining a beam of particles or radiation such as X-rays. O

**Collins miner**
A type of remote-controlled continuous miner for thin seam extraction. D

**collision**
A close approach of two or more particles, photons, atoms or nuclei, during which such quantities as energy, momentum and charge may be exchanged. O

**colloid**
A substance composed of extremely small particles, ranging from 0.2 micron to 0.005 micron which when mixed with a liquid will not settle, but remain permanently suspended; the colloidal suspension thus formed has properties that are quite different from the simple, solid-liquid mixture of a solution. D

**colloidal fuel**
A mixture of finely pulverized coal and fuel oil, which remains homogeneous in storage. It has a high calorific value and is used in oil-fired boilers as substitute for fuel oil alone. D

**combination utility**
Utility which supplies more than one type of utility service (electricity, water, gas, etc.). G

**combined cycle**
Designation of a power plant which uses two technologies to generate electricity. Usually applied to gas-turbine and steam-turbine combinations. Z

**combustibility**
An assessment of the speed of combustion of a coal under specified conditions. D

**combustible schist**
Another name for carbonaceous or bituminous shale. D

**combustion**
(1) The process of burning. (2) Rapid oxidation caused by the union of the oxygen of the air with a material. M

**commodity price adjustment clause**
Clause in a rate schedule that provides for an adjustment of the customer's bill if the price of commodities or index of commodity prices varies from a specified standard. L

**common banded coal**
The common variety of bituminous and subbituminous coal. D

**common carriers**
Legal classification of oil pipelines engaged in interstate commerce. Interstate pipelines are required to carry the oil of any company, provided it meets the conditions of the pipeline's tariffs. More commonly used with regard to truckers engaged in interstate carriage. M

**common plant**
Utility plant used by a utility company which renders more than one utility service, such as electric and gas, to such an extent and in such a manner as to render segregation impractical, as would be the case of a garage housing electric and gas utility trucks. K

**community noise equivalent level**
Community noise equivalent level (CNEL) is a scale which accounts for all the A-weighted acoustic energy received at a point, from all noise events causing noise levels above some prescribed value. Weighting factors are included which place greater importance upon noise events occurring during the evening hours (7:00 p.m. to 10:00 p.m.) and even greater importance upon noise events at night (10:00 p.m. to 6:00 a.m.). E

**company use**
Kilowatt hours used by an electric utility company or the electric department of a combination utility company in the operation of its business, exclusive of station use and energy lost and unaccounted for. K

**composting**
A controlled process of degrading organic matter by microorganisms. (1) Mechanical—a method in which the compost is continuously and mechanically mixed and aerated. (2) Ventilated cell—compost is mixed and aerated by being dropped through a vertical series of ventilated cells. (3) Windrow—an open-air method in which compostable material is placed in windrows, piles, or ventilated bins or pits and occasionally turned or mixed. The process may be anaerobic or aerobic. E

**compressed air**
Air compressed in volume. D

**compression**
In general, the act of increasing the pressure on a fluid. It is usually attended by a reduction in volume. M

**compression pressure**
The pressure of the gases in the cylinder of an internal-combustion engine at the end of the compression stroke. M

**compression process**
The recovery of natural gasoline from "wet natural gas" by compression and condensation rather than by absorption. M

**compressor**
A machine which draws in air or other gas, compresses it, and discharges it at a higher pressure. M

**compressor station**
(1) Any permanent combination of facilities which supplies the energy to move gas in transmission lines or into storage by increasing the pressure. (2) For reporting to A.G.A., includes field (other than temporary), line, and storage compressor stations of producing and transmission companies; distribution companies include underground storage compressor stations and stations used for transporting gas between local distribution systems but not booster or pumping stations used within a local distribution system. DD

**Compton effect**
Elastic scattering of photons (X-rays or gamma rays) by electrons. In each such process the electron gains energy and recoils, and the photon loses energy. This is one of three ways photons lose energy upon interacting with matter. It is named for A. H. Compton, American physicist, who discovered it in 1923. O

**concentrating collectors**
Collectors where large amounts of solar radiation are concentrated upon a relatively small collection area whereby temperatures higher than those attainable by flat-plate collectors are reached. Conventional concentrating collectors must track the same during its daily motion. The principal use of concentrating collectors has been in the production of steam or high temperature fluids for use in refrigeration or power generation. R

**concentrator**
A device or structure that concentrates windstreams or solar rays. U

**condensate**
(1) The liquid resulting when a vapor is subjected to cooling and/or pressure reduction. (2) Liquid hydrocarbons condensed from gas and oil wells. DD

**condensation**
(1) In physics and engineering, the act or process of changing a vapor to a liquid, or a lighter liquid to another and denser form, by depression of temperature or increase of pressure. (2) In chemistry, a reaction involving union between atoms in the same or different molecules (often with elimination of water or other unimportant byproduct) to form a new compound of greater complexity and, frequently, greater molecular weight or density. M

**condensation nucleus**
A particle on which condensation of water vapor begins in the free atmosphere, where it invariably takes place on hydroscopic dust or hydroscopic gases. The common sources of the latter are sea salt, products of combustion, and the dust blown from the earth's surface. E

**condenser**
Ordinarily, a water-cooled heat exchanger used for cooling and liquefying oil vapors. Where the cooling medium used is air, the condenser is called an air condenser. M

**conduction**
The transfer of heat through matter by the communication of kinetic energy from particle to particle rather than by a flow of heated material. D

**conduit bank**
A conduit bank is a length of one or more underground conduits or ducts (whether or not enclosed in concrete) designed to contain underground

cables. A gallery or cable tunnel for power cables is generally treated as a conduit bank for property reporting purposes. K

**confinement time**
The time (in fusion) during which the reacting materials (deuterium and tritium, for instance) are physically confined at proper density to react. V

**connected load**
Sum of the ratings of the electric power consuming apparatus connected to the system, or part of the system, under consideration. C

**connection charge**
An amount to be paid by a customer in a lump sum, or in installments, for connecting the customer's facilities to the supplier's facilities. L

**conservation**
(1) Conserving, preserving, guarding, or protecting; keeping in a safe or entire state. (2) Using in an effective manner or holding for necessary uses, as mineral resources. D

**conservation of energy**
The total energy of an isolated system remains constant irrespective of whatever internal changes may take place, energy disappearing in one form and reappearing in another. D

**constant dollars**
(1) Dollars of constant purchasing power. (2) The value of a dollar during a specified year. C

**consumer price index (BLS)**
This index issued by the U.S. Department of Labor, Bureau of Statistics, as a measure of average changes in the retail prices of goods and services usually bought by the families of wage earners and clerical workers living in cities. It was formerly entitled "Consumer's Price Index for Moderate-Income Families in Large Cities Combined." K

**consumption charge**
The portion of a utility charge based on energy actually consumed as distinguished from the demand charge. D

**containment**
The provision of a gastight shell or other enclosure around a nuclear reactor to confine fission products that otherwise might be released to the atmosphere in the event of an accident. O

**containment vessel**
A gastight shell or other enclosure around a nuclear reactor. O

**contiguous**
Touching without fusion, applicable whether the parts are alike or unlike. D

**Continental shelf**
The extension of the continental land mass into the oceans, under relatively shallow seas, as opposed to the deeper basins. G

**continuous miner**
A mining machine designed to remove coal from the face and to load that coal into cars or conveyors without the use of cutting machines, drills, or explosives. D

**contour mining**
Strip mining in mountainous terrain with steep slopes. Land is extremely difficult to restore, if not impossible, in such regions.

**contract drilling**
Drilling work done for a company or person by a person or company furnishing the drilling equipment and labor for a specified cost, which is based on the amount and type of work. D

**control rod**
A rod, plate, or tube containing a material that readily absorbs neutrons (hafnium, boron, etc.), used to control

the power of a nuclear reactor. By absorbing neutrons, a control rod prevents the neutrons from causing further fission. O

**controlled thermonuclear reaction**
Controlled fusion, that is, fusion produced under research conditions, or for production of useful power. O

**CONUS**
Continental United States.

**convection**
The transfer of heat by means of the upward motion of the particles of a liquid or a gas which is heated from beneath. D

**conventional fuels**
The fossil fuels: coal, oil or gas. K

**conventional mining**
The cycle of operations which includes cutting the coal, drilling the shot holes, charging and shooting the holes, loading the broken coal, and installing roof support. Also known as cyclic mining. D

**conversion efficiency**
The percentage of total thermal energy that is actually converted into electricity by an electric generating plant. X

**conversion factor**
A numerical constant by which a quantity with its value expressed in units of one kind is multiplied to express the value in units of another kind. E

**conversion of energy**
The principle that the total amount of energy in an isolated system remains unchanged while internal changes of any kind occur. E

**conversion processes**
Processes by which refiners are able to produce gasoline from groups of hydrocarbons that are not normally in the gasoline range. The basic conversion processes are thermal cracking, catalytic cracking, and polymerization. M

**converter**
(1) An apparatus for transforming the quality or quantity of electrical energy. (2) A term formerly applied to the transformer, but now restricted to a machine utilizing mechanical rotation. D

**converter reactor**
(1) A reactor that produces some fissionable material, but less than it consumes. (2) In some usages, a reactor that produces a fissionable material different from the fuel burned, regardless of the ratio. (3) In both usages the process is known as conversion. O

**coolant**
A substance circulated through a nuclear reactor to remove or transfer heat. Common coolants are water, air, carbon dioxide, liquid sodium and sodium-potassium alloy (NaK). O

**cooling tower**
A device to remove excess heat from water used in industrial operations, notably in electric power generation. E

**COP**
Coefficient of perfomance.

**core**
(1) The central portion of a nuclear reactor containing the fuel elements and usually the moderator, but not the reflector. (2) In mining, cylindrical sample of rock obtained in core drilling. O

**core oil**
An oil compound having specific properties rendering it suitable for use in making foundry cores. M

**Coriolis effect (Coriolis force)**
The diverting force of the earth's rotation which causes horizontally moving water or air particles to be diverted towards the right in the northern hemisphere and towards the left in the southern hemisphere. E

**corrosion**
The gradual eating away of metallic surfaces as the result of oxidation or other chemical reaction. It is caused by acids or other corrosive agents. M

**cosmic rays**
Radiation of many sorts but mostly atomic nuclei (protons) with very high energies, originating outside the earth's atmosphere. Cosmic radiation is part of the natural background radiation. O

**Cottrell precipitator**
A piece of equipment designed for the removal of dusts or mists from gases by passing them through an electrostatic field. M

**coulomb**
The coulomb is the quantity of electricity transported in 1 second by a current of 1 ampere. E

**counter**
A general designation applied to radiation detection instruments or survey meters that detect and measure radiation in terms of individual ionizations, displaying them either as the accumulated total or their rate of occurrence. O

**course stacking**
The method of shovel operation in which no ground is hauled away. The shovel simply stacks the ground on the opposite side from the working cut, or it may turn entirely around, dumping the spoil on a bank behind. D

**CPA**
Canadian Petroleum Association, or certified public accountant.

**CPI**
Consumer price index.

**cpm**
Cycles per minute.

**cps**
(1) Cycles per second. (2) Conventional power station.

**CPSC**
Consumer Product Safety Commission.

**cracking**
A process in which relatively heavy hydrocarbons (such as fuel oils and naphthas from petroleum) are broken up into lighter products (such as gasoline and ethylene) by means of heat, and usually pressure, and sometimes catalysts. D

**cracking plant**
An oil refinery. G

**crankshaft**
The entire engine shaft that converts reciprocating motion and force of pistons and connecting rods to rotary motion and torque. D

**CRBR**
Clinch River breeder reactor.

**creep**
A slow movement of rock debris or soil due to gravity, down a slope for instance. It is usually imperceptible except to observations over a long period. D

**creosote**
A colorless to yellowish oily liquid compound consisting of a mixture of phenols distilled from wood, and having a smoky odor and burning taste. D

**CRG**
Catalytic rich gas.

**cribbing**
The construction of cribs, or timbers laid at right angles to each other, sometimes filled with earth, as a roof support in a mine or as a support for machinery. D

**critical**
Capable of sustaining a nuclear chain reaction. O

**critical area**
In prospecting work, an area found to be favorable, from geological age and structural considerations. D

**critical area of extraction**
The area of coal required to be worked to cause a surface point to suffer all the subsidence possible from the extract of a given seam. D

**critical coefficient**
The ratio of the critical temperature to the critical pressure. D

**critical compression pressure**
The highest possible pressure in a fuel-air mixture before spontaneous ignition takes place. M

**critical compression ratio**
The ratio of the volume or fuel-air mixture at critical compression pressure to the maximum volume of the mixture in the cylinder before compression. M

**critical density**
The density of a substance at its critical temperature and under its critical pressure. D

**critical depth**
A given quantity of water in an open conduit may flow at two depths having the same energy head. When these depths coincide, the energy head is a minimum and the corresponding depth is Belanger's critical depth. D

**critical flow**
(1) A condition of flow in which the mean velocity is at one of the critical values, ordinarily at Belanger's critical depth and velocity. (2) Used in reference to Reynold's critical velocities which define the point at which the flow changes from streamline or nonturbulent flow. (3) The maximum discharge of a conduit which has a free outlet or getaway and has the water ponded at the inlet. E

**critical mass**
The smallest mass of fissionable material that will support a self-sustaining chain reaction under stated conditions. O

**critical point**
The point at which the properties of a liquid and its vapor become indistinguishable. It is generally synonymous with critical temperature. D

**critical potential**
A potential which produces a sudden change in magnitude of the current. D

**critical pressure**
The pressure necessary to condense a gas at the critical temperature, above which temperature the gas cannot be liquified, no matter what pressure is applied. M

**critical slope**
The maximum angle with the horizontal at which a sloped bank of soil or given height of soil will stand unsupported. D

**critical state**
That particular pressure and temperature at which liquid or gaseous phases reverse at the slightest change in conditions. M

**critical temperature**
The maximum temperature at which a gas can be liquefied by pressure (critical pressure); above this temperature the gas cannot be liquefied, no matter what pressure is applied. M

**critical velocity**
Reynold's critical velocity is that at which the flow changes from laminar to turbulent, and where friction ceases to be proportional to the first power of the velocity and becomes proportional to a higher power—practically the square. D

**critical viscosity**
The viscosity at which measurable friction begins an abrupt increase as a result of changing conditions, e.g., pressure and speed. M

**critical volume**
The specific volume of a substance in its critical state. D

**criticality**
The state of a nuclear reactor when it is sustaining a chain reaction. O

**crop coal**
Coal of inferior quality near the surface. D

**cropping**
Coal cutting beyond the normal cutting plane. D

**cross section**
(1) A profile portraying an interpretation of a vertical section of the earth explored by geophysical and/or geological methods. (2) In the nuclear field, a measure of the probability that a nuclear reaction will occur. Usually measured in barns, it is the apparent (or effective) area presented by a target nucleus (or particle) to an on-coming particle or other nuclear radiation, such as a photon or gamma radiation. O

**cross-wind horizontal-axis rotors**
Windmills in which the axis of rotation is both horizontal to the surface of the earth and perpendicular to the direction of the windstream, somewhat like a water wheel. U

**CRT**
Cathode ray tube.

**crude**
(1) A substance in its natural unprocessed state, crude ore or crude oil, for example. (2) In a natural state; not cooked or prepared by fire or heat; not altered or prepared for use by any process; not refined. Synonym for raw. See crude oil. D

**crude oil**
(1) Petroleum in its natural unprocessed, unrefined state. (2) A mixture of thousands of different hydrocarbons—compounds of hydrogen and carbon. The mixture varies widely from one oil field to another. Crude oil's basic unit is a molecule of one carbon atom linked with four hydrogen atoms. This is the molecule of methane. Theoretically, millions of variations on this are possible, and millions of different hydrocarbon compounds can be formed. M

**crude ore**
The unconcentrated ore as it leaves the mine. D

**crude shale oil**
The oil obtained as a distillate by the destructive distillation of oil shale. D

**cryogenics**
The study and production of very low temperatures and their associated phenomena. G

**crystallization**
A separation process of oil refining in which cooling the mixture causes some of its compounds to solidify or crystallize and separate out of the liquid. M

**CSC**
Civil Service Commission.

**CSG**
Consolidated synthetic gas.

**CSRS**
Cooperative State Research Service.

**CTR**
Controlled thermonuclear reactor.

**cu**
Cubic.

**cubic**
(1) Having the form of a cube, as a cubic crystal. (2) Referring to directions parallel to the faces of a cube, as cubic cleavage. D

**cubic foot**
(1) The most common unit of measurement of gas volume. (2) The amount of gas required to fill a volume of one cubic foot under stated conditions of temperature, pressure, and water vapor. DD

**cubic foot per second**
(1) A unit of measurement of flowing liquids, equal to a flow of 1 cubic foot per second past a given section of channel. Also, called "second foot." (2) Known as "cusec" in Great Britain and other countries. E

**culm**
In anthracite terminology, the waste accumulation of coal, bone and rock from old dry breakers. G

**curie [ Symbol c ]**
(1) The basic unit to describe the intensity of radioactivity in a sample of material. The curie is equal to 37 billion disintegrations per second, which is approximately the rate of decay of 1 gram of radium. (2) A quantity of any nuclide having a curie of radioactivity. Named for Marie and Pierre Curie, who discovered radium in 1898. O

**curie point; curie temperature**
(1) The temperature at which there is a transition in a substance from one phase to another of markedly different magnetic properties. (2) Specifically, the temperature at which there is a transition between the ferro-magnetic and paramagnetic phases. D

**current**
(1) The part of a fluid body (as water or air) moving continuously in a certain direction. (2) Commonly used to express the flow of electricity in a wire.

**current density**
The amount of current per unit area of electrode. D

**current dollars**
Dollars that have not been reduced to a common basis of constant purchasing power but instead reflect, e.g., anticipated future inflation. They cannot be used for computations unless the assumed inflation rate is stated. C

**current efficiency**
(1) The proportion of current used in a given process to accomplish a desired result. (2) In electroplating, the proportion used in depositing or dissolving metal. D

**current gas (top gas)**
The total volume of gas in a storage reservoir which is in excess of the cushion gas. DD

**current meter**
A device for determining the velocity of moving water or fluid. E

**cushion gas**
The total volume of gas which will maintain the required rate of delivery during an output cycle. DD

**customer charge**
An amount to be paid periodically by a customer without regard to demand or energy consumption. L

**cut**
(1) To intersect a vein or working. (2) To excavate coal. D

**cutie pie**
A common radiation survey meter used to determine exposure levels or to locate possible radiation hazards. O

**cut-in speed**
The wind speed at which a wind machine is activated, as the wind speed increases. U

**CVCS**
Chemical and volume control system.

**CWA**
Clean Water Act.

**cycle**
(1) In the oil industry, a series of events (or processes) which repeat themselves. (2) In discontinuous chemical processes, total time from the start of one reaction period to the start of the next reaction period. Includes reaction period, regeneration period, and the periods required for purging and valve changes. (3) In the electrical industry, in one cycle of alternating electric current the current goes from zero potential or voltage to a maximum in one direction, back to zero, then to a maximum in the other direction and then back again to zero. The number of such complete cycles made each second determines the frequency of the current. Direct current does not fluctuate from positive to negative and hence, cycles or frequency can apply only to alternating current. M

**cyclotron**
A particle accelerator in which charged particles receive repeated synchronized accelerations by electrical fields as the particles spiral outward from their source. The particles are kept in the spiral by a powerful magnetic field. O

**cylinder**
The tubular barrel portion of a piston engine within which the piston moves. M

**cylinder oil**
A viscous lubricating stock, used for lubricating the cylinders and valves of steam engines. M

# D

**d**
Abbreviation for density, specific volume, dyne, deuteron, or differential of. D

**DAF**
Dry ash-free.

**Dalton's Law**
In a mixture of gases, the total pressure is equal to the sum of the pressures that the gases would exert separately. D

**dam**
(1) A barrier to keep foul air or water, from mine workings. (2) A retaining wall or bank for water. D

**damp**
Any mine gas, or mixture of gases, particularly those deficient in oxygen. D

**damping**
(1) In seismology, a resistance, contrary to friction, independent of the nature of the contacting surface. (2) A force opposing vibration, damping acts to decrease the amplitudes of successive free vibrations. Damping may result from internal friction within the system, from air resistance, or from mechanical or magnetic absorbers. D

**Darcy's Law**
A statement in fluid dynamics: the velocity of flow of a liquid through a porous medium due to difference in pressure is proportional to the pressure gradient in the direction of flow. D

**data processing**
A series of planned actions motivated by input signals for a defined purpose. D

**datum**
A point, line or surface with reference to which positions (as elevations) are measured or indicated. D

**daughter**
A nuclide formed by the radioactive decay of another nuclide, which in this context is called the parent. W

**db**
Decibel. The unit for measuring sound intensity.

**dc**
Direct current.

**DCA**
Defense Communications Agency.

**DCF ROR**
Discounted cash flow rate of return.

**DDC**
Direct digital control.

**dead oil**
An oil with a density greater than water, which is distilled from tar. M

**dead-weight tons (dwt)**
The total lifting capacity of a ship expressed in long tons (2,240 lbs.). For example, the oil tanker Universe Ireland is listed as 312,000 dwt. which means it can carry 312,000 tons of oil or about 1.9 million barrels. G

**debris**
(1) Rock fragments, sand, earth, and sometimes organic matter, in a heterogeneous mass, as at the foot of a cliff. (2) The silt, sand, and gravel that flow from hydraulic mines; referred to by miners as tailings, slums, and sometimes slickens. D

**debutanization**
Distillation to separate butane and lighter components of the feed from pentane and heavier components. M

**decay chain**
A radioactive series. W

**decay heat**
The heat produced by the decay of radioactive nuclides. W

**decay, radioactive**
The spontaneous transformation of one nuclide into a different nuclide or into a different energy state of the same nuclide. The process results in a decrease, with time, of the number of the original radioactive atoms in a sample. It involves the emission from the nucleus of alpha particles, beta particles (or electrons) or gamma rays; or the nuclear capture or ejection of orbital electrons; or fission. Also called radioactive disintegration. O

**decibel**
The unit on a logarithmic scale for measuring sound intensity. Named in honor of Alexander Graham Bell (1847-1922). When sound or noise is created it gives off energy which is measured in decibels. D

**declination**
The angle which the magnetic needle makes with the geographic meridian. It is said to be east or west, depending on which way the north end of the needle points, east or west of the geographic meridian.

**decoking**
The removal of petroleum coke from equipment such as coking drums. Hydraulic decoking uses high-velocity water streams as the cutting means. M

**decommissioning**
Process of removing a facility or area from operation and decontaminating and/or disposing of it or placing it in a condition of standby with appropriate controls and safeguards. P

**decomposition potential**
The minimum potential difference necessary to decompose the electrolyte of the cell. D

**decomposition value**
Minimum voltage at which continuous electric current flows through an electrolytic solution of normal strength. D

**decontamination**
The removal of radioactive contaminants from surfaces or equipment, as by cleaning and washing with chemicals. O

**de-energize**
To disconnect any circuit or device from the source of power. D

**deep mining**
The exploitation of coal or mineral deposits at depths exceeding about 3,000 feet. It would appear that the deepest coal mine in the world is the Rieu de Coeur colliery at Quaregnan,

Belgium (4,462 ft.) with a rock temperature of 126° F, and it is planned to go even lower to 4,650 feet. D

**definite proportions law**
A chemical compound always contains the same elements in the same proportions by weight. D

**deflagrate**
(1) To burn, burst into flame. (2) To burn rapidly with a sudden evolution of flame and vapor. D

**deflagration**
An explosive combustion reaction that moves through a mixture of reactants at a speed less than that of sound in the mixture; when unconfined, a deflagration may or may not produce significant overpressure. D

**deflation**
Removal of loose material by the wind, leaving the rocks bare to the continuous attack of the weather. D

**deflocculant**
Any organic or inorganic material which is used as an electrolyte to disperse nonmetallic or metallic particles in a liquid, (that is, basic materials such as calgonate, sodium silicate, soda ash, etc., are used as deflocculants in clay slips). D

**deformed point**
The temperature observed during the measurement of expansivity by the interferometer method at which viscous flow exactly counteracts thermal expansion. D

**deg**
Degree (angular measurement).

**degasification**
Progressive loss of gases in a substance leading to the formation of a more condensed product. Applied primarily to the formation of solid bitumens from liquid bitumens, but also used in connection with coal formation. D

**degradation**
The excessive crushing of coal during cutting, loading, and transportation. D

**degree**
A unit space or a unit division marked on various instruments, as thermometers and astronomical instruments. D

**degree-day**
A unit measuring the extent to which the outdoor mean (average of maximum and minimum) daily dry-bulb temperature falls below (in the case of heating) or rises above (in the case of cooling) an assumed base. The base is normally taken as 65° F for heating and for cooling unless otherwise designated. One degree-day is counted for each degree of deficiency below (for heating) or excess over (for cooling) the assumed base, for each calendar day on which such deficiency or excess occurs. K

**degree Engler**
A measure of viscosity. The ratio of the time of flow of 200 ml of the liquid tested, through the viscosimeter devised by Engler, to the time required for the flow of the same volume of water gives the number of degrees Engler. M

**degree-hour**
The product of 1 hour and (usually) the number of °F the hourly mean temperature is above a base point, usually 85°F. The degree hour is used in measuring, roughly, the cooling load in summer for cases where process heat, heat from human beings, and humidity are relatively unimportant as compared with dry-bulb temperature. D

**degrees Kelvin**
(1) Absolute temperature on the centigrade scale. (2) Degrees C plus 273.16. D

**degrees Rankine**
(1) Absolute temperature on the Fahrenheit scale. (2) Degrees F plus 459.6. D

**deionization**
Removal of ions from solution by chemical means. D

**DEMA**
Diesel Engine Manufacturers Association.

**demand**
The rate at which electric energy is delivered to or by a system, part of a system, or a piece of equipment expressed in kilowatts, kilovolt-amperes or other suitable unit at a given instant or averaged over any designated period of time. The primary source of "demand" is the power-consuming equipment of the customers. In the gas industry, the rate at which gas is delivered to or by a system, part of a system, or a piece of equipment, expressed in cubic feet or therms or multiples thereof, for a designated period of time called the demand interval. K

**demand charge**
That part of a utility service charged for on the basis of the possible demand as distinguished from the energy actually consumed. D

**demand factor**
The ratio of the maximum demand on an electric generating and distributing system to the total connected load of such a system; usually expressed in percentage. E

**demand interval**
The period of time during which the electric energy flow is averaged in determining demand, such as 60-minute, 30-minute, 15-minute, or instantaneous. K

**demand power; peak power**
The maximum amount of energy consumed in any consecutive number of minutes, for example, 15 or 30 minutes, during the month. Demand is measured in kilowatts and is the average rate of consumed energy during the peak period. D

**demethanization**
The process of distillation in which methane is separated from the heavier components. Also called demethanation. M

**denatured alcohol**
Ethyl alcohol to which has been added a small amount of some compound which renders the alcohol unfit for human consumption and, hence, free from the internal revenue tax. M

**density**
The amount of matter per unit volume, expressed in pounds per cubic foot. M

**density, dry**
The weight (dry) of a substance per unit volume at a stated temperature.

**density of gases**
The vapor density of a gas, or its density relative to hydrogen, is the number of times a volume of the gas is heavier than the same volume of hydrogen, the volume of both gases being at the same temperature and pressure. D

**DEPA**
Defense Electric Power Administration.

**dependable capacity**
The load-carrying ability for the time interval and period specified when related to the characteristics of the load to be supplied. Dependable capacity of a station is determined by such factors as capability, operating power factor, and portion of the load which the station is to supply. K

**depleted uranium**
Uranium having less uranium-235 atoms than found in nature, which is

0.71 percent. Depleted uranium is a byproduct of the enrichment process. G

**depletion allowance**
A tax allowance extended to the owner of exhaustible resources based on an estimate of the permanent reduction in value caused by the removal of the resource. G

**depolymerization**
A change of long molecules, such as in coal, into single molecules such as aromatics, usually accompanied by substitution of hydrogen or oxygen done under heating processes. Y

**deposit**
(1) Anything laid down. Formerly applied only to matter left by the agency of water, but now includes mineral matter in any form that is precipitated by chemical or other agencies, as the ores in veins. (2) Mineral deposit or ore deposit is used to designate a natural occurrence of a useful mineral, or an ore, in sufficient extent and degree of concentration to invite exploitation. D

**depth of stratum**
The vertical distance from the surface of the earth to a stratum. D

**derived fuel**
A fuel obtained from a raw fuel by some process of preparation for use; for example, coke, charcoal, benzol, and petrol. D

**desalting; desalinizing**
(1) Any process for making potable water from sea water or other salt waters. Distillation is the oldest method. Electrodialysis is an inherently good method because the energy is used to remove the small proportion of salt from the relatively large quantity of water instead of removing the water from the salt. Other methods are freezing by direct contact of refrigerant with sea water; foam separation; liquid-liquid extraction; various nonelectric membrane processes; and ion exchange. (2) In the oil industry, removal of mineral salts (mostly chlorides) from crude oils. D

**desiccation**
(1) A drying out; used in connection with sediments that lose water. (2) Applied to the process of evaporation from bodies in arid regions, thus producing evaporites. D

**design day**
A 24-hour period of the greatest theoretical gas demand, used as basis for designing purchase contracts, and/or production facilities, and/or delivery capacity. DD

**design horsepower**
(1) The specified horsepower for a chain drive multiplied by a service factor. It is the value used to select the chain size for the drive. (2) Used to designate the horsepower for which a piece of equipment is designed. D

**design load**
The load generally taken as the worst combination of forces and loads which a structure is calculated to sustain. The term is similarly applied to such projects as air conditioning. D

**design temperature**
The temperature which an apparatus or system is designed to maintain or operate against under the most extreme conditions. The former is the inside design temperature; the latter, the outside design temperature. D

**design voltage**
The nominal voltage for which a line or piece of equipment is designed. This is a reference level of voltage for identification and not necessarily the precise level at which it operates. K

**desorption**
(1) The reverse process of adsorption whereby absorbed matter is removed

from the adsorbent. (2) The reverse process of absorption. D

**destressed area; destressed zone**
In strata control, a term used to describe an area where the force is much less than would be expected after considering the depth and type of strata. D

**destressing**
In deep mining, relief of pressure on abutments of excavation. D

**destructive distillation**
The heating process of fossil fuels, hydrocarbons, or carbohydrates or other solids accompanied by their decomposition. The destructive distillation of coal yields coke, tar, ammonia, and gas. The destructive distillation of biomass yields char and gas. Water present is vaporized. Y

**desulfurization**
The removal of sulfur or sulfur compounds. M

**desulfurize**
(1) To free from sulfur. (2) To remove the sulfur from an ore or mineral by some suitable process, as by roasting. D

**detector**
Material or a device that is sensitive to radiation and can produce a response signal suitable for measurement or analysis. A radiation detection instrument. There are numerous types of detectors that are sensitive to various stimuli. O

**detergency**
The ability of a substance to clean and to wash away undesirable substances. Detergents may either be oil-soluble or water-soluble. Soap and synthetic detergents help to wet, disperse, and deflocculate solid particles. Oil-soluble detergents are used in motor oils to disperse, loosen, and remove carbon, dirt, etc. from interior surfaces of internal-combustion engines. M

**detergent oil**
A lubricating oil possessing special sludge-dispersing properties for use in internal-combustion engines. These properties are usually conferred on the oil by the incorporation of special additives. Detergent oils hold sludge particles in suspension and thus promote engine cleanliness. M

**detergents**
Gasoline additives which remove and prevent deposits in fuel systems and carburetors. M

**detonation**
(1) Sharp explosion. (2) The term used to describe the knock-producing type of combustion in spark-ignition, internal-combustion engines. M

**detrital**
Descriptive of minerals occurring in sedimentary rocks that were derived from pre-existing igneous, sedimentary, or metamorphic rocks. Synonym for clastic; allogenic. D

**Deuterium [Symbol $H^2$ or D]**
An isotope of hydrogen whose nucleus contains one neutron and one proton and is therefore about twice as heavy as the nucleus of normal hydrogen, which is only a single proton. Deuterium is often referred to as heavy hydrogen; it occurs in nature as 1 atom to 6500 atoms of normal hydrogen. It is nonradioactive. O

**deuteron**
The nucleus of deuterium. It contains one proton and one neutron. O

**development wells**
Wells drilled within an area that has already proved productive of oil or natural gas. M

**device, nuclear**
A nuclear explosive used for peaceful purposes, tests or experiments. The term is used to distinguish these explosives from nuclear weapons, which are packaged units ready for transportation or use by military forces. O

**dewatering coal**
The removal of moisture from coal after it has passed through the washer. D

**dewpoint**
(1) The temperature to which air must be cooled, at constant pressure and constant water vapor content, in order for saturation or condensation to occur. (2) The temperature at which the saturation pressure is the same as the existing vapor pressure. Also called saturation point. D

**dewpoint hygrometer**
An instrument for determining the dewpoint; a type of hygrometer. D

**dialysis**
The separation of substances in solution by means of semipermeable membranes (as of parchment, cellophane, or living cells) through which the smaller molecules and ions diffuse readily but through which the larger molecules and colloidal particles diffuse very slowly or not at all. Such separations are important in nature. D

**diastrophism**
The process of deformation that produces in the earth's crust its continents and ocean basins, plateaus and mountains, folds of strata, and faults. D

**diatomaceous earth**
(1) A fine siliceous material resembling chalk used in waste water treatment plants to filter sewage effluent to remove solids. (2) Also used as inactive ingredients in pesticide formulations applied as dust or powder; diatomite. E

**diatomic**
Consisting of two atoms; having two atoms in a molecule. D

**dielectric**
A material which offers relatively high resistance to the passage of an electric current but through which magnetic or electrostatic lines of force may pass. Most insulating materials, for example, air, porcelain, mica, and glass, are dielectrics and a perfect vacuum would constitute a perfect dielectric. D

**diesel engine**
Developed in the late 1800's by Dr. Rudolph Diesel, the diesel engine is fundamentally different from gasoline engines. The ignition is caused by the heat of compressed air in a cylinder, not by a spark plug as in gasoline engines. Modern diesel fuels are carefully manufactured in several grades, ranging from heavy oils to light kerosene-type oils. M

**dieseling**
In a compressor, explosions of mixtures of air and lubricating oil in the compression chambers or other parts of the air system. D

**difference of potential**
The difference in electrical pressure existing between any two points in an electrical system or between any point of such a system and the earth. Determined by a voltmeter. D

**diffuser**
A device or structure that diffuses a windstream. U

**diffuse radiation**
The solar radiation which strikes an object after the radiation has been

scattered and diverted by particles in the atmosphere (non-directional in nature). H

**diffusion**
The permeation of one substance through another, such as gas through gas, liquid, or solid; solute through solvent; liquid through liquid or solid; and finally solid through solid. The pressure corresponding to that exerted by dissolved material in its diffusions from a more concentrated to a less concentrated part of a solution is called osmotic pressure. D

**digital computer**
Machine which makes mathematical computations by methods in which digits are added or subtracted in accordance with the coding signals to which the machine is sensitive. D

**diluvium**
Sand, gravel, clay, etc., in surfical deposits. Formerly, according to some authors, alluvium was the effect of the ordinary, and diluvium was the effect of the extraordinary action of water. D

**dipole**
Coordinate valence link, between two atoms. Electrical symmetry of a molecule. When a molecule is formed by sharing of two electrons between a donor atom and an acceptor, it is more positive at the donor end and more negative at the acceptor end, and has a dipole moment. D

**direct current (D-C) (dc)**
Electricity that flows continuously in one direction as contrasted with alternating current. K

**direct-cycle reactor system**
A nuclear power plant system in which the coolant or heat transfer fluid circulates first through the reactor and then directly to a turbine. O

**direct energy conversion**
The generation of electricity from an energy source in a manner that does not involve transference of energy to a working fluid. Direct conversion methods have no moving parts and usually produce direct current. Some methods include thermoelectric conversion, thermionic conversion and magnetohydrodynamic conversion. G

**directional drilling**
Rotary drilling techniques perfected to deflect the drill from the vertical in a gradual curve as the hole goes deeper. Such controlled directional drilling can tap reserves inaccessible by vertical drilling. For example, deposits lying beneath the Pacific Ocean off California have been reached by wells drilled directionally from shore. M

**direct radiation**
Radiation which passes through the atmosphere without deflection or scattering due to molecules in the air. H

**discretionary income**
The amount of a person's income that remains after he meets all his basic needs and commitments, such as food, clothing, shelter, taxes and debt payments.

**disintegration**
The breaking up and crumbling away of a rock, caused by the action of moisture, heat, frost, air, and the internal chemical reaction of the component parts of rocks when acted upon by these surface influents. D

**disintermediation**
Withdrawal of deposits from savings and loan associations and savings banks to take advantage of higher interest rates in money markets, resulting in a shortage of mortgage credit for housing.

**dispersant**
A chemical agent used to break up concentrations of organic material. In cleaning oil spills, dispersants are used to disperse oil from the water surface. E

**dispersing agent**
A material that increases the stability of a suspension of particles in a liquid medium by deflocculation of primary particles. D

**displacement**
A general term for the change in position of any point on one side of a fault plane relative to any corresponding point.

**displacement power**
Power from one generating source which displaces power from another generating source. Usually this permits power from the latter source to be transmitted to more distant loads. L

**disposal**
Emplacement or release of waste material without the intention of retrieval. Disposal may be totally irreversible or retrieval may be possible; but it is the absence of an intention to retrieve which defines disposal. In general, disposal does not require continuing surveillance, although surveillance may be appropriate for limited time periods. Z

**dissection**
The effect of erosion in destroying the continuity of a relatively even surface by cutting ravines or valleys into it. D

**dissolved gas**
Natural gas in solution in crude oil in the reservoir. G

**dissolved gas drive**
In some oil accumulations, the gas does not form a cap but remains dissolved in the oil. When the formation is opened, the gas expands and drives the mixture to the surface. This type of drive is a type of generation in flush production. M

**dissolved oxygen**
The oxygen dissolved in water or sewage. Adequately dissolved oxygen (DO) is necessary for the life of fish and other aquatic organisms and for the prevention of offensive odors. Low dissolved oxygen concentrations generally are due to discharge of excessive organic solids having high BOD, the result of inadequate waste treatment. E

**distillate**
That portion of a liquid which is removed as a vapor and condensed during a distillation process. M

**distillates**
A classification of fuel oils, distillates are the lighter oils, some of which are used in space heating, water heating, and cooking. The major market for distillates is in the automatic central heating of homes and smaller apartment houses and buildings. M

**distillation**
Vaporization of a liquid and its subsequent condensation in a different chamber. The separation of one group of petroleum constituents from another by means of volatilization in some form of closed apparatus, such as a still, by the aid of heat. Condensation of gases can be done at different temperatures to provide different substances. Y

**distribution**
The act or process of distributing electric energy from convenient points on the transmission or bulk power system to the consumers. K

**distribution line**
One or more circuits of a distribution system on the same line of poles or supporting structures, operating at relatively low voltage as compared with transmission lines. K

**distribution system**
That portion of an electric system used to deliver electric energy from points on the transmission or bulk power system to the consumers. L

**diurnal variation**
The daily variation in the earth's magnetic field. D

**diversity**
That characteristic of variety of electric loads whereby individual maximum demands usually occur at different times. Diversity among customer's loads results in diversity among the loads of distribution transformers, feeders, and substations, as well as between entire systems. K

**diversity factor**
The ratio of the sum of the non-coincident maximum demands of two or more loads to their coincident maximum demand for the same period. K

**DOC**
Department of Commerce.

**DOD**
Department of Defense.

**DOE**
Department of Energy.

**DOI**
Department of the Interior.

**DOL**
Department of Labor.

**domestic coke**
Domestic coke is normally a byproduct of coal-gas plants and commercial byproduct plants. The general characteristics of the coal, therefore, are fixed by the requirements for gas and coking coals. D

**dope**
Slang expression for additive in the oil industry. M

**Doppler effect**
A shift in the measured frequency of a wave pattern caused by movement of the receiving device or the wave source. The moving receiver will intercept more or fewer waves per unit time, depending on whether it is moving toward or away from the source of the waves. D

**dose equivalent**
(1) A term used to express the amount of effective radiation when modifying factors have been considered. (2) The product of absorbed dose multiplied by a quality factor multiplied by a distribution factor. It is expressed numerically in rems. W

**dose, permissible**
The amount of radiation which may be received by an individual within a specified period with expectation of no harmful result to himself. D

**dose rate**
The radiation dose delivered per unit time and measured, for instance, in rems per hour. O

**dosimeter**
A device that measures radiation dose, such as a film badge or ionization chamber. O

**DOT**
Department of Transportation.

**doubling dose**
Radiation dose which would eventually cause a doubling of gene mutations. W

**downwind**
On the opposite side from the direction from which the wind is blowing.

**dragline**
A type of excavating equipment which casts a rope-hung bucket a considerable distance, collects the dug material by pulling the bucket toward itself on the ground with a second rope, elevates the bucket and dumps

the material on a spoil bank, in a hopper, or on a pile. D

**draw**
(1) Strictly speaking, the distance on the surface to which the subsidence or creep extends beyond the workings. (2) The effect of creep upon the pillars of a mine. D

**drawdown**
Distance that the water surface of a reservoir is lowered from a given elevation as the result of the withdrawal of water. L

**dredge**
(1) Large floating contrivance utilized in underwater excavation for the purpose of developing and maintaining water depths in canals, rivers, and harbors. (2) Raising the level of lowland areas and improving drainage; constructing dams and dikes; removing overburden from submerged ore bodies prior to open-pit mining; or to recover subaqueous deposits having commercial value. D

**drift mine**
A coal mine which is entered directly through a horizontal opening. G

**drift mining**
The working of relatively shallow coal seams by drifts from the surface. The drifts are generally inclined and may be driven in rock or in a seam. Drift mining may be viewed as intermediate between opencase coal mining and shaft or deep mining. D

**dropping point**
The temperature at which grease passes from a semisolid to a liquid state under the conditions of ASTM Method D 566. M

**dross**
Small coal which is inferior or worthless, and often mixed with dirt. D

**dry-bulb temperature**
Temperature of air as indicated by a standard thermometer, as contrasted with wet-bulb temperature dependent upon atmospheric humidity. D

**dry-bulb thermometer**
A thermometer with an uncovered bulb, used with a wet-bulb thermometer to determine atmospheric humidity. The two thermometers constitute the essential parts of a psychrometer. D

**dry cell**
A primary cell which does away with the liquid electrolyte so that it may be used in any position. D

**drycleaned coal**
Coal from which impurities have been removed mechanically without the use of liquid media. D

**dry hole**
A drilled well which does not yield gas and/or oil in quantities or condition to support commercial production. G

**dry limestone process**
A method of controlling air pollution caused by sulfur oxides. The polluted gases are exposed to limestone which combines with oxides of sulfur to form manageable residues. E

**dry coal**
Coal containing little hydrogen. D

**dry gas**
A gas which does not contain fractions that may easily condense under normal atmospheric conditions. M

**dry point**
In a distillation test, the temperature at which the last drop of petroleum fluid evaporates. M

**dry process**
A process which dehydrates grease during the initial manufacturing stage. The exact amount of necessary water is added later. M

**dry steam**
An energy source obtained when hot water boils in an underground reservoir, steam rises, some of it condensing on surrounding rock and dry steam approaches the surface to be tapped and used in a turbine.

**dual-cycle reactor system**
A reactor-turbine system in which part of the steam fed to the turbine is generated directly in the reactor and part in a separate heat exchanger. A combination of direct-cycle and indirect-cycle reactor systems. O

**dual-fuel engine**
A diesel engine which may be operated as an oil diesel, a gas diesel, or a combination of both, as it is equipped with controls or parts to permit operating as one or the other. M

**dull coal; dulls**
Any coal which absorbs the greater part of the incident light instead of reflecting it. D

**ductile**
In mineralogy, capable of considerable deformation, especially stretching, without breaking; said of several native metals and occasionally said of some tellurides and sulfides. D

**DWT, dwt**
Dead weight tons.

**dynamic**
Forces tending to produce motion. D

**dynamic meter**
The specific work unit, $10^5$ dyne centimeters per gram, necessary to lift the unit mass 1 meter against the force of gravity. D

**dynamics**
(1) Mathematics concerned with forces not in equilibrium and, therefore, exhibiting free or potential energy. (2) Electrodynamics has to do with electrons; thermodynamics, with atoms and molecules. (3) Particle dynamics is that of moving masses. D

**dynamite**
An industrial explosive which is detonated by blasting caps. The principal explosive ingredient is nitroglycerin or specially sensitized ammonium nitrate. D

**dynamo**
A machine for converting mechanical energy into electrical energy by magneto-electric induction. A dynamo may also be used as a motor. D

**dynamometer**
An apparatus for measuring force or power, especially the power developed by a motor. It commonly embodies a spring to be compressed or a weight to be sustained by the force, combined with an index or automatic recorder to show the work performed. M

**dyne**
The absolute centimeter-gram-second unit of force, defined as that force which will impart to a free mass of one gram an acceleration of one centimeter per second per second. E

# E

**e**
Symbol for electron or sometimes for energy.

**E**
Symbol for energy.

**earth coal**
A name sometimes given to lignite. An earthy brown coal. D

**earth current**
A light electric current apparently traversing the earth's surface but which in reality exists in a wire grounded at both ends, due to small potential differences between the two points at which the wire is grounded. D

**earth oil**
Same as petroleum. D

**earthquake**
A local trembling, shaking, undulating, or sudden shock of the surface of the earth, sometimes accompanied by fissuring or by permanent change of level. Earthquakes are most common in volcanic regions, but often occur elsewhere. D

**earth's magnetic poles**
Areas in the higher latitudes where the lines of magnetic force converge. D

**earth tremor**
A slight earthquake. D

**easement**
(1) An incorporated right existing distinct from the ownership of the soil, consisting of a liberty privilege, or use of another's land without profit or compensation. (2) A right-of-way. D

**ebb current**
The movement of the tidal current away from shore or down a tidal stream. D

**EBR**
Experimental breeder reactor.

**ebulated**
A boiling bed involved in the heating of hydrocarbon or carbohydrates for gasification or liquefaction processes. Y

**ebullition**
The boiling point of a liquid. There is a rapid formation of bubbles of saturated vapor in the interior of the liquid which rise to the surface and escape. E

**ECC**
Environmental Control Council.

**eccentric**
A device for converting continuous circular motion into reciprocating rectilinear motion, consisting of a disk mounted out of center on a driving shaft, and surrounded by a collar or strap connected with a rod. Rotation of the driving shaft gives the rod a back-and-forth motion. D

**ECCS**
Emergency core cooling system.

**ECE**
Economic Commission for Europe.

**echo ranging**
Locating underwater objects by sending sound pulses into water. Target range is derived by measuring transit time of sound pulse. D

**ecological impact**
The total effect of an environmental change, either natural or man-made, on the ecology of the area. E

**ecology**
The interrelationships of living things to one another and to their environment or the study of such interrelationships. E

**economic coal reserves**
The reserves in coal seams which are believed to be workable with regard to thickness and depth. In most cases, a maximum depth of about 4000 feet is taken, and a minimum thickness of about two feet. The minimum economic thickness varies according to quality and workability. D

**economic geology**
The science of locating and processing ores. D

**ecosystem**
A complex of the community of living things and the environment forming a functioning whole in nature. W

**ECPA**
Energy Conservation and Production Act.

**ECS**
Environmental control systems.

**eddy**
A circular movement of water. Eddies may be formed where currents pass obstructions or between two adjacent currents flowing counter to each other. D

**eddy loss**
Energy lost by eddies as distinct from that lost by friction. D

**edge well**
(1) A well located at the edge of an oil or gas accumulation or at the edge of a lensed reservoir. (2) A well at or near the contact of oil and/or gas and water. D

**edging**
In forming, reducing the flange radius by retracting the forming punch a small amount after the stroke but prior to releasing the pressure. D

**EDP**
Environmental development plan.

**EDS**
Exxon Donor Solvent process.

**EEI**
Edison Electric Institute. K

**EER**
Energy efficiency ratio.

**effective capacity**
The maximum load which a machine, apparatus, device, plant or system is capable of carrying under existing service conditions. DD

**effective grounding**
In mining, effective grounding means that the path to ground from circuits, equipment, or conductor enclosures is

permanent and continuous and has carrying capacity ample to conduct safely any currents liable to be imposed upon it. D

**effective horsepower**
The amount of useful energy that can be delivered by an engine. D

**effective stack height**
The sum of three terms: (1) actual height, (2) the rise caused by the velocity of the stack gases, and (3) the rise attributable to the density difference between the stack gases and the atmosphere. E

**effervescence**
Evolution of gas in bubbles from a liquid. D

**efficiency**
That percentage of the total energy content of a fuel which is converted into power. The remaining energy is lost to the environment as heat. W

**efflorescence**
In geology, the formation of crystals by the evaporation of water from solutions brought to the surface by capillarity. D

**effluent**
A liquid, solid, or gaseous product, frequently waste, discharged or emerging from a process. D

**effusion**
That property of gases which allows them to pass through porous bodies, that is, the flow of gases through larger holes than those to which diffusion is strictly applicable. D

**EGD**
Evironmental guidance document.

**EGR**
Enhanced gas recovery.

**egress**
The provision of two or more exits from a confined space containing machinery to minimize the risk of a person being trapped in the event of an outbreak of fire or escape of steam or noxious gases, The same applies to mine workings. D

**EH & S**
Environmental health and safety.

**EHS**
Environmental Health Services, one of the Federal National Institutes of Health in the Department of Health and Human Services.

**EHV**
(1) Extra-high voltage. (2) Electric hybrid vehicle.

**EIA**
(1) Energy Information Administration, DOE. (2) Environmental impact assessment.

**Einstein equation**
See mass-energy equation.

**EIS**
Environmental impact statement.

**elastic**
(1) Capable of sustaining stress without permanent deformation. (2) The term is also used to denote conformity to the law of stress-strain proportionality. D

**elasticity**
The property or quality of being elastic, that is, an elastic body returns to its original form or condition after a displacing force is removed. D

**elastomer**
A substance which can be stretched and, after having been stretched and the stress removed, will return with force to approximately its original length. M

**electrical conductivity**
The numerical equal of the reciprocal of resistivity. The unit of conductivity is the mho. D

**electrical energy**
The energy of moving electrons. D

**electrical heat**
When a current flows in a circuit which contains resistance, heat is produced and the resistance and conductors of the circuit are raised in temperature. D

**electrical system**
A system in which all the conductors and apparatus are electrically connected to a common source of electromotive force. D

**electric charge**
A property of matter resulting from an imbalance between the number of protons and the number of electrons in a given piece of matter. The electron has a negative charge; the proton, a positive charge. Like charges repel each other; unlike, attract. D

**electric desalting**
A continuous process to remove inorganic salts and other impurities from crude oil by settling out in an electrostatic field. M

**electric energy**
As commonly used in the electric utility industry, electric energy means kilowatt hours. K

**electric generation**
This term refers to the act or process of transforming other forms of energy into electric energy, or to the amount of electric energy so produced, expressed in kilowatt hours. K

**electric generator**
A machine which transforms mechanical energy into electric energy. K

**electricity**
A material agency which, when in motion, exhibits magnetic, chemical and thermal effects and which, whether in motion or at rest, is of such a nature that when it is present in two or more localities, within certain limits of association, a mutual interaction of force between such localities is observed. D

**electric rate**
Unit prices and the quantities to which they apply as specified in an electric rate schedule or sales contract. L

**electric rate schedule**
Statement of the electric rate and the terms and conditions governing its application. L

**electric space heating**
Space heating of a dwelling or business establishment or other structure using permanently installed electric heating as the principal source of space heating throughout the entire premises. K

**electric system**
Physically connected electric generating, transmission, and distribution facilities operated as a unit under one control. L

**electric system loss**
Total electric energy loss in the electric system. It consists of transmission, transformation, and distribution losses, and unaccounted-for energy losses between sources of supply and points of delivery. L

**electrochemical**
Chemical action employing a current of electricity to cause or to sustain the action. D

**electrochemical cell (battery)**
A battery consists of two or more cells connected together in ways suited to delivering the needed energy at a suitable electrical pressure, or voltage. Each cell contains two conducting electrodes, one electrically positive and the other negative, made of dissimilar materials, usually metals or their salts. These electrodes are immersed in chemical solutions called electrolytes,

whose function is mainly to carry positive ions from the negative pole to the positive. To neutralize the charge that results, a current of electrons will then flow through external conducting wires from one pole to the other, and this can be made to give up energy if sent through a proper energy-converting device, such as an electric motor. CC

**electrode**
Conducting body that is brought in conducting contact with the ground. D

**electrodialysis**
A process that uses electrical current and an arrangement of permeable membranes to separate soluble minerals from water. Often used to desalinize salt or brackish water. E

**electrofining**
A process for contacting a light hydrocarbon stream with a treating agent (acid, caustic, doctor, etc.), then assisting the action of separation of the chemical phase from the hydrocarbon phase by an electrostatic field. M

**electrolysis**
Process of breaking a compound into its chemical components by passing an electric current through an electrolyte. D

**electrolyte**
A nonmetallic electric conductor (as a solution, liquid, or fused solid) in which current is carried by the movement of ions instead of electrons with the liberation of matter at the electrodes; a liquid ionic conductor. D

**electrolytes**
Materials which when placed in solution make the solution conductive to electrical currents. E

**electrolytic cell**
An assembly, consisting of a vessel, electrodes, and an electrolyte in which electrolysis can be carried out. D

**electrolytic process**
A process employing the electric current for separating and depositing metals from solution. D

**electromagnet**
A core of magnetic metal (as soft iron) that is surrounded wholly or in part by a coil or wire, that is magnetized when an electric current is passed through the wire, and that retains its power of attraction only while the current is flowing. D

**electromagnetic radiation**
Radiation consisting of associated and interacting electric and magnetic waves that travel at the speed of light. Examples: light, radio waves, gamma rays, X- rays. All can be transmitted through a vacuum. O

**electromagnetism**
Every electric current generates a magnetic field which is in a plane perpendicular to the current. The strength of the field is proportional to the current and in the case of a long, straight wire is inversely proportional to the distance from the wire. This principle is important in magnetic prospecting insofar as it forms the basis for certain types of geomagnetic instruments. D

**electromotive force**
Something that moves or tends to move electricity. The amount of energy derived from an electrical source per unit quantity of electricity passing through the source (as a cell or a generator). D

**electron [Symbol $e^-$]**
An elementary particle with a unit negative electrical charge and a mass 1/1837 that of the proton. Electrons surround the positively charged nucleus and determine the chemical properties of the atom. Positive electrons, or positrons, also exist. O

**electron capture (abbreviation EC)**
A mode of radioactive decay of a nuclide in which an orbital electron is captured by and merges with the nucleus, thus forming a new nuclide with the mass number unchanged but the atomic number decreased by 1. O

**electro-negative**
Descriptive of an element or group which ionizes negatively, or acquires electrons and, therefore, becomes a negatively charged anion. In electrolysis, moves to anode. D

**electron volt (abbreviation ev or eV )**
The amount of kinetic energy gained by an electron when it is accelerated through an electric potential difference of 1 volt. It is equivalent to 1.603 X $10^{-12}$ erg. It is a unit of energy, or work, not of voltage. O

**electropositive**
Positive charge; having more protons than electrons. An electropositive ion in a cation. D

**electrostatic precipitator**
(1) The most efficient of the dust samplers, the electrostatic precipitator is a medium-volume instrument. Air is drawn through a metal tube serving as a collecting surface (the anode) in which a platinum wire mounted axially acts as the ionizing and precipitating electrode (the cathode). A potential of about 10,000 volts direct current is maintained across the tube and wire. (2) A pollution control device that removes particles in gas by an electrical charge. The particles are accumulated on an electrode and rinsed away. D

**electrostatics**
Science of electric charges captured by bodies which then acquire special characteristics due to their retention of such charges. D

**element**
One of the 103 chemical substances that cannot be divided into simpler substances by chemical means. A substance whose atoms all have the same atomic number. Examples: hydrogen, lead, uranium. Not to be confused with fuel element. O

**elementary particles**
The simplest particles of matter and radiation. Most are shortlived and do not exist under normal conditions (exceptions are electrons, neutrons, protons and neutrinos). Originally this term was applied to any particle that could not be subdivided, or to constituents of atoms; now it is applied to nucleons (protons and neutrons), electrons, mesons, muons, baryons, strange particles, and the anti-particles of each of these, and to photons, but not to alpha particles or deuterons. Also called fundamental particles. O

**elkerite**
A variety of bitumen formed through a slow oxidation of petroleum. D

**elutriation**
Purification or sizing by washing and pouring off the lighter or finer matter suspended in water, leaving the heavier or coarser portions behind. D

**eluvial**
Formed by the rotting of rock in place to a greater or lesser depth. D

**eluvium**
Atmospheric accumulations in situ, or at least only shifted by wind, in distinction to alluvium, which requires the action of water. D

**emanation**
The escape of radioactive gases from the materials in which they are formed; for example, radon from radium and krypton and xenon from a substance undergoing fission. D

**EMARS**
Energy minerals allocation recommendation system.

**EMB**
Energy Mobilization Board.

**emergency core cooling system (ECCS)**
A safety system in a nuclear reactor, the function of which is to prevent the fuel in the reactor from melting should a sudden loss of normal coolant occur. X

**emission factor**
The average amount of a pollutant emitted from each type of polluting source in relation to a specific amount of material processed. For example, an emission factor for a blast furnace (used to make iron) would be a number of pounds of particulates per ton of raw materials. E

**emissions**
The total of substances discharged into the air from a stack, vent or other discrete source. E

**emissivity**
The proportion of the incident energy that is radiated, not reflected, from a surface. H

**EMS**
Energy management system.

**emulsification**
The process of emulsifying or dispersing one liquid in a second liquid in which it is completely or partly immiscible. M

**encroachment**
(1) To work coal or mineral beyond the boundary which divides one mine area from another. (2) The advancement of water, replacing withdrawn oil or gas in a reservoir. D

**endogenetic**
Pertaining to rocks resulting from physical and chemical reactions with their origin being due to forces within the material. D

**endothermic**
A chemical reaction requiring an input of heat, so that the temperature of the reacting bodies is lowered. Y

**end user**
Any ultimate consumer of any type of fossil fuel (petroleum, coal, natural gas) or electricity whether generated by fossil fuel or other energy sources. End users are classified in six sectors of the economy—agricultural, commercial, residential, industrial, government, and transportation. Refiners and electric generating stations are generally not included in this definition. F

**energized facilities**
Facilities that are under load (supplying energy to load) or carrying rated voltage and frequency but not supplying load. K

**energy**
The capability of doing work. There are several forms of energy, including kinetic, potential, thermal, and electromagnetic. One form of energy may be changed to another, such as burned coal to produce steam to drive a turbine which produces electricity. Except for some hydroelectric and nuclear power, most of the world's energy comes from energy in the form of fossil fuels, which are burned to produce heat. G

**energy absorption**
(1) Generally, it refers to the energy absorbed by any material subjected to loading. (2) Specifically, it is a measure of toughness or impact strength of a material; the energy needed to fracture a specimen in an impact test. It is the difference in kinetic energy of the striker before and after impact, expressed as total energy (foot-pound or inch-pound) for metals and ceramics and energy per

inch for plastics and electrical insulating materials. D

**energy band**
Every spectrum of valence electrons in a polyatomic material. Conduction is not significant if the energy band is filled. D

**energy charge**
That portion of the charge for electric service based upon the electric energy consumed or billed. L

**energy conversion efficiency**
A measure of how complete a process converts the energy contained in a particular resource (e.g., primary energy) into usable energy (e.g., secondary energy; end-use energy); usually expressed as a percentage of the input energy. Z

**energy dissipation**
The transformation of mechanical energy into heat energy. In fluids, this is accomplished by viscous shear. The rate of energy dissipation in flowing fluids varies with the scale and the degree of the turbulence. Baffles and other damping methods are used to dissipate energy. E

**energy, dump**
Energy generated by water power that cannot be stored or conserved when such energy is beyond the immediate needs of the producing system. K

**energy, economy**
Energy produced and supplied from a more economical source in one system, substituted for that being produced or capable of being produced by a less economical source in another system. K

**energy efficiency**
The product of the current efficiency multiplied by the voltage efficiency.

**energy gap**
Forbidden part of the energy spectrum of valence electrons. If the lower energy band is filled, electrons must be activated across this gap before electronic conductance is realized. D

**energy gradient**
The slope of the energy line of a body of flowing water with reference to a datum plane. E

**energy intensiveness (EI)**
A measure of energy utilization per unit of output. For passenger transport, for example, it is a measure of calories used per passenger mile. V

**energy, interchange**
Kilowatt hours delivered to or received by one electric utility system from another. They may be returned in kind at a later time or may be accumulated as energy balances until the end of a stated period. Settlement may be by payment or on a pooling basis. K

**energy level**
The distance from an atomic nucleus at which electrons can have orbits. May be thought of as a shell surrounding the nucleus. D

**energy loss**
Difference between energy input and output. L

**energy plantations**
(1) The growing of plant material for its fuel value. (2) A renewable source of energy-rich, fixed carbon produced by photosynthesis. Q

**engine horsepower**
A standard unit of power equal to 746 watts in the United States and nearly equivalent to the English gravitational unit which equals 550 foot pounds of work per second, or 33,000 foot pounds per minute. D

**engineered storage**
Storage of radioactive wastes, usually in suitable sealed containers, in any of a variety of structures especially designed to protect them from water

and weather and to prevent leakage to the biosphere by accident or sabotage. This storage may also provide for extracting heat of radioactive decay from the waste. P

**Engler degree**
See degree Engler.

**enriched material**
Material in which the percentage of a given isotope present in a material has been artificially increased, so that it is higher than the percentage of that isotope naturally found in the material. Enriched uranium contains more of the fissionable isotope uranium-235 than the naturally occurring percentage (0.7%). O

**enriching**
Increasing the heat content of a gas by mixing with it a gas of higher Btu content. For nuclear field, see enriched material. DD

**enrichment**
A nuclear energy term applied to isotopic enrichment. O

**ENS**
European Nuclear Society.

**enthalpy**
(1) Total heat content of air. (2) The sum of the enthalpies of dry air and water vapor, per unit weight of dry air. Measured in British thermal units per pound. D

**entrained bed**
Combustion process in which pulverized coal is carried in a gas stream. Z

**entropy**
A measure of the unavailable energy in a system, that is, energy that cannot be converted into another form of energy. D

**entry driver**
A combination mining machine designed and built to work in entries and other narrow places, and to load coal as it is broken down. D

**environment**
The sum of all external conditions and influences affecting the life, development, and ultimately, the survival of an organism. E

**environmental impact statement**
A document prepared by a Federal agency on the environmental impact of its proposals for legislation and other major action significantly affecting the quality of the human environment. Environmental impact statements are used as tools for decision-making and are required by the National Environmental Policy Act. E

**EOR**
Enhanced oil recovery.

**eolation**
(1) The process by which wind modifies land surfaces, both directly by transportation of dust and sand and by the work of sandblasts, and indirectly by wave action on shores. (2) Eeolic gradation. D

**eolian**
(1) Of, relating to, formed by, or deposited from the wind or currents of air. (2) Applied to sand dunes which have been accumulated by the wind. Eolian was formerly spelled aeolian. D

**EPA**
Environmental Protection Agency.

**EPAA**
Emergency Petroleum Allocation Act.

**EPCA**
Energy Policy and Conservation Act.

**epi-**
A prefix meaning alteration. D

**epidermal deformation**
Deformations affecting the outer cover of the earth's crust, such as slumping, volcanotectonic collapse, sliding, compressive settling, etc. D

**epigene**
Formed, originating, or taking place on or not far below the surface of the earth. D

**epithermal reactor**
An intermediate reactor. O

**epoch**
(1) A geologic time unit corresponding to a series. (2) A subdivision of a period, formerly used for other smaller divisions of geologic time. D

**EPRI**
Electric Power Research Institute, Palo Alto, California.

**equation of motion**
(1) The Newtonian law of motion states that the product of mass and acceleration equals the vector sum of the forces. (2) In meterological and oceanographic use, both sides of the equation of motion are divided by mass to give force per unit mass. The forces considered in ocean currents are gravity, Coriolis force, pressure gradient force, and frictional forces. D

**equation of time**
The measure, in minutes, of the extent by which solar time, as told by the sundial, runs faster or slower than civil or mean time, as determined by a clock running at a uniform time. R

**equilibrium**
When two or more forces act upon a body in such a manner that the body remains at rest the forces are said to be in equilibrium, or in a perfect balance. D

**ERA**
Economic Regulatory Administration, DOE.

**ERAB**
Energy Research Advisory Board, DOE.

**ERDA**
Energy Research and Development Administration, absorbed into DOE.

**erg**
(1) A unit of work done by a force of one dyne acting through a distance of one centimeter. (2) Unit of energy which can exert a force of one dyne through a distance of one centimeter. E

**erode**
(1) To wear away, as land, by the action of water. (2) To produce or to form by erosion, or by wearing away, as glaciers. D

**EROS**
Earth resources observation systems.

**erosion**
The wearing away of the land surface by wind or water. Erosion occurs naturally from weather or run-off but is often intensified by man's land-clearing practices. E

**ERS**
Economic Research Service in the Federal Department of Agriculture.

**ERTS**
Earth Resource Technology Satellite.

**ESA**
Environmentally sensitive areas.

**ESCOE**
Engineering Societies Commission on Energy, Inc.

**ESECA**
Energy Supplies and Environmental Coordination Act.

**ester**
A compound formed by the action of alcohol and acid; e.g., ethyl alcohol reacts with acetic acid to produce

ethyl acetate (an ester) and water. This reaction is called esterification. M

**estuary**
The mouth of a river where the tide meets the river current.

**ethanol**
Ethyl alcohol, $C_2H_5OH$, also known as grain alcohol. Can be used as fuel or mixed with gasoline as a fuel. Y

**ethyl fluid**
A gasoline antiknock compound manufactured by Ethyl Corporation. M

**ethylene**
Colorless olefinic gas, $C_2H_4$, with characteristic sweet odor and taste, derived from the cracking of petroleum. M

**EURATOM**
European Atomic Energy Community.

**eutrophic**
Applied to a lake which is rich in dissolved nutrients, but is frequently shallow and has seasonal oxygen deficiency in the stagnant bottom waters. M

**ev**
(1) Electron volt. (2) Electric vehicle.

**evaporation**
The conversion of a liquid into vapor, usually by means of heat. M

**excavator**
A large number of power-operated digging and loading machines. These are used increasingly in opencast mining and in quarrying. D

**excess air**
In practice, complete combustion cannot be obtained without slightly more air than is theoretically necessary. The amount of excess air varies with the design and mechanical condition of the appliance, but ranges from 15 percent upwards. D

**excess reactivity**
More reactivity than that needed to achieve criticality. Excess reactivity is built into a reactor (by using extra fuel) in order to compensate for fuel burnup and the accumulation of fission-product poisons during operations. O

**excitation (generally)**
The addition of energy to a system, thereby transferring it from its ground state to an excited state. D

**excitation (in the electrical industry)**
The power required to energize the magnetic field of generators in an electric generating station. K

**excited state**
The state of an atom or nucleus when it possesses more than its normal energy. The excess energy is usually released eventually as a gamma ray or photon. D

**exciter**
An auxiliary generator that supplies energy for the field excitation of another electric machine. D

**exclusion area**
An area immediately surrounding a nuclear reactor where all human habitation is prohibited to assure safety in the event of an accident. W

**excursion**
A sudden, very rapid rise in the power level of a reactor caused by supercriticality. Excursions are usually quickly suppressed by the negative temperature coefficient of the reactor and/or by automatic control rods. W

**exergy**
(1) Concept related to thermodynamically available energy and useful work. (2) A term commonly used in Europe and associated with technical capability to perform work. Z

**exhaustion**
In mining, the complete removal of ore reserves. D

**exine**
The outer of the two layers forming the wall of certain spores. D

**exinoid**
A coal constituent similar to material derived from plant exines. D

**exosphere**
Space beyond the earth's atmosphere. It begins at a height of about 1,000 kilometers. D

**exothermic**
A chemical reaction which produces heat, so that the temperature of the reacting bodies is raised. Y

**experimental reactor**
A reactor to test the design of new reactors. O

**exploitation**
The process of winning or producing from the earth the oil, gas, minerals, or rocks which have been found as the result of exploration. D

**exploration drilling**
Drilling boreholes by the rotary, diamond, percussive, or any other method of drilling for geologic information or in search of a mineral deposit. D

**exploratory wells**
(1) Wells classified as those drilled to find the limits of an oil or gas-bearing formation, that is partly developed. (2) Those drilled in search of a new productive formation in an area that is already productive. (2) Those drilled in an area where neither oil nor natural gas has ever been found. The last are known as new-field wildcats. M

**explosive strength**
A measure of the amount of energy released by an explosive on detonation and its capacity to do useful work. D

**explosive stripping**
A method, encouraged by the introduction of low-cost AN-FO explosives, in which by using an excess of explosives in the strip mine bench, up to about 40 percent of the overburden can be removed from the coal seam by the energy of the explosive, thereby requiring no excavation. D

**extensometer**
Instrument used for measuring small deformations, deflection, or displacements. D

**extraction**
(1) In coal mining, the process of mining and removal of coal or ore from a mine. D (2) In the oil industry, the process of separating a material, by means of a solvent, into a fraction soluble in the solvent (extract) and an insoluble residue. (3) The separation of a metal or valuable mineral from an ore or concentrate. M

**extraction-turbine**
A steam-turbine from which steam for process work or heating is tapped at a suitable stage in the expansion. Z

**extra high voltage (EHV)**
A term applied to voltage levels of transmission lines which are higher than the voltage levels commonly used. At present, the electric utility industry generally considers EHV to be any voltage of 345,000 volts or higher. K

**extraneous gas**
That volume of gas which is not indigenous to the storage area. DD

**extravastion**
The eruption of molten or liquid material from the earth, as lava from a vent, water from a geyser, etc. D

# F

**f.**
Fusion, freezing, or frequency. Symbol for coefficient of friction. D

**F.**
Chemical symbol for fluorine. Also abbreviation for Fahrenheit, or force.

**FAA**
Federal Aviation Administration.

**fabric**
The special arrangement and orientation of rock components, whether crystals or sedimentary particles, as determined by their sizes, shapes, etc. D

**fabric filters**
A device for removing dust and particulate matter from industrial emissions much like a home vacuum cleaner bag. The most common use of fabric filters is the baghouse. E

**face**
The solid surface of the unbroken portion of the coalbed at the advancing end of the working place. D

**face timbering**
The placing of safety posts at the working face to support the roof of the mine. D

**facies**
The aspect belonging to a geologic unit of sedimentation, including mineral composition, type of bedding, fossil content, etc. Sedimentary facies are really segregated parts of differing nature belonging to any genetically related body of sedimentary deposit. D

**facilities charge**
Amount to be paid by the customer as lump sum, or periodically, as reimbursement for facilities furnished. The charge may include operation and maintenance as well as fixed costs. L

**facing**
The main vertical joints often seen in coal seams; they may be confined to the coal, or continue into the adjoining rocks. D

**factor**
One of several elements, circumstances, or influences which tend to the production of a given result. D

**Fahrenheit**
The commonly used thermometer scale in which the freezing point of water is 32° and the boiling point is 212°. Named after Gabriel Fahrenheit, a German physicist (1686-1736). D

**fall**
A mass of roof or side which has fallen in any subterranean working or gallery, resulting from any cause. D

**FAO**
Food and Agricultural Organization of the United Nations.

**farad**
(1) Unit of electric capacitance. (2) It is the capacitance of a capacitor between the plates of which there appears a difference of potential of 1 volt when it is charged by a quantity of electricity equal to 1 coulomb. E

**faraday**
The quantity of electricity transferred in electrolysis per equivalent weight of any element or ion, being equal to about 96,500 coulombs per gram equivalent. D

**fast breeder reactor**
A reactor that operates with fast neutrons and produces more fissionable material than it consumes. O

**Fast Flux Test Facility (FFTF)**
An experimental liquid metal fast breeder reactor which will be used to test various fuels and reactor components. X

**fast neutron**
A neutron with kinetic energy greater than approximately 100,000 electron volts. W

**fast reactor**
A reactor in which the fission chain reaction is sustained primarily by fast neutrons rather than by thermal or intermediate neutrons. Fast reactors contain little or no moderator to slow down the neutrons from the speeds at which they are ejected from fissioning nuclei. D

**fathom**
(1) A unit of linear measurement that equals 6 feet or 1.828 meters. (2) A measure used for sea depths and sometimes for shaft and rope lengths. D

**fathometer**
An instrument used in measuring the depth of water by the time required for a sound wave to travel from the surface to bottom and for its echo to be returned. It may also be used for measuring the rise and fall of the tides in offshore localities. E

**fat oil**
The bottoms or enriched oil drawn from the absorber, as opposed to lean oil. M

**fault**
A fracture or a fracture zone along which there has been displacement of the two sides relative to one another parallel to the fracture. The displacement may be a few inches or many miles. D

**fauna**
All the invertebrate and vertebrate animals of any given age or region. All the plants are similarly called flora. D

**FBC**
Fluidized bed combustion.

**FBR**
Fast breeder reactor.

**FCC**
Federal Communications Commission.

**FCG**
Fuel cell generation.

**FCR**
Full core reserve.

**FDA**
Food and Drug Administration, part of the Federal Department of Health and Human Services.

**FDPC**
Federal data processing centers.

**featheredge**
The thin end of a wedge-shaped piece of rock or coal. D

**feedback**
In automatic control of a process, control of an earlier stage by means of variance registered at a later stage. D

**feeder**
Very small fissures or cracks through which methane escapes from the coal. As working faces are advanced, fresh feeders are encountered in each fall of coal. D

**feeder (electrical industry)**
An electric line for supplying electric energy within an electric service area or sub-area. K

**feedometer**
Device which weighs a passing stream of ore. D

**feed rate**
Rate at which a drilling bit is advanced into or penetrates the rock formation being drilled expressed in inches per minute, inch per bit revolution, number of bit revolutions per inch of advance, or feet per hour.

**feedstock**
Fossil fuels used for their chemical properties, rather than their values as fuel; e.g., oil used to produce elastics and synthetic fabrics.

**FEO**
Federal Energy Office.

**FERC**
Federal Energy Regulatory Commission.

**fermentation**
The breakdown of organic materials by metabolism of microorganisms in which energy is produced.

**ferric**
Of, pertaining to, or containing iron in the trivalent state, for example, ferric chloride. D

**ferride**
A member of a group of elements that are related to iron. The group includes chromium, cobalt, manganese, nickel, titanium, and vanadium, besides iron. D

**ferrite**
Pure, or nearly pure, metallic iron.

**fertile material**
Material, not itself fissionable by thermal neutrons, that can be converted into a fissile material by irradiation in a reactor. There are two basic fertile materials, uranium-238 and thorium-232. When these fertile materials capture neutrons, they are partially converted into fissile plutonium-239 and uranium-255, respectively. P

**FFB**
Fuels from biomass.

**FFTF**
Fast flux test facility.

**FGD**
Fluid gas desulfurization.

**fiber**
The smallest single strands of asbestos or other fibrous materials. D

**field**
A section of land containing, yielding, or worked for a natural resource, for example, a coalfield, an oilfield, or a diamond field. D

**fiery mine**
A mine in which the seam or seams of coal being worked give off a large amount of methane. D

**fill**
Any sediment deposited by an agent to fill or to fill partly a valley, a sink, or other depression. D

**film badge**
A package of photographic film worn like a badge by workers in the nuclear industry to measure exposure to ionizing radiation. The absorbed dose can be calculated by the degree of film darkening caused by the irradiation. D

**film coefficient**
(1) The heat transferred by convection per unit area per degree temperature difference between the surface and the fluid. (2) Also called unit convection and conductance, surface coefficient. D

**filter aid**
A low-density, inert, fibrous, or fine granular material used to increase the rate and improve the quality of filtration. D

**fines**
Very small material produced in breaking up large lumps, as of ore or coal. D

**FIRE**
Forest industry renewable energy.

**fire**
(1) The manifestation of rapid combustion, or combination of materials with oxygen. (2) To blast with gunpowder or other explosives. D

**fire classification**
The following explains the National Fire Protection Association classifications. Class A fires are defined as those in ordinary solid, combustible materials, such as coal, wood, rubber, textiles, paper and rubbish. Class B fires are defined as those in flammable liquids, such as fuel or lubricating oils, grease, paint, varnish, and lacquer. Class C fires are defined as those in (live) electric equipment, such as oil-filled transformers, generators, motors, switch panels, circuit breakers, insulated electrical conductors, and other electrical devices. D

**firedamp**
(1) A combustible gas that is formed in mines by decomposition of coal or other carbonaceous matter, and that consists chiefly of methane. (2) The explosive mixture formed by this gas with air. D

**fire flooding (in situ combustion)**
A second thermal recovery technique in which oil-moving energy is created by injecting air or oxygen-bearing gas into the reservoir and then igniting a portion of the reservoir oil. This starts a narrow burning area that travels from the injection well toward the producing well, pushing the oil ahead of it. This system does not have to be fast, for heat is more important than speed. To get the oil moving, temperatures may go as high as 1200°F. Careful regulation of injected air keeps the entire process under close control. M

**firewall**
A wall to prevent the spread of fire usually made of noncombustible materials. D

**firing**
The process of initiating the action of an explosive charge or the operation of a mechanism which results in a blasting action.

**First Law of Thermodynamics**
States that energy (measured by its quantity or enthalphy) is never created or destroyed, but only changed in form and quality. First Law efficiency measures the fraction of the energy supplied to a device or process that it delivers in its output (having no regard to any degradation of energy quality). Thus a furnace with a First Law efficiency of 70

percent delivers 70 percent of the energy present in its fuel as useful heat to the building, while the other 30 percent escapes up the flue or to other useless places. C

**fiscal policy**
Management of government spending, taxation and the budget deficit or surplus to stimulate or restrain the economy.

**Fischer-Tropsch process**
(1) A process of synthesizing hydrocarbons and oxygenated chemicals from a mixture of hydrogen and carbon monoxide. (2) A coal gasification process wherein the carbon in coal is gasified with oxygen and/or steam to form a synthesis gas. The hydrogen content of the gas can be increased by reacting CO with steam by the water-gas shift reaction. In the presence of a catalyst, the synthesis gas mixture reacts to form a mixture of hydrocarbons. Temperature, pressure, and reactant ratios determine the reaction; depending upon relative concentrations, either liquid or gaseous products can be reduced. M

**fissile**
Capable of being split, as schist, slate, and shale. D

**fissile material**
While sometimes used as a synonym for fissionable material, this term has also acquired a more restricted meaning, namely, any material fissionable by neutrons of all energies, including (and especially) thermal (slow) neutrons as well as fast neutrons; for example, uranium-235 and plutonium-239. O

**fission**
The splitting of a heavy nucleus into two approximately equal parts (which are nuclei of lighter elements), accompanied by the release of a relatively large amount of energy and generally one or more neutrons. Fission can occur spontaneously, but usually is caused by nuclear absorption of gamma rays, neutrons or other particles. O

**fissionable material**
Commonly used as a synonym for fissile material. The meaning of this term also has been extended to include material that can be fissioned by fast neutrons only, such as uranium-238. Used in reactor operations to mean fuel. O

**fission fragments**
The two or more nuclei which are formed by the fission of a nucleus. Also referred to as primary fission products. They are of medium atomic weight and are radioactive. W

**fission products**
The nuclei (fission fragments) formed by the fission of heavy elements, plus the nuclides formed by the fission fragments' radioactive decay. W

**fission yield**
(1) The amount of energy released by fission in a thermonuclear (fusion) explosion as distinct from that released by fusion. (2) The amount (percentage) of a given nuclide produced by fission. O

**fissure**
An extensive crack, break, or fracture in the rocks. A mere joint or crack persisting only for a few inches or a few feet is not usually termed a fissure by geologists or miners, although in a strict physical sense it is one. Where there are well-defined boundaries, very slight evidence of ore within such boundaries is sufficient to prove the existence of a lode. Such boundaries constitute the sides of a fissure. D

**fixation**
The act or process by which a fluid or a gas becomes or is rendered firm or

stable in consistency, and evaporation or volatilization is prevented. D

**fixed-bed operation**
A type of operation in which the catalyst remains stationary in the reactor. The catalyst may be regenerated in place periodically. To be contrasted with fluid-bed operation. M

**fixed costs**
Cost associated with investment in plant. L

**fixed gas**
Gas which will not condense under pressure and temperature conditions available in the process under consideration. M

**fixed operation costs**
Costs other than those associated with investment in plant, which do not vary or fluctuate with changes in operation or utilization of plant. L

**fixed solids**
The residue remaining after ignition of suspended or dissolved matter according to standard methods. E

**flame**
A burning mixture of a combustible gas (or vapor) and air. Solid fuels burn with a glow, but with little flame. Principle types of flames are luminous, nonluminous, long (lazy) flames, and short flames. D

**flame test**
The use of the characteristic coloration imparted to a flame to detect the presence of certain elements.

**flare gas**
(1) Unutilized natural gas burned in flares at an oil field. (2) Waste gas. G

**flashpoint**
The minimum temperature at which sufficient vapor is released by a liquid or solid to form a flammable vapor-air mixture at atmospheric pressure. D

**flat and meter rate schedule**
Rate schedule of two components, the first of which is a customer or service charge, and the second of which is a price for the energy consumed. L

**flat-plate collector**
A device used to absorb solar radiation for the purposes of heating water, for example. They generally consist of five components: (1) glazing, which may be one or more sheets of glass or a diathermanous (radiation-transmitting) plastic film or sheet; (2) tubes or fins for conducting or directing the heat-transfer fluid from the inlet duct or header to the outlet; (3) plate, generally metallic, which may be flat, corrugated, or grooved, to which the tubes or the fins are attached in some manner which produces a good thermal bond; (4) insulation, which minimizes downward heat loss from the plate; and (5) container or casing, which surrounds the foregoing components and keeps them free from dust, moisture, etc. R

**flat rate schedule**
A rate schedule that provides for a specified charge irrespective of the energy used or demand. L

**flexible pipe**
A drilling technique being tested which runs as a continuous unit, and which is designed to permit raising and lowering the pipe without having to couple and uncouple joints. M

**float coal**
Small, irregularly shaped isolated deposits of coal imbedded in sandstone or in siltstone. D

**flooding**
The drowning out of a well by water that sometimes results from drilling too deeply into the sand. D

**flood plain**
The flat ground along a stream, covered by water at the flood stage. D

**flood tide**
The flow, or rising toward the shore, is called the flood tide, and the falling away, ebb tide. D

**flow rate**
Weight of dry air flowing per unit time, measured in pounds per hour. D

**flue gas**
Gas from the combustion of fuel, the heating value of which has been substantially spent and which is, therefore, discarded to the flue or stack. M

**fluid-bed operation**
A type of operation based on the tendency of finely divided powders to develop high concentrations (i.e., settle) in a gas stream of low velocity. The fluidized-powder technique involves suspending the finely divided powder in an upwardly flowing stream of gas. Each particle is supported by the gas, and this suspension has most of the characteristics associated with a true liquid. Thus, it can flow through pipes and valves, and it will develop hydrostatic heads which allow it to flow from one vessel to another. M

**fluid catalyst**
The finely divided particles used as the catalyst in a fluid-bed catalytic process. Microspherical or powdered catalysts of either natural or synthetic origin may be used. M

**fluid fuel reactor**
A type of reactor (for example, a fused-salt reactor) whose fuel is in fluid form. O

**fluidized bed**
A fluidized bed results when a fluid, usually a gas, flows upward through a bed of suitably sized solid particles at a velocity high enough to buoy the particles, to overcome the influence of gravity, and to impart to them an appearance of great turbulence. G

**fluidized bed reactor**
A reactor design in which the fuel ranges in size from small particles to pellets. Although the fuel particles are solid, their entire mass behaves like a fluid because a stream of liquid or gas coolant keeps them moving. O

**flume**
An inclined channel, usually of wood and often supported on a trestle, for conveying water from a distance to be utilized for power, transportation, etc., as in placer mining, logging, etc. D

**fluorescence**
Many substances can absorb energy (as from X-rays, ultraviolet light, or radioactive particles), and immediately emit this energy as an electromagnetic photon, often of visible light. This emission is fluorescence. The emitting substances are said to be fluorescent. O

**fluorides**
Gaseous, solid or dissolved compounds containing fluorine, emitted into the air or water from a number of industrial processes. Fluorides in the air are a cause of vegetation damage and, indirectly, of livestock damage. E

**flushing**
A drilling method in which water or some other thicker liquid, for instance a mixture of water and clay, is driven into the borehole, through the rod and bit. The water rises along the rod on its outer side, that is between the walls of the borehole and the rod, and with such a velocity that the broken rock fragments are carried up by this water current (direct flushing); or water enters the borehole around the rod and issues upwards through the rod (indirect flushing). D

**flush production**
The phase during which the rate of oil flow is governed by natural pressure within the reservoir, usually the first stage in an oil well's life. It occurs when the drill taps an oilbearing formation that has enough natural pressure to enable the petroleum to flow by itself. There are three types of "drives" which can generate this force: gas cap drive, dissolved gas drive and water drive. M

**fluvial**
Of or pertaining to streams and rivers; produced by stream or river action, as a fluvial plane. D

**flux**
(1) The rate of flow of energy across a given area perpendicular to the direction of flow. (2) A flowing or discharge of fluid. (3) A substance used to promote fusion (as by removing impurities) of metals or minerals; the rate of transfer of fluid, particles, or energy (as radiant energy) across a given surface. E

**fly ash**
The finely divided particles of ash entrained in the flue gases arising from the combustion of fuel. The particles of ash may contain incompletely burned fuel. The term has been applied predominantly to the gas-born ash from boilers with spreader stoker, underfeed stoker, and pulverized fuel (coal) firing. E

**FMC**
Federal Maritime Commission.

**fob**
Free on board.

**focus**
In seismology, the source of a given set of elastic waves. The true center of an earthquake, within which the strain energy is first converted to elastic wave energy. D

**focusing collector**
Focusing collectors utilize optical systems and/or concentrating reflector surfaces to increase the intensity of solar radiation on the energy-absorbing surface. See parabolic reflector. H

**food chain**
The pathways by which any material (such as radioactive material from fallout) passes from the first absorbing organism through plants and animals to humans. W

**footage**
(1) The payment of miners by the running foot of work. (2) The sum given. (3) The number of feet of borehole drilled per unit of time, or that required to complete a specific project or contract. D

**footing**
(1) The characteristics of the material directly beneath the base of a drill tripod, a derrick, or mast uprights. (2) The material placed under such members to produce a firm base on which they may be set. D

**foot pound**
The amount of work done in raising one pound one foot, or in overcoming a pressure of one pound through a distance of one foot. BB

**force**
Influence which, when brought to bear on a body, changes its rate of momentum. Types of force are attractive, accelerating, repulsive. Measured in dynes or poundals. D

**force vibration**
Vibration of a structure, generally caused by engines or machines, sometimes by wind. H

**forced circulation**
A solar system which utilizes a pump or fan to circulate a liquid or air

(respectively), as distinguished from a system relying on natural circulation. H

**foreshaft sinking**
The first 150 feet or so of shaft sinking from the surface, during which time the plant and services for the main shaft sinking are installed. D

**formation**
(1) As defined and used by the U. S. Geological Survey, the ordinary unit of geologic mapping consisting of a large and persistent stratum of some one kind of rock. (2) Also it is loosely employed for any local and more or less related group of rocks. D

**formation drilling**
Boreholes drilled primarily to determine the structural, petrologic, and geologic characteristics of the overburden and rock strata penetrated. Also called formation testing. D

**form energy**
The potentiality of minerals to develop crystal form within a solid medium, such as rock. D

**formula**
The formula of a compound is arrived at by: (1) writing the symbols for the elements making up the compound; and (2) following the symbols with the appropriate figure, showing how many atoms of each element are in one molecule of the compound. Also called molecular formula. D

**fossil**
Originally, a rock, mineral, or other substance dug out of the earth. Now, any remains, impression, or trace of an animal or plant of past geologic ages that have been preserved in the earth's crust. D

**fossil fuel**
Naturally occurring substances derived from plants and animals which lived in ages past. The bodies of these long-dead organisms have become our recoverable fuels which can be burned, such as coal, crude oil, or natural gas and other hydrocarbon types of fuels such as shale, lignite, etc. Y

**fossil-fuel plant**
Electric power plant utilizing fossil fuel: coal, lignite, oil, or natural gas as its source of energy. L

**foul**
A condition of the atmosphere of a mine, so contaminated by gases as to be unfit for respiration. Impure. D

**foundation**
(1) The shafts, machinery, building, railways, workshop, etc., of a mine, commonly called a plant. (2) The ground upon which a substructure is supported. D

**FPC**
Federal Power Commission.

**fpm**
Feet per minute.

**FPR**
Federal Procurement Regulations.

**fps**
Feet per second.

**FR**
Federal Register.

**fractional distillation**
The separation of the components of a liquid mixture by vaporizing and collecting the fractions, or cuts, which condense in different temperature ranges. M

**fractional horsepower motor**
An electric motor rated at less than one horsepower. D

**fractionating column**
A column arranged to separate various fractions of petroleum by a single

distillation. The column may be tapped at different points along its length to separate various fractions in the order of their condensing temperatures (boiling points). M

**fractionation**
Separation in successive stages, each removing from a mixture some proportion of one of the substances. The operation may be precipitation, crystallization, distillation, etc. M

**fracturing**
A process of opening up underground channels for trapped hydrocarbons in oil wells, but by force rather than by a chemical action such as in acidizing. A specially treated fluid is pumped into the producing well under high pressure, thus splitting open fissures in the reservoir rock. The pressure is released and the fluid flows back out of the well. M

**fragmentation**
The breaking of coal, ore, or rock by blasting so that the bulk of the material is small enough to load, handle and transport. D

**Frasch process**
A process formerly used for removing sulfur by distilling oil in the presence of copper oxide. M

**FRC**
Future Requirements Committee.

**free energy**
That part of the energy of a system which is the maximum available for doing work. D

**free moisture**
Moisture in coal that can be removed by ordinary air drying. D

**free radical**
A complex of abnormal valence which possesses additive properties but does not carry an electrical charge and is not a free ion. M

**free water**
Water that is free to move through a soil mass under the influence of gravity. Also called gravitation water; ground water; phreatic water. D

**freezing index**
The number of degree days between the highest and lowest points on the cumulative degree days-time curve for one freezing season. It is a measure of the combined duration and magnitude of below-freezing temperatures occurring during any given freezing season. D

**freezing point**
The temperature at which a substance freezes (e.g., at which a liquid solidifies). M

**frequency**
(1) The number of repetitions of a periodic process in a unit of time. (2) In the electrical industry, the number of cycles through which an alternating current passes per second. Frequency has been generally standardized in the United States electric utility industry at 60 cycles per second (60 hertz). D

**fresh air**
Air free from the presence of deleterious gases. D

**friction**
(1) Resistance to the motion of one surface against another. (2) A widespread force which slows down movement and causes heat. M

**friction head**
The additional pressure that the pump must develop to overcome the frictional resistance offered by the pipe, by bends or turns in the pipeline, by changes in the pipe diameter, by valves, and by couplings. D

**front**
(1) A designation for the mouth or collar of a borehole. (2) The working

attachment of a shovel, as dragline, hoe, or dipper stick. D

**front-end loader**
A tractor loader with a digging bucket mounted and operated at the front end of the tractor. D

**ft**
Foot.

**FTC**
Federal Trade Commission.

**ft-lb**
Foot-pound.

**FUA**
Fuel Use Act.

**fuel**
(1) Any substance which will produce heat upon combustion or in a case of food will produce heat as a result of respiration. The energy content of common fuels is as follows:

1 barrel (bbl.) of crude oil= 5,800,000 Btu.

1 cubic foot (cf) of natural gas= 1,032 Btu.

1 ton of coal= 24,000,000 to 27,000,000 Btu.

Two trillion Btu's per year are about equal to 1,000 barrels of crude oil per day. (2) As applied to atomic energy, fuel is defined as fissionable material used or usable to produce energy in a reactor. (3) Also applied to a mixture, such as natural uranium, in which only part of the atoms are readily fissionable, if the mixture can be made to sustain a chain reaction. D

**fuel adjustment clause**
A clause in a rate schedule that provides for an adjustment of the customer's bill if the cost of fuel varies from a specified unit cost. L

**fuel-air ratio**
(1) The ratio of the weights of fuel and air supplied to an engine during equal operating intervals. (2) In ASTM methods for testing the knocking characteristics of motor fuels, D 357 and D 908, this ratio is adjusted for maximum knock at each compression ratio and for each fuel tested. M

**fuel cell**
An electrochemical device to convert chemical energy directly into electricity. It is similar in some respects to a storage battery or a dry cell. Like a battery, the fuel cell produces electricity by a chemical reaction. Unlike a storage battery, however, the fuel cell continues to produce electicity as long as fuel is added. In this respect, a fuel cell operates like an engine. M

**fuel cell battery**
A chemical battery, with perhaps sodium, chlorine, and mercury among ingredients, which will have self-generating tendencies. D

**fuel costs (electric utility companies)**
(1) Cents per million Btu consumed. Since coal is purchased on the basis of its heat content, its cost is measured by computing the "cents per million Btu" of the fuel consumed. (2) It is the total cost of fuel consumed divided by its total Btu content, and the answer is multiplied by one million. K

**fuel costs (coal)**
Average cost per (short) ton ($ per ton)—includes bituminous and anthracite coal and relatively small amounts of coke, lignite, and wood. K

**fuel costs (gas)**
(1) Average cost per Mcf (cents per thousand cubic feet)—includes natural, manufactured, mixed, and waste gas. (2) Frequently expressed as cost per therm (100,000 Btu). K

**fuel costs (nuclear)**
Nuclear fuel costs can be given on a fuel cycle basis. A fuel cycle consists of all the steps associated with

procurement, use and disposal of nuclear fuel. Accounting for the cost of each step in the fuel cycle including interest charges, nuclear fuel costs can be given in cents per million Btu or mills per kilowatt hour for the cycle lifetime of the fuel which is normally five to six years. K

**fuel costs (oil)**
Average cost per bbl—42 gallons ($ per barrel)—includes fuel oil, crude and diesel oil, and small amounts of tar and gasoline. K

**fuel cycle**
The series of steps involved in supplying fuel for nuclear power reactors. It includes mining, refining, the original fabrication of fuel elements, their use in a reactor, chemical processing to recover the fissionable material remaining in the spent fuel, re-enrichment of the fuel material, and refabrication into new fuel elements. O

**fuel efficiency**
The ratio of the heat produced by a fuel for doing work to the available heat of the fuel. D

**fuel element**
A rod, tube, plate or other mechanical shape or form into which nuclear fuel is fabricated for use in a reactor. O

**fuel for electric generation**
This includes all types of fuel—solid, liquid, gaseous, and nuclear—used exclusively for the production of electric energy. Fuel for other purposes, such as building heating or steam sales, is excluded. K

**fuel gas**
Any gas used for heating. M

**fuel oil**
Any liquid or liquefiable petroleum product burned for the generation of heat in a furnace or firebox, or for the generation of power in an engine, exclusive of oils with a flash point below 100°F (Tag closed-cup tester) and oils burned in cotton-or wool-wick burners. M

**fuel oil, classifications of**
Domestic fuel oils are classed as Nos. 1, 2, and 3; industrial fuel oils as Nos. 4, 5, and 6. Occasionally, fuels are referred to as light, medium, and heavy domestic oils; and light, medium, and heavy industrial oils. BB

**fuel oil equivalent**
The term used in indicating the No. 6 fuel oil equivalent of a given fuel on a thermal value basis. M

**fuel rate**
The amount of fuel needed to generate one kilowatt hour of electricity. In 1969, the rates were 0.88 pounds of coal, average, in the United States electricity industry, 0.076 gallons of oil and 10.4 cubic feet of natural gas. G

**fuel reprocessing**
The processing of reactor fuel to recover the unused fissionable material. L

**full employment**
State of the economy when the unemployment rate is relatively low (conventionally about 4 percent of the labor force) and production is close to capacity.

**full-seam mining**
A mining system, brought on by the advent of mechanical loading and mechanical coal cleaning, in which the entire section is dislodged together and the coal separated from the rock outside of the mine by the cleaning plant. D

**fume**
The gas and smoke (more especially the noxious or poisonous gases) given off by the explosion or detonation of blasting powder or dynamite. The

character of the fume is influenced largely by the completeness of detonation. The degree of confinement of the charge and the size of the detonator has a great influence on the character of the fumes produced. D

**furnace**

A structure in which, with the aid of heat so produced, the operation of roasting, reduction, fusion, steam-generation, desiccation, etc., is carried on, or, as in some mines, the air current is heated, to facilitate its ascent and thus aid ventilation. D

**fusain**

This term was introduced by Grand'Eury in 1882. This substance is recognized macroscopically by its black or gray-black color, its silky luster, its fibrous structure and its extreme friability. It is the only constituent of coal which marks and blackens objects with which it comes in contact. Fusain may include a high proportion of mineral material, which strengthens it and reduces its friability; it retains, however, its silky luster. In macroscopic description of seams, only those bands having a thickness of several millimeters are recorded. Microscopic examination shows that fusain consists in the main of fusite. Fusain occurs as wide bands and lenses in almost all humic coals. It is widely distributed, but not abundant. The term was first used in the United States by J. J. Stevenson as a synonym for mineral charcoal. In the Thiessen-U. S. Bureau of Mines system, fusain is a component with a minimum band width of 37 microns. Microscopically, fusain closely resembles wood charcoal, usually being soft, friable, and black, and disintegrating readily into a black powder when roughly handled. A hard variety exists, impregnated with mineral matter. Examined microscopically in thin sections, fusain is usually opaque (black) although in very thin sections it may be slightly translucent and dark red.

**fused-salt reactor**

A type of reactor that uses molten salts of uranium for both fuel and coolant. O

**fusinization**

The process in coalification which results in the formation of fusain.

**fusion**

The formation of a heavier nucleus from two lighter ones (such as hydrogen isotopes), with the attendant release of energy (as in a hydrogen bomb). O

**fusion point**

The temperature at which melting takes place. Most refractory materials have no definite melting points, but soften gradually over a range of temperatures. D

**fusite**

In 1955 the Nomenclature Subcommittee of the International Committee for Coal Petrology resolved to use this term for the microlithotype consisting principally of the macerals fusinite, semifusinite and sclerotinite. D

**FWPCA**

Federal Water Pollution Control Act.

# G

**g.**
Abbreviation for gram, or gravity.

**gad**
A steel wedge used in mining. D

**gage**
(1) An instrument for measuring, indicating, or regulating the capacity, quantity, dimensions, power, amount, properties, etc., of anything, hence, a standard of comparison. (2) Spacing of tracks or wheels. M

**gal**
Gallon.

**gale**
Wind blowing from 30 to 60 miles per hour. Different sources define gale with different windspeeds. All agree that the average is between 40 and 50 mph.

**gallatin**
The heavy oil of coal tar used in the Bethell process for the preservation of timber. Also called dead oil. D

**gallery**
A horizontal or nearly horizontal underground passage, either natural or artificial. D

**gallon**
A unit measure of volume. A U. S. Gallon contains 231 cu. in. or 3.785 liters; it is 0.83268 times the imperial gallon. One U. S. gallon of water weighs 8.3374 lb. at 60° F. M

**galvanic cell**
A cell in which chemical change is the source of electrical energy. It usually consists of two dissimilar conductors in contact with each other and with an electrolyte, or of two similar conductors in contact with each other and with dissimilar electrolytes. D

**galvanize**
To coat with zinc. D

**galvanometer**
An instrument for measuring a small electric current or for detecting its presence or its direction by means of the movements of a magnetic needle or of a coil in a magnetic field that registers usually on a scale or by a moving beam of light reflected from a mirror attached to the needle or the coil. D

**GAMA**
Gas Appliance Manufacturers Association.

**gamma radiation**
Emission by radioactive substances of quanta of energy corresponding to X-rays and visible light but with a much shorter wavelength than light. They may be detected by gamma-ray Geiger counters. D

**gamma rays [Symbol (gamma)]**
High-energy, short-wave length electromagnetic radiation. Gamma radiation frequently accompanies alpha and beta emissions and always accompanies fission. Gamma rays are very penetrating and are best stopped or shielded against by dense materials, such as lead or depleted uranium. Gamma rays are essentially similar to X-rays, but are usually more energetic, and are nuclear in origin. O

**gangue**
Undesired minerals associated with ore, mostly nonmetallic. D

**gangway**
A main haulage road underground. D

**gantry**
A bridge or platform carrying a traveling crane or winch and supported by a pair of towers, trestles, or side frames running on parallel tracks. D

**Gantt chart**
Construction program for major engineering works, set out in graphic form. Down the vertical axis in sequence are set out the items concerned. The abscissa shows the period covered in days, or weeks, and the period allowed for each item marked by a horizontal line. The chart displays the inter-relation between the items, and aids in ensuring that no item is so delayed as to impede progress on a later one which depends on it. D

**GAO**
General Accounting Office.

**garnet**
A group of silicate minerals including several species with related chemical structure. D

**gas**
Any aeriform liquid other than atmospheric air, such as carbon dioxide, carbon monoxide, methane and natural gas. Gasoline is commonly, but improperly, referred to simply as "gas." M

**gas cap drive**
Often there is a considerable cap of gas trapped above the oil in a formation. When the rock is penetrated, this gas expands and exerts enough pressure on the oil to move it toward the only escape available--the well bore leading to the surface. M

**gas, casinghead**
Unprocessed natural gas produced from a reservoir containing oil. Sometimes called Bradenhead Gas. DD

**Gas cooled fast breeder reactor (GCBR)**
A fast breeder reactor which is cooled by a gas, usually helium, under pressure. G

**gas cycling**
A process of increasing the production of natural gas liquids from a gas field that is rich in these substances. It is neither a pressure maintenance nor secondary recovery process. "Wet" gas from the producing wells is sent to a special processing plant where liquids are removed and the remaining "dry" gas is returned to the gas reservoir through injection wells. Each time the dry gas circulates through the reservoir it picks up additional liquids which are, in turn, extracted when it reaches the processing plant again. When all of the recoverable liquids have been removed from the reservoir, the remaining gas is sold as dry gas. M

**gas detector**
A device to show the presence of firedamp, etc., in a mine. D

**gas, dry**
(1) Gas whose water content has been reduced by a dehydration process. (2) Gas containing little or no hydrocarbons, commercially recoverable as liquid product. Gas in this second definition preferably should be called "lean gas." Specified small quantities of liquids are permitted by varying statutory definitions in certain states. DD

**gaseous diffusion (plant)**
A method of isotopic separation based on the fact that gas atoms or molecules with different masses will diffuse through a porous barrier (or membrane) at different rates. The method used to separate uranium-235 from uranium-238; it requires large gaseous-diffusion plants and enormous amounts of electric power. O

**gaseous fuel**
Includes natural gas and the prepared varieties, such as coal gas, oil gas, iron blast furnace gas, as well as producer gas. D

**gasification**
(1) The conversion of a solid or a liquid to a gas that is suitable for use as a fuel which may have low, medium or high Btu content. (2) The formation of gas from carbon material to gaseous products which might include reaction with an oxygen, steam, carbon dioxide with products CO, $CO_2$, H, $CH_4$. Y

**gas impurities**
Undesirable matter in gas, such as dust, excessive water vapor, hydrogen sulfide, tar and ammonia. G

**gas oil**
A petroleum distillate with a viscosity and boiling range between those of kerosene and lubricating oil. M

**gas-oil ratio**
The relative quantity of gas in cubic feet to a barrel of oil produced. M

**gasoline**
A refined petroleum product which, by its composition, is suitable for use as a fuel in internal-combustion engines. M

**gasometers**
Tall metal chambers (gas holders) fitted with a roof which rises or falls with the entry or removal of gas. The roof is weighted so that the gas is compressed, thus enabling it to pass along gas mains and pipes at a suitable pressure (about 9 inch water gage) for efficient combustion. D

**gasoscope**
An apparatus for detecting the presence of dangerous gas escaping into a coal mine or a dwelling house. D

**gassy**
A coal mine is rated gassy by the U. S. Bureau of Mines if an ignition occurs or if a methane content exceeding 0.25 percent can be detected, and work must be halted if the methane exceeds 1.5 percent in a return airway. D

**gas well**
A well that produces chiefly natural gas. D

**gas works**
A plant for manufacturing gas. D

**GATE**
Group to advance total energy.

**gather**
To assemble loaded cars from several production points and deliver them to main haulage for transport to the surface or pit bottom. D

**gauss**
The unit of magnetic field intensity equal to 1 dyne per unit pole. The

preferred term for this unit is oersted. One oersted equals $10^5$ gammas. Gauss was used before the official adoption of the oersted in 1932. D

**gauze**
The wire mesh used to prevent the passage of flame from a flame safety lamp to the external atmosphere. D

**Gay-Lussac's law**
When gases react, they do so in volumes which bear a simple ratio to one another, and to the volumes of their products if these are gaseous, temperature and pressure remaining constant. Also called law of gaseous volume. D

**GCBR, GCFR (s)**
Gas-cooled fast breeder reactor.

**GCHWR**
Gas-cooled heavy-water modulated reactor.

**GCR**
Gas-cooled reactor.

**GDP**
Gross domestic product.

**Geiger-Muller counter (Geiger-Muller tube)**
A radiation detection and measuring instrument. It consists of a gas-filled (Geiger-Muller) tube containing electrodes, between which there is an electrical voltage but no current flowing. When ionizing radiation passes through the tube, a short, intense pulse of current passes from the negative electrode to the positive electrode and is measured or counted. The number of pulses per second measures the intensity of radiation. It is also often known as the Geiger counter; it was named for Hans Geiger and W. Muller who invented it in the 1920's. O

**Geissler tube**
A sealed and partly evacuated glass tube containing electrodes, used for the study of electric discharges through gases. D

**gel**
A jelly-like material representing the coagulation of a colloidal subtance. M

**general geology**
The branch of geology treating the problems of dynamic geology in relation to the geologic history of the earth. D

**general soil survey**
A general investigation of superficial deposits. The sampling procedure may include augers, boreholes, and trial pits; and tests are made to cover soil identification. This type of survey aims at establishing soil profiles and locating areas requiring special investigation. D

**generating station (generating plant or power plant)**
A station at which are located prime movers, electric generators, and auxiliary equipment for converting mechanical, chemical, and/or nuclear energy into electric energy. K

**generation, electric**
The process of transforming other forms of energy into electric energy. G

**genetically significant dose**
A population-averaged dose which estimates the potential genetic effects of radiation on future generations. It takes into consideration the number of people in various age groups, the average dose to the reproductive organs to which people in these groups are exposed, and their expected number of future children. W

**genetic effects of radiation**
(1) Radiation effects that can be transferred from parent to offspring. (2) Any radiation-caused changes in the genetic material of sex cells.

**genter filter**
A filter utilized in coal-washing plants for the recovery of fine coal particles. D

**geochemical exploration**
Exploration or prospecting methods depending on chemical analysis of the rocks or soil, or of soil gas, or of plants. D

**geochemistry**
(1) The study of the relative and absolute abundances of the elements and of the atomic species (isotopes) in the earth. (2) The distribution and migration of the individual elements in the various parts of the earth (the atmosphere, hydrosphere, crust, etc.), and in minerals and rocks, with the object of discovering principles governing this distribution and migration. D

**geofault**
A large fault directly affecting the relief of the earth's surface, on land or beneath the sea. D

**geognosy**
A branch of geology that deals with the materials of the earth and its general interior and exterior constitution. D

**Geological Survey**
A bureau of the Department of Interior established in 1879. The objectives of the Survey are to "perform surveys, investigations, and research covering topography, geology, and the mineral and water resources of the United States; classify land as to mineral character and water and power resources; enforce departmental regulations applicable to oil, gas, and other mining leases, permits, licenses, development contracts; and publish and disseminate data relative to the foregoing activities." G

**geophone**
Instrument for detecting seismic signals. M

**geophysics**
The science of the earth with respect to its structure, composition, and development. D

**geopressured resources**
Resources in regions where deep sedimentary basins filled with sand and clay or shale of Tertiary Age (less than 80 million years) are generally undercompacted below depths of 2 to 3 kilometers and, therefore, the interstitial fluid pressure carries a part of the overburden load. Waters in geopressured zones are not meteoric but are produced by compaction and dehydration of the marine sediments themselves. Geopressured deposits are hotter than normally pressured deposits. They occur in continuous belts, are commonly bounded by regional faults, and extend for hundreds of miles. J

**geopressured zones**
In the deep sedimentary deposits of the gulf coast, heat, water, and natural gas are trapped between impermeable shale beds. Pressures in these formations are unusually high so geopressured zones have three sources of energy—the intrinsic heat, the mechanical energy from the high pressures and the natural gas. The zones seem to have high potential as energy sources, but so little is presently known about them that major questions must be answered about how to utilize them.

**geotechnics**
The engineering behavior of all cuttings and slopes in the ground. This term is gradually replacing the term "soil mechanics." D

**geothermal; geothermic**
Of or relating to the heat of the earth's interior. G

**geothermal energy**
Heat from the interior of the earth. It has high potential for electric power production, residential space heating, desalinization, refrigeration, and air conditioning. Geothermal resources fall into three categories: steam, hot water and hot rocks. B

**geothermal energy resources**
Resources consisting of the thermal energy and the fluids (water plus dissolved minerals) which are found in the earth where the temperature of the formation and fluids significantly exceeds that which is to be anticipated on the basis of normal vertical temperature gradients and normal vertical heat flow. J

**geothermal gradient**
(1) The change in temperature of the earth with depth, expressed either in degrees per unit depth, or in units of depth per degree. (2) The mean rate of increase in temperature with depth in areas that are not adjacent to volcanic regions is about 1 degree F in about 55 feet, corresponding to about 100 degrees F per mile of depth. G

**geothermal power hazards**
Steam often smells sulphuric and makes a lot of noise at well sites. It usually contains hydrogen sulfide which can be hazardous if oxidized into sulfuric acid and discharged into the air or water. Other trace elements include nitrogen, boron, fluorides and arsenic. Withdrawal of steam or water can drop the local water table and release excess brine which must be reinjected to prevent earth subsidence. Since most geothermal resources are far from most population centers, large transmission line systems would have to be developed at great expense. D

**geothermal steam**
Steam drawn from deep within the earth. There are about 90 known places in the continental United States where geothermal steam could be harnessed for power. These are in California, Idaho, Nevada, and Oregon. G

**geothermometer**
(1) A thermometer designed to measure temperatures in deep-sea deposits or in boreholes deep below the surface of the earth. (2) A geologic thermometer. D

**geyser**
A volcano in miniature, from which hot water and steam are erupted periodically instead of lava and ashes during the waning phase of volcanic activity. Named from the Great Geyser in Iceland, though the most familiar example is probably Old Faithful in Yellowstone Park, Wyoming. D

**gib**
A temporary support at the face to prevent coal from falling before the cut is complete, either by hand or by machine. D

**Girbotal process**
A wet scrubbing process for removing hydrogen sulfide from fuel gases in which aqueous solutions of aliphatic amines dissolve $H_2S$ and $CO_2$ from gases. D

**Glauber's salts**
Salts in sealed cans to store heat. They are crystalline until they reach 90°F, when they melt, and, in doing so, absorb large amounts of heat, which is then held between the molecules of the liquid salts as latent heat of fusion. When the temperature falls again, the salts recrystallize and give out the heat they have been holding, which can be used to warm air circulated through the house. About 20 tons of such salts are needed to store

enough heat to keep a moderately-sized home warm for about two weeks, and the chemicals required are somewhat expensive. CC

**glazing materials on solar collectors**
The purpose of the glazing is to admit as much solar radiation as possible and to reduce the upward loss of heat to the lowest attainable value. In addition to serving as a heat-trap by admitting short-wave solar radiation and retaining longwave thermal radiation, the glazing also reduces heat loss by convection. Glass is the principal material used to glaze solar collectors, but plastic films and sheets are also used. They possess the advantage of being able to withstand hail and other stones, and, in the form of thin films, they are completely flexible. R

**glove box**
A sealed box in which workers, using gloves attached to and passing through openings in the box, can handle radioactive materials safely from the outside. O

**glow**
(1) The incandescence of a heated substance. (2) The light from such a substance. D

**GMT**
Greenwich mean time. D

**GNP**
Gross national product.

**goaf; gob**
That part of a mine from which the coal has been worked away and the space more or less filled up. D

**gobbing**
The act of stowing waste in a mine. D

**gouge**
A layer of soft material along the wall of a vein, favoring the miner, by enabling him, after gouging it out with a pick, to attack the solid vein from the side. D

**governor**
(1) A device for regulating the speed of an engine or motor under varying conditions of load and pressure. (2) A device for regulating the flow or pressure of a fluid, as gas or water. D

**gpd**
Gallons per day.

**gph**
Gallons per hour.

**gpm**
Gallons per minute.

**GPO**
Government Printing Office.

**gps**
Gallons per second.

**grade of coal**
A term to indicate the nature of coal mainly as determined by the amount and nature of the ash and the sulfur content. The term grade is sometimes used as a synonym for rank. D

**gradienter**
A surveyor's instrument consisting of a small telescope mounted on a tripod and fitted with a spirit level and a graduated vertical arc; used for determining grades, etc. D

**grading**
The degree of mixing of size classes in sedimentary material; well graded implies more or less uniform distribution from coarse to fine, poorly graded implies uniformity in size or lack of continuous distribution. D

**graduator**
An apparatus for evaporating a liquid by causing it to flow over large surfaces while exposed to a current of air. D

**Graham's Law of Diffusion**
The relative rates of diffusion of two gases are inversely proportional to the square roots of their densities. D

**grail**
(1) Gravel or sand. (2) Anything in fine particles. D

**gram-centimenter**
(1) A unit of work. (2) The work done in raising the weight of 1 gram vertically 1 centimeter. (3) 981 ergs. D

**gram-degree**
Same as calorie. D

**Granby cars**
A popular type of automatically dumped car for hand or power-shovel loading. In this type car, a wheel attached to the side of the car body engages an inclined track at the dumping point. D

**graphite (reactor grade)**
A very pure form of carbon used as a moderator in nuclear reactors. W

**graphitization**
The condition existing when, under the combined influences of steam, elevated temperatures, and time, a portion of the combined carbon in iron reverts to graphite. This condition is most likely to occur in the heat-affected zone adjacent to a weld.

**gravimeter; gravity meter**
An instrument for measuring the value of gravity or for measuring variations in the magnitude of the earth's gravitational field. Measurements of gravity are accomplished generally by one of three methods: dropped ball, pendulum or spring gravimeter. D

**gravitational separation**
The separation of oil, gas, and water in a reservoir rock in accordance with their relative gravities. D

**gravity**
The force by which substances are attracted to each other, or fall to earth. D

**gravity API**
The gravity scale developed by the American Petroleum Institute to express the density of liquid petroleum products. In this scale, water has a gravity of 10° API; and liquids lighter than water (such as petroleum oils) have API gravities numerically greater than 10. D

**gravity feed systems**
(1) A system that depends on gravity to operate. (2) In the solar field, a system which utilizes the principle that heated water is less dense than cool water and therefore rises, causing natural circulation. H

**greases**
A type of lubricant used to lubricate hard-to-reach places, made essentially of lubricating oil, soaps, additives, and fillers. Greases have been developed in hundreds of consistencies to meet innumerable performance characteristics. M

**Great Coal Age**
Another name for the Coal Measures or the Pennsylvanian. So called because the greatest coal deposits of the world are found in formations of this age. D

**greenhouse effect**
The heating effect of the atmosphere upon the earth. Light waves from the sun pass through the air and are absorbed by the earth. The earth then re-radiates this energy as heat waves that are absorbed by the air, specifically by carbon dioxide. The air thus behaves like glass in a greenhouse, allowing the passage of light but not of heat. Thus many scientists theorize that an increase in the atmospheric concentration of $CO_2$ can eventually cause an increase in the earth's surface temperature. E

**Gresham's Law**
When two or more coins are equal for the purpose of discharging a debt, but

unequal in intrinsic value, the one with the lowest intrinsic value will be circulated and the others hoarded. Similarly, cheap imitations tend to replace goods held to a standard of quality which are costly to maintain. D

**GRI**
Gas Research Institute.

**grit**
Sand, especially coarse sand.

**gross calorific value**
In the case of solid fuels and liquid fuels of low volatility, the heat produced by combustion of unit quantity, at constant volume, in an oxygen-bomb calorimeter under specified conditions. D

**gross energy consumption**
Total energy inputs into the economy, including coal, petroleum, natural gas, and the electricity generated by hydroelectric, nuclear, and geothermal power plants. Includes feedstocks and raw materials from fossil fuels such as asphalt, lubricants, chemical plant feedstocks, coke. Gross consumption includes conversion, transmission and distribution losses by the electric power sector. G

**gross generation**
The total amount of electric energy produced by the generating units in a generating station or stations. K

**Gross National Product (GNP)**
This is a comprehensive measure of aggregate economic activity. GNP measures final goods and services, eliminating intermediate products consumed in the production of end products. Because the use of energy is so intimately bound to economic activity, a forecast of GNP is necessary for forecasting energy demand.

**gross system capability**
The net generating station capability of a system at a stated period of time (usually at the time of the system's maximum load), plus capability available at such time from other sources through firm power contracts. K

**grounds**
Ground coal. D

**ground state**
The state of a nucleus, atom or molecule at its lowest (normal) energy level. O

**ground water**
The water which permeates, in an unbroken sheet, the rock masses of the earth, filling their pores and fissures. D

**GSA**
General Services Administration.

**gusher**
(1) An oil well with a strong natural outflow. (2) Sometimes is applied to a geyser. D

**GW**
Gigawatt, 1 million kilowatts.

**GWe**
Gigawatt (s) electric.

**GWt**
Gigawatt (s) thermal.

**GWyr**
Gigawatt years.

**gyro cracking**
A vapor-phase thermal cracking process. M

# H

**H**

Henry. Unit of electrical inductance. One H is equivalent to $10^9$ electromagnetic units.

**hade**

The angle of inclination of a vein measured from the vertical; dip is measured from the horizontal. D

**half-life**

The time in which half the atoms of a particular radioactive substance disintegrate to another nuclear form. Measured half-lives vary from millionths of a second to billions of years. O

**half life, effective**

The time required for a radionuclide contained in a biological system, such as a human or an animal, to reduce its activity by half as a combined result of radioactive decay and biological elimination. Compare biological half life and half life. W

**half mask**

The part of a mine rescue, or oxygen-breathing apparatus which covers the nose and mouth only, and through which the wearer breathes the oxygen furnished by the apparatus. D

**half-thickness**

The thickness of any given absorber that will reduce the intensity of a beam of radiation to one-half its initial value. W

**halite**

Impure common salt, NaCl. D

**halogens**

Members of the family of very active, chemical elements consisting of bromine, chlorine, fluorine, and iodine. Chemically, the halogens resemble each other closely, all being monovalent, non-metallic, and capable of forming negative ions. D

**hammerpick**

A compressed-air-operated hand machine used by miners to break up the harder rocks in a mine. It consists mainly of a pick and a hammer operated by compressed air. D

**hand cleaning**

The removal by hand of impurities from coal, or vice-versa. D

**hang-up**

Underground, blockage of ore pass or chute by rock. D

**hard coal**
All coal of higher rank than lignite. In the United States the term is restricted to anthracite. D

**hardness scale**
Quantitative units by means of which the relative hardness of minerals and metals can be determined, which for convenience is expressed in Mohs', Knoop, or scleroscope units for minerals, and Vickers, Brinell, or Rockwell units for metals. D

**haring cell**
A four-electrode cell for measurement of electrolyte resistance and electrode polarization during electrolysis. D

**harmonic**
A sinusoidal quantity having a frequency which is an integral multiple of the frequency of a periodic quantity to which it is related. D

**hatch**
A door or gate. D

**haul**
(1) The distance from the coal face to pit bottom or surface, in drift mining. (2) The distance quarry or opencast products must be moved to the treatment plant or construction site. (3) The distance from the shaft or opencast pit to spoil dump. D

**haulage**
The drawing or conveying, in cars, or movement of men, supplies, ore and waste both underground and on the surface. D

**haunt**
Coal sold at the pithead. D

**hawser**
(1) Any wire rope used for towing on lake or sea. (2) A large rope, varying from 5 to 24 inches in circumference, of 6 to 9 strands and left-handed twist. D

**HC**
Hydrocarbons.

**HE**
High explosive. D

**head**
(1) Development openings in a coal seam. (2) An advance main roadway driven in solid coal; the top portion of a seam in the coal face. (3) A unit of pressure intensity usually given in inches or feet of a column of fluid. Thus 1 foot of water head is the pressure from a column of water 1 foot high. D

**headbox**
A device for distributing a suspension of solids in water to a machine, or for retarding the rate of flow, as to a topfeed filter or for eliminating by over-flow some of the finest particles.

**header**
An entry-boring machine that bores the entire section of the entry in one operation. D

**heading**
(1) A passage leading from the gangway. The main entry in a coal mine is laid out with the precision of a main avenue in a city. From it at right angles headings run like cross streets, lined on each side with breasts, or chambers. (2) A smaller excavation driven in advance of the full-size section. D

**health physics**
The science concerned with recognition, evaluation, and control of health hazards from ionizing radiation.

**heat**
Form of energy generated or transferred by combustion, chemical reaction, mechanical means or passage of electricity and measurable by its thermal effects. D

**heat budget**
The total amount of the sun's heat received on the earth during any one year must exactly equal the total amount which is lost from the earth by reflection and radiation into space. D

**heat conduction**
The transfer of heat within a substance or from one substance to another without radiation, current, or mixing. AA

**heat content**
(1) A thermodynamic function; e.g., at constant pressure, $H = U + PV$ where $U$ = internal or intrinsic energy, $P$ = pressure, and $V$ = volume. (2) Commonly, the amount of heat in a quantity of matter at some given temperature. M

**heat energy**
Energy in the form of heat. D

**heat exchanger**
Any device that transfers heat from one fluid (liquid or gas) to another or to the environment. O

**heating season**
The coldest months of the year when pollution emissions are higher in some areas because of increased fossil-fuel consumption. E

**heating tendency**
The ability of a coal to fire spontaneously. This phenomenon can occur whenever the heat generated from oxidation reactions in a coal exceeds the heat dissipated. This characteristic varies for different types of coals and even for coals of the same classification but of different origin. D

**heating value**
The amount of heat produced by the complete combustion of a unit quantity of fuel. The gross, higher heating value, is that which is obtained when all of the products of combustion are cooled to the temperature existing before combustion, the water vapor formed during combustion is condensed, and all the necessary corrections have been made. The net, or lower heating value, is obtained by subtracting the latent heat of vaporization of the water vapor formed by the combustion of the hydrogen in the fuel from the gross, or higher heating value. DD

**heat island effect**
An air circulation problem peculiar to cities. Tall buildings, heat from pavements and concentrations of pollutants create a haze dome that prevents rising hot air from being cooled at its normal rate. A self-contained circulation system is put in motion that can be broken by relatively strong winds. If such winds are absent, the heat island can trap high concentrations of pollutants and present a serious health problem. E

**heat loss**
The heat transferred from a hot object by convection, conduction and radiation. H

**heat of combustion**
The heat created when a substance is completely burned in oxygen. The calorific, thermal, or heating value of a fuel is the total amount of heat developed by the complete combustion of a unit quantity of fuel; it is reported as calories per gram or Btu per pound. M

**heat of fusion**
The latent heat required to change a solid to a liquid. D

**heat of linkage**
The energy necessary to break a chemical bond. D

**heat of reaction**
The quantity of heat consumed or liberated in a chemical reaction, as

heat of combustion, heat of neutralization, or heat of formation. D

**heat of vaporization**
The latent heat required to change a liquid to a gas. D

**heat pump**
(1) A reversible heating and cooling mechanism that can produce additional usable heat from the amount stored, such as a mechanical refrigerating system which is used for air cooling in the summer and which, when the evaporator and condenser effects are reversed, can absorb heat from the outside air or water in the winter and raise it to a higher potential so that it can also be used for winter heating. FF (2) A method of gathering solar energy involving the collection of heat stored in the waters of a pond, or in the ground itself. A pump of this sort, a device for digging a thermal hole in one place and piling the heat taken from this up into a thermal hill somewhere else, can be used for either heating or cooling. The extra energy is derived from stored solar power which the pump gathers and delivers. CC

**heat rate**
A measure of generating station thermal efficiency, generally expressed in Btu per net kilowatt hour. It is computed by dividing the total Btu content of fuel burned for electric generation by the resulting net kilowatt hour generation. K

**heat sink**
Anything that absorbs heat, usually part of the environment, such as the air, a river, or outer space. D

**heat transfer coefficient**
The rate at which heat is transferred per hour, per unit surface, per degree of temperature difference. M

**heat transfer media**
A fluid used in the transport of thermal energy. In some systems the transport media is air, while in others, a liquid such as water or a water/antifreeze solution is used.

**heat transport**
The transport of heat by a unit volume of ocean water is a function of the specific heat, density, the temperature and the north-south component of the velocity, if the transport of heat from the equator toward the poles is being considered. Meterological phenomena account for most heat transport. Ocean currents, however, are considered of major importance. D

**heat treatment**
Heating and cooling a solid metal or alloy in such a way as to obtain desired conditions or properties. D

**heat unit**
Measure of amount of heat, such as the Btu or the calorie. M

**heat value**
(1) The energy released by burning a given amount of the substance. (2) Energy equivalent. V

**heaving**
Applied to the rising of the bottom after removal of the coal. D

**heavy oils**
(1) High density oils usually viscous but still mobile at reservoir conditions; oils of "low" API gravity. (2) Feedstocks requiring upgrading before pipelining and conventional refining—in this latter sense oil recovered in-situ by some methods from oil sands such as in Cold Lake, is also referred to as heavy oil. Z

**heavy water [Symbol $D_2O$]**
Water containing significantly more than the natural proportion (one in 6500) of heavy hydrogen (deuterium) atoms to ordinary hydrogen atoms. Heavy water is used as a moderator in some reactors because it slows down neutrons effectively and also has a

low cross section for absorption of neutrons. O

**heavy-water-moderated reactor**
A reactor that uses heavy water as its moderator. Heavy water is an excellent moderator and thus permits the use of inexpensive natural (unenriched) uranium as a fuel. O

**hectare**
A metric unit of land measure equal to 10,000 square meters or 2.471 acres. D

**hecto-**
A prefix used to denote one hundred times. D

**heel**
A small body of coal left under a larger body as a support. Known also as heel of coal. D

**heliochemical process**
A technological process of utilizing solar energy through photosynthesis, for the maintenance of life on this planet. R

**helioelectrical process**
Technological process by which solar energy can be utilized using photovoltaic converters, providing power for all of the communications satellites and ultimately, being very valuable for terrestrial applications. R

**heliostat**
An electromechanical system that orients a mirror so that direct radiation is reflected on a fixed position regardless of the position of the sun. Q

**heliothermal process**
A process by which solar energy can be utilized to provide much of the thermal energy needed for space heating and cooling and domestic water heating. R

**heliothermometer**
An instrument that measures the sun's heat by means of thermoelectric junction. E

**Helmholtz coil**
A pair of similar coaxial coils with their distance apart equal to their radium, which permits an accurate calculation of the magnetic field between the coils. Used in calibration of magnetometers. D

**hemicellulose**
One of the cellulose materials which is a carbohydrate which serves to cement the plant cellulose fibers together which has the chemical formula $C_9 H_{10} O_3$ ) 0 $CH_3$ ) n. Y

**Henry**
Unit of electrical induction. With electromotive force of 1 volt and current of 1 ampere. One Henry (H) = $10^9$ electromagnetic units. Symbol H. D

**Henry's law**
The mass of a gas dissolved by a given volume of liquid at a given temperature is proportional to the pressure. The volume dissolved is independent of pressure. D

**HEPA**
High-efficiency particulate air (filter).

**hermetic**
Made impervious to air and other fluids by fusion. Originally applied to the closing of glass vessels by fusing the ends and by extension applied to any mode of airtight closure. D

**hertz**
Cycles per second. U. S. electrical supply has a frequency of 60 hertz. G

**heterogeneous**
(1) Differing in kind. (2) Having unlike qualities or possessing different characteristics, as opposed to homogeneous. M

**heterogeneous reactor**
A reactor in which the fuel is separate from the moderator and is arranged in discrete bodies, such as fuel elements. Most reactors are heterogeneous. O

**HEU**
Highly enriched uranium.

**HEW**
Formerly Department of Health, Education and Welfare. Now called Department of Health and Human Services.

**HHS**
Department of Health and Human Services.

**HHV**
High heating value.

**high-efficiency particulate air filter (HEPA)**
An air filter capable of removing at least 99.97 percent of the particulate material in an air system. P

**high-energy fuel**
A fuel having higher energy release than that of hydrocarbons with oxygen; e.g., hydrogen, boron compounds; fluorine. M

**highest useful compression ratio**
The optimum value of the compression ratio; it varies with fuel, engine, and operating conditions. M

**high explosive**
An explosive with a nitroglycerin base requiring the use of a detonator to initiate the explosion, which is violent and practically instantaneous. High explosives may be divided into gelatins and dynamites and also a special type known as permitteds (or permitted explosives) for use in gassy or dusty coal mines. They possess much greater concentrated strength than low explosives such as black powder. D

**high frequency**
Electrically, rapid reciprocation as with alternating current or an oscillating circuit. D

**high-level liquid waste**
Fluid materials, disposed of by storage in underground tanks, that are contaminated by greater than 100 microcuries/milliliter of mixed fission products or more than 2 microcuries/milliliter of cesium-137, strontium-90, or long-lived alpha emitters. P

**high potential**
As applied in the U. S. Bureau of Mines Federal Mine Safety Code, means voltages in excess of 650 volts. D

**high pressure**
A liquid or aeriform gas pressurized to more than 150 pounds per square inch. D

**high-rank coals**
Coals containing less than 4 percent of moisture in the air-dried coal or more than 84 percent of carbon (dry ash-free coal). All other coals are considered as low-rank coals. D

**High temperature gas cooled reactor (HTGCR)**
A promising approach to commercial nuclear power which it is claimed would permit more efficient use of uranium and also some use of thorium in its fuel cycle. Offers greater thermal efficiency than light water reactors. G

**high-volatile A bituminous coal**
Nonagglomerating bituminous coal having less than 69 percent of fixed carbon and more than 31 percent of volatile matter and 14,000 or more British thermal units (moist mineral-matter-free). D

**high-volatile B bituminous coal**
Nonagglomerating bituminous coal having 13,000 or more, and less than

14,000 British thermal units (moist, mineral-matter-free). D

**high-volatile C bituminous coal**
Either agglomerating or nonweathering bituminous coal having 11,000 or more, and less than 13,000 British thermal units (moist, mineral-matter-free). D

**high-volatile coals**
Coals containing over 32 percent of volatile matter with a coal rank code No. 400 to 900. D

**high voltage**
(1) A high electrical pressure or electromotive force. (2) That which is greater than 650 volts. Also called high potential. D

**"highwalls"**
A term applied to the great gashes on mountainsides caused by explosive charges, power shovels or mammoth auger drills used to bore out the minerals in strip mining. R

**hitch**
(1) A fault. (2) Fractures and dislocations of strata common in coal measures, accompanied by more or less displacement. D

**hl**
Hectoliter. D

**hm**
Hectometer. D

**hog**
The upward bending or camber of a beam. Prestressed concrete beams are frequently cast with a hog to counteract any downward sag. D

**hog fuel**
(1) Forest wastes or residuals. (2) Bark, shavings, sawdust, low grade lumber and rejects resulting from the operation of pulp saw and plywood mills. Z

**holder, gas**
A gas tight receptacle or container in which gas is stored for future use. There are two general ways of storing gas: (1) at approximately constant pressure (low pressure containers) in which case the volume of the container changes and, (2) in containers of constant volume (usually high pressure containers) in which case the quantity of gas stored varies with the pressure. DD

**hole**
(1) To undercut a seam of coal by hand or machine. (2) A drill hole, borehole or well. D

**holocellulose**
The name describing the combination of cellulose and hemicellulose. Y

**homogeneous**
Of the same kind of nature; consisting of similar parts or of elements of a like nature; uniform in characteristics. M

**homogeneous reactor**
A reactor in which the fuel is mixed with or dissolved in the moderator or coolant. Example: a fused-salt reactor. O

**homogenizing**
Holding at a high temperature to eliminate or decrease segregation by diffusion. D

**hopper**
A vessel into which materials are fed, usually constructed in the form of an inverted pyramid or cone terminating in an opening through which the materials are discharged (not primarily intended for storage).

**hopper car**
A railway car for coal, gravel, etc., shaped like a hopper, with an opening to discharge the contents. G

**horizon mining**
A system of mine development which is suitable for inclined, and perhaps faulted, coal seams. Main stone heading are driven at predetermined levels, from the winding shaft to intersect and gain access to the seams to be developed. The stone headings, or horizons are from 100 to 200 yards vertically apart, depending on the seams available and their inclination. D

**horizontal-axis rotors**
Windmills in which the axis of rotation is parallel to the direction of the windstream; typical of conventional windmills; head-on type. U

**horsepower**
(1) A unit of rate of operation. (2) One mechanical horsepower equals 33,000 ft-lb per min, or 550 ft-lb per sec. M

**horsepower hour**
One horsepower expended for one hour, or the horsepower multiplied by the number of hours. One horsepower hour equals 1,980,000 foot-pounds, 0.745 kilowatt-hours, 2,545 Btu (mean). G

**hot**
Highly radioactive. O

**hot cell**
A heavily shielded enclosure in which radioactive materials can be handled by persons using remote manipulators and viewing the materials through shielded windows or periscopes. O

**hot dry rock**
Possibly the Nation's largest and most widely distributed source of geothermal energy, the development of this resource is difficult. It requires drilling a hole, and cracking the deep rock, either hydraulically as in conventional oil field development or by explosives. Water would then be injected into the fractured rock; after being heated to nearly the temperature of the surrounding rock, the hot water would be withdrawn through another hole. The heat would be extracted at the surface and the water re-injected into the underground system. This potential energy source has never been put into use.

**hot reserve**
Thermal generating capacity maintained at a temperature and condition which will permit it to be placed into service promptly. L

**hot spot**
(1) A surface area of higher-than-average radioactivity. (2) A part of a fuel element surface that has become overheated. P

**hp**
Horsepower.

**HPCS**
High-pressure core spray.

**HPFL**
High performance fuels laboratory.

**hr**
Hour.

**HTGR**
High-temperature gas-cooled reactor.

**humacite**
A group name for bitumens which vary from gelatinous to hard resinous or elastic. Believed to represent an emulsion of highly acidic (humic acids) hydrocarbons with a varying amount of water (as high as 90 percent). Insoluble in organic solvents. D

**humic coals**
(1) A group of coals including the ordinary bituminous varieties, which have been formed from accumulation of vegetable debris that have maintained their morphological organization with little decay. The majority

of them are banded and have a tendency to develop joining or cleat. (2) Chemically, humic coals are characterized by hydrogen varying between 4 and 6 percent. D

**humid heat**
Ratio of the increase in total heat per pound of dry air to the rise in temperature, with constant pressure and humidity ratio. D

**humidifier**
A device for maintaining the correct degree of moisture in the atmosphere. D

**humidity**
The condition of the atmosphere with respect to water vapor. When the word humidity is used without a qualifying adjective the relative humidity is usually meant. Humidity may be expressed in many different ways; for example, absolute humidity, mixing ratio, saturation deficit, and specific humidity. D

**humins**
In coal, amorphous brown to black substances, formed by natural decomposition from vegetable substances, insoluble in alkali carbonates, water, and benzol. D

**hunger**
(1) Dirty, mottled clay, formed from the weathering of shale. (2) Crystalline calcium carbonate found in the joints of coal seams. D

**HVAB**
High-volatile "A" bituminous. D

**HVAC**
Heating, ventilating, and air conditioning.

**HVACs**
High voltage alternating current system.

**HVBB**
High-volatile "B" bituminous. D

**HVCB**
High-volatile "C" bituminous. D

**HVDCs**
High-voltage direct current system.

**HWLWR**
Heavy-water moderated boiling light-water cooled reactor.

**HWMP**
Hazardous Waste Management Program.

**hydrate**
A compound or complex ion formed by the union of water with some other substance and represented as actually containing water. D

**hydration**
The chemical process of combination or union of water with other substances. E

**hydraulic**
Strictly, having to do with water in motion, but term is extended to cover all liquids which convey, store or transfer pressure energy to reactants. D

**hydraulic capacity**
The rating of a hydroelectric generating unit or the sum of such ratings for all units in a station or stations. K

**hydraulic engineer**
One who handles the engineering work of design, erection, and construction of sewage-disposal plants, waterworks, dams, water-operated power plants. D

**hydraulic excavation**
Excavation by means of a high pressure jet of water, the resulting waterborne excavated material being conducted through flumes to the desired dumping point. D

**hydraulic extraction**
A term which has been given to the processes of excavating and transporting coal or other material by water energy. Also called hydroextraction. D

**hydraulic fluid**
A fluid of petroleum or nonpetroleum origin used in hydraulic systems. Low viscosity, low rate of change of viscosity with temperature, and low pour point are desirable characteristics. M

**hydraulic fracturing**
A general term, for which there are numerous trade or service names, for the fracturing of rock in an oil or gas reservoir by pumping a fluid under high pressure into the well. The purpose is to produce artificial openings in the rock in order to increase permeability. G

**hydraulic mining**
Mining by washing sand and dirt away with water which leaves the desired mineral. D

**hydraulic theory**
A theory of oil and gas migration that suggests that migration is caused by the movement of underground water which carries along oil and gas. D

**hydro**
A term used to identify a type of generating station or power or energy output in which the prime mover is driven by water power. K

**hydrocarbon**
(1) A compound that contains only hydrogen and carbon, the principal components of fossil fuels. (2) One of the principle products produced from biomass such as methane. The simplest and lightest forms of hydrocarbon are gaseous. With greater molecular weights they are liquid, while the heaviest are solids. Y

**hydrocracking**
Usually used as an adjunct to catalytic cracking, a process in which hydrogen is added to materials with large molecules, then cracking at pressures of 70 to 150 atmospheres and at moderate temperature from 500 to 750°F. Feedstocks can include crude oil, residue, petroleum tars, and asphalts. Y

**hydrodynamics**
That branch of hydraulics which relates to the flow of liquids over weirs, or through pipes, channels, and openings.

**hydrodynamometer**
An instrument for determining the velocity of a fluid in motion by its pressure. D

**hydroelectric**
Power (electricity) produced by water power. Also called hydropower.

**hydroelectric power station**
A power station in which electricity is generated by the energy of falling water. D

**hydroelectric scheme**
A complete project for water development which will include the design and construction of a dam, tunnels, spillways, power station intakes, and many other constructional works over a wide area. D

**hydrogasification**
A gasification which involves the addition of hydrogen to products of primary gasification to optimize the formation of methane usually as a part of heating process. Y

**hydrogenation**
(1) The chemical reaction in which hydrogen is added as gas to a substance in the presence of a catalyst at high temperatures and high pressure. Hydrogen reacts with certain organic compounds to saturate double and

triple bonds. (2) It is a process used for the production of oil or gas from coal, but it has also been successfully used in laboratory to make these fuels from organic waste. Y

**hydrogeneous**
(1) Formed or produced by water. (2) Applied to rocks formed by the action of water, in contradistinction to pyrogenous rocks, or rocks formed by the action of heat. D

**hydrogeneous coal**
(1) Coal high in volatile matter, for example, gas coal or sapropelic coal. (2) Coals containing a large quantity of moisture; for example, brown coal. D

**hydrologic cycle**
The complete cycle of phenomena through which water passes, commencing as atmospheric water vapor, passing into liquid and solid form as precipitation, thence along or into the ground surface, and finally again returning to the form of atmospheric water vapor through evaporation and transpiration. Also called the water cycle. D

**hydrolysis**
A chemical process that breaks complex organic molecules into simple molecules. For example, starch and cellulose hydrolyzed by acids or enzymes to produce simple sugars which can be fermented to produce alcohol. Y

**hydrometer**
An instrument used for determining the density or specific gravity of fluids, such as drilling mud or oil, by the principle of buoyancy. The instrument is floated in the fluid and sinks to a greater or lesser depth depending on the density of the fluid, the amount of submergence being indicated by graduations or divisions on the instrument. D

**hydrometrograph**
An instrument for determining and recording the quantity of water discharged from a pipe, orifice, etc., in a given time. D

**hydrophilic**
Of, or relating to, having, or denoting a strong affinity for water. D

**hydrosphere**
The aqueous envelope of the earth, including the ocean, all lakes, streams, and underground waters, and the water vapor in the atmosphere. D

**hydrostatic pressure**
The pressure of, or corresponding to the weight of a column of water at rest. D

**hydrotator**
A coal washer of the classifier type whose agitator or rotator consists of hollow arms radiating from a central distributing manifold or center head. D

**hydrothermal reservoir**
One of the forms of geothermal reservoir systems. Consists of naturally circulating hot water or steam ("wet steam") or those which contain mostly vapor ("dry steam"). The latter type of hydrothermal reservoir is the most desirable type with present technology. V

**HYGAS**
A process to produce pipeline quality gas by hydrogasification of coal. G

**hygrometry**
Measurement of atmospheric humidity. D

**hygroscopic**
Having the property of absorbing moisture. M

**hysteresis**
Energy loss amounting to the difference between the work input and the work output in a cycle of extension and retraction, as in the stretching and relaxing of rubber. M

# I

**i**
Symbol for the instantaneous value of electric current in amperes. D

**IAEA**
International Atomic Energy Agency.

**IBG**
Intermediate Btu-gas.

**ICC**
Interstate Commerce Commission.

**ICONS**
Information Center on Nuclear Standards.

**ICOP**
Imported crude oil processing.

**ICRP**
International Commission on Radiological Protection.

**id**
Inside diameter.

**IEA**
International Energy Agency.

**IEEE**
Institute of Electrical and Electronic Engineers.

**ignite**
To heat a gaseous mixture to the temperature at which combustion takes place. D

**ignition control compounds**
Gasoline additives which reduce deposits in the combustion chamber so that they will not preignite the gasoline, and also modify deposits on spark plugs to reduce spark plug misfiring. M

**ignition point**
Of solids and liquids, the minimum temperature at which combustion can occur, but at which point it is not necessarily continuous. The point or temperature at which a substance takes fire. D

**IGT**
Institute of Gas Technology. G

**ihp**
Indicated horsepower.

**IIASA**
International Institute of Applied Systems Analysis.

**IIT**
Illinois Institute of Technology.

**illuminating gas**
(1) Coal and carbureted water gases and their various mixtures. (2) The different classes of oil gas. (3) Acetylene, gasoline gas, and producer gas. The first is the most important for illuminating purposes. Producer gas is the most important for fuel and power gas. D

**image furnace**
Apparatus for the production of a very high temperature in a small space by focusing the radiation from the sun (solar furnace) or from an electric arc (arc-image furnace) by means of mirrors and/or lenses. Such furnaces have been used for the preparation and study of some special ceramics. D

**immiscible**
(1) Not capable of mixing. (2) Tending to form two layers, as to oil and water. M

**imp**
A tremendous impulse of energy shot through the sea for ocean floor mining purposes. D

**impeller**
Rotating member of centrifugal pump, which receives inflowing water or ore pulp at or near its center and accelerates it radially to the periphery, where it is discharged with the kinetic energy needed to carry it through the pumping system. D

**impermeable**
Having a texture that does not permit water to move through it perceptibly under the head differences ordinarily found in subsurface water. Synonym for impervious. D

**impermeable dry rock**
Hot impermeable dry rock systems are those geothermal regions where the heat is contained almost entirely in impermeable rock of very low porosity. The heat generated by interactions of large plates of the earth's surface has been in part responsible for the presence of large, once-molten masses of granite forming the core of the Sierras, for example, and various other once-molten masses called batholiths. It has been estimated that the thermal energy released by cooling one cubic mile of rock from an initial temperature of 350°C to 177°C would be equivalent to that available from 300 million barrels of oil. J

**impervious**
Impassable, applied to impermeable strata, such as clays, shales, etc., which will not permit the penetration of water, petroleum, or natural gas. D

**impregnation**
The treatment of porous castings with a sealing medium to stop pressure leaks. D

**impulse turbine**
A water turbine, such as the Pelton wheel, in which the driving force is provided more by the speed of the water than by a fall in its pressure. D

**in**
Inch.

**in-lb.**
Inch-pound.

**inch-pound**
(1) The work done in raising 1 pound 1 inch. (2) A unit of work. Abbreviation, in-lb. D

**inch ton**
A unit of work equal to that required to lift 1 ton 1 inch high. D

**incineration**
The controlled process by which solid, liquid or gaseous combustible wastes are burned and changed into gases; the residue produced contains little or no combustible material. E

**incoalation**
The process of coal formation which begins after peat formation is completed without there being any sharp boundary between the two processes. D

**incombustible**
Applies to substances that will not burn. D

**incompatible**
Applied to a substance which, for chemical, physical or physiological reasons, cannot be mixed with another without changing its nature or effect. M

**incomplete combustion**
A term applied to combustion in which all of the fuel is not burned; for example, leaving unburned carbon in ashes. M

**incorporation**
A process by which material contributing to coal formation responds to diagenetic and metamorphic agencies of coalification and becomes a part of the coal without undergoing any material modification. See also coalification. Compare vitrinization; fusinization. D

**incremental energy costs**
The additional cost of producing and/or transmitting electric energy above some base cost previously determined. L

**index fossil**
A fossil which, because of its wide geographic distribution and restricted time range, can be used to identify and to date the strata or the succession of strata in which it occurs. Synonym for guide fossil. D

**index of refraction; refractive index**
The ratio of the velocity of light, or of other radiation; in the first, of two media to its velocity; in the second, as it passes from one medium into the other, the first medium usually being taken to be a vacuum or air. Symbol, n. D

**indicated horsepower**
In an engine the power actually given by the steam (or working medium), or the rate at which work is done on the engine piston. M

**indicated reserves**
Reserves based partly on specific measurements, samples, or production data, and partly from projections for a reasonable distance on geological evidence. G

**indigenous**
Originating in a specific place; in situ. D

**indirect-cycle reactor system**
A reactor system in which a heat exchanger transfers heat from the reactor coolant to a second fluid which then drives a turbine. O

**induced radioactivity**
Radioactivity that is created when substances are bombarded with neutrons as from a nuclear explosion or in a reactor, or with charged particles and photons produced by accelerators. W

**induction**
The production of magnetization or electrification in a body by the mere proximity of magnetized or electrified bodies, or of an electric current in a conductor by the variation of the magnetic field in its vicinity. D

**induction meter**
An instrument for measuring a.c. circuits that will measure both power and energy (kilowatt hours). It is the type used for household and other consumer electricity meters. D

**industrial air conditioning**
Air conditioning in industrial plants where (usually) the objective is the furtherance of a manufacturing

process rather than the comfort of human beings. D

**industrial calorific value**
The calorific value obtained when coal is burned under a boiler. D

**industrial degree-day**
A degree-day unit based (usually) on a 45°F or 55°F mean daily temperature so as to be applicable to industrial buildings maintained at relatively low temperatures. D

**inert gas**
A gas that does not react with other substances under ordinary conditions. E

**inertial confinement**
One of two major techniques used in nuclear fusion experimentation. (See "Magnetic Confinement.") A frozen pellet of deuterium and tritium is bombarded from all sides by an energy source—a laser beam of charged particles. The resulting implosion of the pellet results in high temperature and density which allows ignition of the fusion reaction and the pellet explodes. V

**inertinite**
This term was proposed to simplify the nomenclature of coal petrography by combining, in a single term, the group of the following macerals: micrinite, semifusinite, fusinite, sclerotinite. This grouping is based on certain similarities in the technological properties of the four macerals. D

**INFCE**
International nuclear fuel cycle evaluation.

**inferred reserves**
Reserves based upon broad geologic knowledge for which quantitative measurements are not available. Such reserves are estimated to be recoverable in the future as a result of extensions, revisions of estimates, and deeper drilling in known fields. G

**infiltration**
The deposition of a mineral among the mineral grains or in the pores of a rock by the permeation of the percolation of water carrying it in solution. D

**influent stream**
A stream, or the reach of a stream is influent with respect to ground water if it contributes water to the zone of saturation. D

**infracrustal rock**
Rock that originated at great depth, either by consolidation from magma or by granitization.

**infrared**
Pertaining to the micron wavelength region of the electromagnetic spectrum from approximately 0.78 $\mu$ to 300 $\mu$. M

**infusible**
Not transformable by heating from solid to liquid state under specified conditions of pressure, temperature, and time. D

**ingate**
The point of entrance from a shaft to a level in a coal mine. D

**inhibitor**
A substance, the presence of which, in small amounts, in a petroleum product prevents or retards undesirable chemical changes from taking place in the product, or in the condition of the equipment in which the product is used. In general, the essential function of inhibitors is to prevent or retard oxidation or corrosion. D

**initial boiling point**
According to ASTM Method D 86, the recorded temperature when the first drop of liquid falls from the end of the condenser. M

**initiation**
The process of causing a high explosive to detonate. The initiation of an explosive charge requires an initiating point, which is usually a primer and electric detonator, or a primer and a detonating cord or fuse. D

**inject**
To introduce, under pressure, a liquid or plastic material into cracks, cavities, or pores in a rock formation. D

**inorganic**
Applied to all substances that do not contain carbon as a constituent, also to a few others in which carbon is present in an unimportant sense, for example, metallic carbonates. Metals, rocks, minerals and a variety of earths are all inorganic. D

**insequent**
(1) Developed on the present surface but not consequent on or controlled by the structure. (2) Descriptive of a certain type of stream, drainage, and dissection. D

**in-situ**
In the natural or original position. Applied to a rock, soil or fossil when occurring in the situation in which it was originally formed or deposited. G

**in-situ coal gasification**
Coal gasified underground by pumping hot steam-oxygen mixtures through coal seams.

**in-situ combustion**
An experimental means of recovery of oil of low gravity and high viscosity which is unrecoverable by other methods. The essence of the method is to heat the oil to increase its mobility by decreasing its viscosity. Heat is applied by igniting the oil sand and keeping the fire alive by the injection of air. The heat breaks the oil down into coke and lighter oils and the coke catches fire. As the combustion front advances, the lighter oils move ahead of the fire into the bore of a producing well. D

**in-situ origin theory**
The theory of the origin of coal that holds that a coal was formed at the place where the plants from which it was derived grew.

**in-situ recovery**
Refers to methods to extract the fuel component of a deposit without removing the deposit from its bed. Extracted gas or oil is pumped or forced out by pressure of displacing fluids. G

**insolation**
(1) Solar radiation which is delivered to any place on the surface of the earth directly from the sun. (2) The rate of such radiation per unit of surface. E

**inspissation**
Drying up. An inspissated oil deposit is one from which the gases and lighter fractions have escaped, and only the heavier oils and asphalt remain. D

**installed capacity**
The total of the capacities as shown by the nameplates of similar kinds of apparatus such as generating units, turbines, synchronous condensers, transformers, or other equipment in a station or system. L

**instantaneous detonator**
A detonator in which there is no designed delay period between the passage of an electric current through the detonator and its explosion. D

**Institution of Mining and Metallurgy: I.M.M.**
The London Institution of Mining and Metallurgy is the central British organization for regulating the professional affairs of suitably qualified

mining engineers engaged in production and treatment of non-ferrous metals and rare earths. D

**insufflator**
An injector for forcing air into a furnace. D

**insulate**
To separate or to shield (a conductor) from conducting bodies by means of nonconductors, so as to prevent transfer of electricity, heat, or sound. D

**insulation**
The prevention of the transfer of energy between two conductors by separation of the conductors with a non-conducting material; or, the nonconducting material itself. If electric, separation of a conductor or charged body from earth or from other conductors is by means of a nonconducting barrier. If thermal, prevention of passage to or from a body of external heat is by use of nonconducting envelope. E

**intake**
The passage by which the ventilating current enters a mine; intake for an adit or entry. See also downcast, which is more appropriate for a shaft. D

**integrated gasoline companies**
Gigantic corporations involved in most steps from production and refining to marketing. B

**intensity**
(1) The energy or the number of photons or particles of any radiation incident upon a unit area or flowing though a unit of solid material per unit of time. (2) In connection with radioactivity, the number of atoms disintegrating per unit of time. O

**intensity of radiation**
At a given place, the energy per unit time entering a sphere of unit cross-sectional area centered at that place. The unit of intensity is the erg per square centimeter second or the watt per square centimeter. D

**interbedded**
Occurring between beds, or lying in a bed parallel to other beds of a different material; same as interstratified. D

**intercalated**
Descriptive of a body of rock interbedded or interlaminated with another body of different rock. D

**intercolline**
Placed between hills; specifically applied to depressions between the cols and crateriform hillocks of volcanic regions. D

**interconnection**
A tie permitting a flow of energy between the facilities of two electric systems. L

**interfacial energy**
Tension at interfaces between the various phases of a system which may include solid, liquid and gas interfaces, varying in their combinations and qualities. D

**interior coalfields**
U. S. includes: Eastern interior field, Illinois, Indiana, Kentucky; Western interior field, Great Plains States from Iowa to Arkansas; Southwestern field, Texas; and Northern field, Michigan. D

**intermediate (epithermal) neutron**
A neutron having energy greater than that of a thermal neutron but less than that of a fast neutron. The range is generally considered to be between about 0.5 and 100,000 electron volts. O

**intermediate (epithermal) reactor**
A reactor in which the chain reaction is sustained mainly by intermediate neutrons. O

**internal-combustion engine**
An engine which operates by means of combustion of a fuel within its cylinder. M

**internal energy**
Of a gas, the total heat energy stored in a unit mass of a gas due to the motion and position of the molecules of the gas. This energy is measured by a thermometer. D

**interruptible load**
Electric power load which may be curtailed at the supplier's discretion, or in accordance with a contractual agreement. L

**interstate waters**
According to law, waters defined as: (1) rivers, lakes, and other waters that flow across or form a part of state or international boundaries; (2) waters of the Great Lakes; (3) coastal waters—whose scope has been defined to include ocean waters seaward to the territorial limits and waters along the coastline (including inland streams) influenced by the tide. E

**interstice**
An opening in anything or between things; especially, a narrow space between the parts of a body or things close together; a crack, a crevice, a chink, or a cranny. D

**interstitial**
Descriptive of void spaces; interstitial or connate water occupies part of the void spaces in the reservoir rock. D

**in-the-seam mining**
The conventional system of mining in which the development headings are driven in the coal seam. D

**intratelluric**
Formed or occurring within the earth; said of the constituents of an effusive rock formed before its appearance on the surface, or of the period of their formation. Also called intratellural. D

**intrinsically safe**
Apparatus that is so constructed that, when installed and operated under the conditions specified by the certifying authority, any electrical sparking that may occur in normal working, either in the apparatus or in the circuit associated therewith, is incapable of causing an ignition of the prescribed flammable gas or vapor. D

**intrusion**
(1) In geology, a mass of igneous rock which, while molten, was forced into or between other rocks. (2) A mass of sedimentary rock occurring in a coal seam. D

**intumesce**
To enlarge, to expand, to swell, or to bubble up (as from being heated). D

**inversion**
An atmospheric condition where a layer of cool air is trapped by a layer of warm air so that it cannot rise. Inversions spread polluted air horizontally rather than vertically so that contaminating substances cannot be widely dispersed. An inversion of several days can cause an air pollution episode.

**ion**
An atom or molecule that has lost or gained one or more electrons. By this ionization it becomes electrically charged. Examples: an alpha particle, which is a helium atom minus two electrons; a proton, which is a hydrogen atom minus its electron. O

**ion engine**
An engine which provides thrust by expelling accelerated or high velocity ions. Ion engines using energy provided by nuclear reactors are proposed for space vehicles. O

**ion exchange**
A chemical process involving the reversible interchange of various ions between a solution and a solid material, usually a plastic or resin. It is

used to separate and purify chemicals, such as fission products, rare earths, etc., in solutions. O

**ionization**
The process of adding one or more electrons to, or removing one or more electrons from, atoms or molecules, thereby creating ions. High temperatures, electrical discharges, or nuclear radiations can cause ionization. O

**ionization chamber**
An instrument that detects and measures ionizing radiation by measuring the electrical current that flows when radiation ionizes gas in a chamber, making the gas a conductor of the electricity. W

**ionization event**
An occurrence in which an ion or group of ions is produced; for example, by passage of a charged particle through matter. W

**ionization potential**
Energy, in volts, needed to remove an electron from a normal atom and to leave it positively charged. To ionize is to dissociate a molecule or a compound into ions of opposite charge. D

**ionized gas**
A gas that is capable of carrying an electric current. G

**ionizing radiation**
Any radiation capable of displacing electrons from atoms or molecules, thereby producing ions. Examples: alpha, beta, gamma radiation, short-wave ultraviolet light. Ionizing radiation may produce severe skin or tissue damage. W

**IP**
Institute of Petroleum (London).

**IPAA**
Independent Petroleum Association of America.

**i.p.s.**
Inches per second.

**ironshot**
Of a certain mineral, streaked or speckled with iron, or an iron ore. D

**irradiated fuel**
Nuclear fuel that has been used in a nuclear reactor. X

**irradiation**
Exposure to radiation, as in a nuclear reactor. O

**irrespirable**
Not respirable; not fit to be breathed. Said of mine gases. D

**ISF**
Intermediate storage facility.

**ISO**
International Organization for Standardization.

**isobar**
An imaginary line or line on a map or chart connecting or marking places on the surface where the height of the barometer reduced to sea level is the same either at a given time or for a certain period. D

**isocals**
Lines of equal calorific value in coal drawn on the map or on a diagram. D

**isomer**
One of two or more nuclides with the same numbers of neutrons and protons in their nuclei, but with different energies; a nuclide in the excited state and a similar nuclide in the ground state are isomers. O

**isomerization**
A refining method which alters the structure of hydrocarbons, rearranging

petroleum molecules to form high-octane products. M

**isotherm**
A line or graph on a diagram, representing constant temperature. M

**isotope**
One of two or more atoms with the same atomic number (the same chemical element) but with different atomic weights. An equivalent statement is that the nuclei of isotopes have the same number of protons but different numbers of neutrons. Isotopes usually have very nearly the same chemical properties, but somewhat different physical properties. See "long-lived isotope." O

**isotope separation**
The process of separating isotopes from one another, or changing their relative abundances, as by gaseous diffusion or electromagnetic separation. Isotope separation is a step in the isotopic enrichment process. W

**isotopic enrichment**
A process by which the relative abundances of the isotopes of a given element are altered, thus producing a form of the element which has been enriched in one particular isotope and depleted in its other isotopic forms. W

**IUPAC**
International Union of Pure and Applied Science.

# J-K

**j**
Joule.

**jack arch**
An arch having horizontal or nearly horizontal upper and lower surfaces; may be called flat arch or straight arch. The term is also used for any arch roughly built. D

**jar block**
Synonym for drive hammer. D

**jackline man**
In petroleum production, one who pumps several oil wells from a central power plant (jack plant), engaging or disengaging rods or cables by which power is transmitted to separate wells. D

**jamb**
(1) A vein or bed of earth or stone, which prevents the miners from following a vein of ore. (2) A large block. D

**JCAE**
Joint Committee on Atomic Energy.

**Jeppe's tables**
A series of tables especially compiled for mining work that includes tables of density, vapor pressure and absolute humidity. D

**jet engine**
An engine which converts fuel and air into a fast-moving stream of hot gases which effect propulsion of the device of which the engine is a part. M

**jet fuel**
Fuel meeting the required properties for use in jet engines and aircraft turbine engines. M

**jig**
(1) A device which separates coal from foreign matter by a means of their difference in specific gravity in a water medium. (2) A kind of shaker conveyor. (3) A guide used in shaping pieces of wood or metal. D

**jimmy**
A shot crowbar. D

**jink**
A coupling between two mine tubs or trains in a set or journey. D

**jinny road**
Underground gravity plane. D

**jobbers**
Wholesalers who buy from refineries and sell to service stations.

**Johannsen's classification**
A mineralogical classification of igneous rocks in which a rock is char-

acterized by a number, the Johannsen number, consisting of three or four digits, each of which has a specific mineralogical significance. D

**joint**
(1) A line of cleavage in a coal seam. (2) The location where two or more members are to be or have been fastened together either mechanically or by brazing or welding. D

**joule**
A unit of energy or work which is equivalent to one watt per second or 0.737 foot-pounds. G

**Joule's Law**
Either of two statements: (1) The rate at which heat is produced by a steady current in any part of an electric circuit is jointly proportional to the resistance and to the square of the current, or (2) the internal energy of an ideal gas depends only upon its temperature irrespective of volume and pressure. D

**Joy double-ended miner**
A cutter loader for continuous mining on a longwall face. It consists of two cutting heads fixed at each end of a caterpillar-mounted chasis. The heads are pivoted and controlled hydraulically for vertical movement. D

**Joy microdyne**
A wet-type dust collector for use at the return end of tunnels or hard headings. It wets and traps dust as it passes through the appliance, and releases it in the form of a slurry which is removed by a pump. D

**Joy miner**
A continuous miner mainly for use in coal headings and extraction of coal pillars. It weighs about 15 tons and comprises (1) a turntable mounted on caterpillars, (2) a ripper bar, and (3) a discharge boom conveyor. The ripper bar has six cutter chains with picks running vertically to the plane of the seam. An intermediate conveyor behind the ripper bar delivers the coal into a small hopper and a discharge conveyor takes it to the outbye end of the machine. D

**Joy transloader**
A rubber-tired self-propelled loading, transporting, and dumping machine. D

**Joy walking miner**
A continuous miner with a walking mechanism instead of caterpillar tracks. The walking mechanism was adopted to make the machine suitable for thin seams. D

**jp fuel**
Jet propulsion fuel.

**JPL**
Jet Propulsion Laboratory.

**jug**
A colloquial equivalent of detector, geophone, etc. D

**junking**
The process of cutting a passage through a pillar of coal. D

**k**
Kilo (thousand); also indicates that the basic number that follows is multiplied by 1000 or $10^3$.

**K**
Kelvin (degree). Also a common symbol for a numerical constant.

**kaolin**
A clay, mainly hydrous aluminum silicate, from which porcelain may be made. Also called china clay; porcelain clay. D

**kata-**
A prefix used with metamorphic names to indicate an origin in the deepest zone of metamorphism. D

**katamorphism**
The breaking down processes of metamorphism, as contrasted with the building up processes of anamorphism. D

**kc**
Kilocycle.

**keg**
(1) A cylindrical container made of steel or some other substance, which contains 25 pounds of blasting powder or gunpowder. (2) Any small cask or barrel having a capacity of 5 to 10 gallons. D

**Kelvin scale**
A fundamental temperature scale in which the temperature measure is based on the behavior of an ideal gas. The zero on this scale is at -273.16°C. Degrees Kelvin, therefore, equals degrees centigrade +273.16. Also known as the absolute thermodynamic scale. M

**kennel coal**
A coal that can be ignited with a match to burn with a bright flame. It is also known as candle coal, and this latter name is probably the origin of the term cannel coal. D

**kerf**
The undercut usually made in the coal to facilitate its fall. D

**kerogen**
A bituminous substance which yields an oily substance when heated in the absence of air. A major component of oil shale. Y

**kerosene**
Any jet fuel, diesel fuel, fuel oil or other petroleum oils derived by refining or processing crude oil or unfinished oils, in whatever type of plant such refining or processing may occur, which has a boiling range in atmospheric pressure from 400 degrees to 550 degrees F. G

**kerosine coal**
Another name for oil shale. D

**KeV**
Kiloelectron volt(s).

**keystone**
A symmetrically tapered piece at the center or crown of an arch. D

**kg**
Kilogram.

**KGRA**
Known geothermal resources areas.

**kill**
As applied to an oil or gas well, this means to shut off the flow of oil or gas temporarily or to destroy the well entirely so that neither oil nor gas can flow. D

**kiln**
A large furnace used for baking, drying, or burning firebrick or refractories, or for calcining ores or other substances. D

**kilo-**
A prefix that multiplies a basic unit by 1000.

**kilocalorie**
A large calorie. Abbreviation, kcal. D

**kilocycle**
One thousand cycles. Abbreviation, kc. D

**kilo electron volt**
One thousand electron volts. Abbreviation, KeV. D

**kilogram**
The unit of weight in the metric system equal to 1,000 g or 2.2046 lb. avoirdupois. Abbreviation, kg. M

**kiloton energy**
The energy of a nuclear explosion which is equivalent to that of an explosion of 1,000 tons of TNT. O

**kilovar (KVAR)**
1,000 reactive voltamperes. K

**kilovoltampere (kVA)**
An electrical term that indicates the energy in an alternating current circuit. It is the product of voltage expressed in kilovolts and current expressed in amperes.. G

**kilowatt**
A unit of power equal to 1,000 watts.

**kilowatt-hour**
A unit of work or energy equal to that expended by one kilowatt in one hour. It is equivalent to 3,412 Btu of heat energy. G

**kinematic viscosity**
The ratio of the absolute viscosity to the density at the temperature of the viscosity measurement. The metric units of kinematic viscosity are the stoke and centistoke, which correspond to the poise and centipoise of absolute viscosity. M

**kinetic energy**
Energy available as a result of motion. M

**kinetic friction**
The condition of friction-bearing surfaces when in motion; commonly called friction of motion. M

**kirwanite**
A variety of anthracite with a metallic luster. D

**km**
Kilometer.

**knapping**
The act of breaking stone. D

**knock**
(1) The noise associated with self-ignition of a portion of the fuel-air mixture ahead of the advancing flame front. (2) In mining, to examine a mine roof for safety. M

**knot**
(1) The unit of speed used in navigation, equal to one nautical mile (6,080.20 ft.)/ hr. (2) An imperfection. E

**kohm**
Kilohm. D

**krypton-85**
A fission product and radioactive isotope of the element krypton, an inert gas, with a half-life of about 10 years. X

**kton**
Kiloton. D

**kv**
Kilovolt.

**kw**
Kilowatt.

**KWe**
Kilowatt(s) electric.

**kwh**
Kilowatt-hour.

**kWh(e)/yr**
Kilowatt hours of electricity per year.

**kwt**
Kilowatt(s) thermal.

**kWyr**
Kilowatt year.

# L

**l**
Liter.

**L**
Abbreviations for: liter, Lambert, lumen, liquid. Also symbol for inductance in Henry's.

**labeled atom**
One rendered radioactive and thus traceable through a chemical process of a flow line. Also called tagged atom. D

**labile**
Applied to particles in a rock that decomposes easily. D

**labor**
(1) A shaft, cavity or other part of a mine from which ore is being, or had been, extracted. (2) A working, as a labor in a quicksilver mine. D

**labor turnover**
The ratio of the new men taken on at a mine to the average total number on the books during the year. At a settled mine, this ratio may be less than 20%, while in remote areas it is considerably higher. D

**lacing**
The timber or other material placed behind and around the main supports. D

**lacustrine**
Produced by or belonging to lakes. D

**lade**
A watercourse, ditch, or drain. D

**ladies**
Building slates, size 16 X 8 inches. D

**LAER**
Lowest achievable emission rates.

**lag**
(1) A flattish piece of wood or other material to wedge the timber or steel supports against the ground and to secure the area between the supports. (2) A lid; a wedge.

**lag of ignition**
The time which elapses during the preflame period. D

**LASL**
Los Alamos Scientific Laboratory.

**lamellar**
(1) Composed of thin layers, plates, scales or lamellae. (2) Disposed in layers like the leaves of a book. D

**laminated coal**
Thinly bedded coal. D

**lamination**
Stratification on a fine scale, each thin stratum, or lamina, being a small fraction of an inch in thickness, typically exhibited by shales and fine-grained sandstones. D

**lamings**
Partings in coal seams. D

**lampblack**
A black or gray pigment made by burning low-grade heavy oils or similar carbonaceous materials with insufficient air, and in a closed system so that the soot can be collected in settling chambers. Properties are markedly different from carbon black. Used as a black pigment for cements and ceramic ware, an ingredient in liquid-air explosives, in lubricating composition, and as a reagent in the cementation of steel. D

**landed cost**
The cost of imported crude oil equal to actual cost of crude at point of origin plus transportation cost to the United States. G

**landing**
Level stage in a shaft, at which cages are loaded and discharged. D

**land reclamation**
Generally means restoring in major or total part land which has been strip mined. B

**land subsidence**
The sinking of a land surface as the result of the withdrawal of underground material. It results from underground mining and is a hazard of the development of geothermal fields. V

**langley**
A unit of solar energy, commonly employed in radiation theory and equal to one gram calorie per square centimeter of irradiated surface. E

**lantern coal**
Another name for cannel coal.

**large coal**
One of the three main size groups by which coal is sold by the National Coal Board of Great Britain. Large coal has no upper size limit and has a lower size limit of 1 to 2 inches.

**laser**
(1) A convenient contracted version of the phrase "light amplification by stimulated emission of radiation." (2) A device making use of a new technique for obtaining exceedingly intense and coherent beams of visible radiation, by making use of the fluorescent properties of a large number of compounds. D

**laser isotope separation**
An isotope enrichment process now in the development stage. Atoms of uranium-235, or molecules containing them, would be selectively ionized or excited by lasers, allowing physical or chemical separation of one of the isotopes. X

**latent energy**
The energy required to cause a change of state at constant temperature, as in the melting of ice or the vaporization of water. E

**latent heat**
That quantity of heat which disappears or becomes concealed in a body while producing some change in it other than the rise of temperature. BB

**lateral development**
Any system of development in coal seams or thick ore bodies in which headings are driven horizontally across the coal or ore and connected to main haulage drifts, entries or shafts. There are many variations and modifications depending on the thickness, shape, and inclination of the deposit. D

**lateral support**
Means whereby walls are braced either vertically or horizontally by columns, pilasters, or crosswalls or by floor or roof constructions, respectively. D

**laterlog**
The electrical resistivity of coal appears to decrease with ash content. The laterlog measures what is virtually the true resistivity of the coal and may ultimately provide information on seam quality. D

**law of definite proportions**
This states that in every sample of any one compound substance, the proportions by weight of the constituent elements are constant. D

**law of equivalent proportion**
When elements combine (or replace one another) to form compounds they do so in weights which are proportional to their equivalents. D

**law of gravitation**
The law, discovered by Sir Isaac Newton, that every particle of matter attracts every other portion of matter, and the stress between them is proportional to the product of their masses divided by the square of their distance apart. D

**law of indestructibility of matter**
Matter is neither created nor destroyed in the course of chemical action. Also called the law of conservation of matter. D

**law of mass action**
The rate of a chemical reaction is directly proportional to the molecular concentrations of the reacting substances. D

**law of motion**
A statement in dynamics, a body at rest remains at rest and a body in motion remains in uniform motion in a straight line unless acted upon by an external force. The acceleration of a body is directly proportional to the applied force and is in the direction of the straight line in which the force acts. For every force and there is an equal and opposite force or reaction. D

**Lawson criterion**
A rough measure of success in fusion. For a self-sustaining fusion reaction to take place, the product of the confinement time (in seconds), and the particle density (in particles per $cm^3$) must be about $10^{14}$. V

**layer**
A bed or stratum of rock. D

**lb**
Pound.

**lb-ft**
Pound-foot.

**LBG**
Low Btu-gas.

**LCRM**
Low cost residual material.

**LDC**
Less developed countries.

**LEA**
Low excess air.

**leach**
To wash or to drain by percolation. D

**leachate**
The liquid that has percolated through the soil or other medium. D

**lead**
Petroleum industry parlance for the motor fuel antiknock additive compound tetraethyllead, or for other organometallic lead antiknock compounds. M

**leaded gasoline**
Refers to gasoline containing tetraethyllead or other organometallic lead antiknock compounds. M

**league**
(1) A unit of linear measure. A land league = 3 statute miles or 15,840 feet. A nautical league = 3 geographical miles or 18,240.78 feet. (2) Also an area that contains 4,428.40 acres or 6.92 square miles. The term is used in Texas land descriptions. D

**leakage**
In nuclear engineering, the escape of neutrons from a reactor core. Leakage lowers a reactor's reactivity. O

**leak vibroscope**
An instrument which detects leaks in water, oil, gas, stream, and air lines by amplifying the sound produced by the escaping fluid. D

**lean gas**
(1) The residual gas from the absorber after the condensable gasoline has been removed from the "wet" gas. (2) In the coal industry, a term used in several European countries for coal with a low volatile matter. M

**lean oil**
Absorption oil from which gasoline fractions have been removed. Oil leaving the stripper in a natural gasoline plant. M

**lease**
Contract between landowner and another granting the latter right to search for and produce oil or mineral substances upon payment of an agreed rental, bonus, and/or royalty. D

**leg**
In mine timbering, a prop or upright member of a set or frame. D

**legend**
The explanation of the symbols and patterns shown on a map or diagram.

**leg wire**
One of the two wires atttached to and forming a part of an electric blasting cap or squib. D

**Lehmann process**
A process for treating coal by disintegration and separation of the petrographic constituents, (fusain, durain, and vitrain). D

**LEL**
Lower explosive limit.

**lens**
A body of ore or of rock thick in the middle and thin at the edges, similar to a double convex lens. D

**lenticle**
A rock stream or rock bed, which, from being thin at the edges, is more or less lens-shaped. Nearly all undeformed strata are lenticles. D

**Lenz's law**
When an electromotive force is induced in a conductor by any change in the relation between the conductor and the magnetic field, the direction of the electromotive force is such as to produce a current whose magnetic field will oppose the change. D

**LeRC**
Lewis Research Center.

**lessee**
One who leases mineral lands, including oil, gas, sulphur and potash; incorrectly leaser. D

**lessor**
One who transfers a mineral lease, including oil and gas leases. D

**LET**
Linear energy transfer.

**lethal dose**
A dose of ionizing radiation sufficient to cause death. Median lethal dose (MLD or LD-50) is the dose required

to kill within a specified period of time (usually within 30 days) half of the individuals in a large group or organisms similarly exposed. The median lethal dose (LD-50/30) for man is about 400-450 roentgens. O

**leveling**
In surveying, measurement of rises and falls, heights and contour lines in engineering projects or map making. D

**levigation**
A means of classifying a material as to particle size by the rate of settling from a suspension.

**LGR**
Light-water cooled, graphite-moderated reactor.

**LHV**
Lower heating value.

**licensed material**
Source material, special nuclear material, or byproduct material received, possessed, used or transferred under a general or special license issued by the NRC or a state. O

**life cycle**
The phases, changes or stages an organism passes through during its lifetime. E

**life cycle cost**
A measure of what a thing will cost totally, not only to buy but also to operate over its lifespan. Often computed as capital cost plus present value of running costs. C

**life of mine**
May be defined as the time in which, through the employment of the available capital, the ore reserves, or such reasonable extension of the ore reserves as conservative geological analysis may justify, will be extracted. D

**lig.**
Lignite. D

**light**
A form of radiant energy which gives rise to the sensation of sight. Light travels through space with a velocity of 186,285 miles per second, in common with radio and other waves of a similar physical nature but different wavelength. D

**lightning arrester**
A lightning arrester is a protective device for limiting surge voltages on equipment by discharging or bypassing surge current; it prevents continued flow of current to ground, and is capable of repeating these functions as specified. D

**light oil**
Any of the products distilled or processed from crude oil up to, but not including, the first lubricating oil distillate. Any of the products beginning with and following the first lubricating oil distillate is referred to as a "heavy oil." M

**light water**
Ordinary water ($H_2O$), as distinguished from heavy water or deuterium oxide ($D_2O$). O

**Light-Water Reactor (LWR)**
A nuclear reactor in which water is the primary coolant moderator with slightly enriched uranium fuel. There are two commercial light-water reactor types--the boiling water reactor (BWR) and the pressurized water reactor (PWR). G

**lignin**
A carbohydrate in the fibrous portion of wood and vascular plants formula $C_9H_{10}O_3(OCH_3)_n$. Y

**lignite**
A brownish-black coal in which the alteration of vegetal material has proceeded further than in peat but not

so far as subbituminous coal. Lignite contains about 30 to 40% water and has a heat value of about 5500-8300 Btu per pound; usually it has a high ash content but generally contains little sulphur, less than 1%. D

**lignitous coal**
Coal containing 75 percent to 84 percent of carbon (ashless, dry basis). D

**lignocellulose**
A carbohydrate made up primarily of lignin and cellulose. Y

**lime coal**
Inferior variety of coal suitable only for lime burning and similar purposes. D

**liming**
The brick, concrete, cast iron or steel casing placed around a tunnel or shaft as a support. D

**liminic coal basin**
A coal basin formed inland from the seacoasts, as opposed to a paralic coal basin. D

**limit switch**
A control to limit some function. Examples are pressure limit switches which shut off the fuel burner when the stream pressure reaches a predetermined point, and temperature limit switches for hot and warm air. D

**limnology**
The study of the physical, chemical, meteorological and biological aspects of fresh waters. E

**linear energy transfer (LET)**
The density of ionization events along the path of a nuclear particle. Beta and gamma radiations have low LET whereas alpha particles and neutrons have high LET. X

**line drop**
Loss in voltage owing to the resistance of conductors conveying electricity from a power station to the consumer. D

**line loss**
(1) Energy loss and power loss on a transmission or distribution line. (2) Kilowatt hours and kilowatts lost in transmission and distribution lines under specified conditions. L

**line pack, gas delivered from**
That volume of gas delivered to the markets supplied by the net change in pressure in the regular system of mains, transmission and/or distribution. For example, the change in the content of a pipeline brought about by the deviation from steady flow condition. DD

**line-sawed**
Said of oil-well casing when worn by the drill rope or cable. D

**lip screen**
A common term applied to stationary screens installed in the loading chutes over which the coal flows as it is loaded into railroad cars for market. D

**liquation**
The process of separating by heat a fusible substance from a substance less fusible. D

**liquefaction**
The process of making or becoming liquid. Conversion of a solid into a liquid by heat, or of a gas into a liquid by cold or by pressure. D

**liquefied natural gas (LNG)**
Natural gas cooled to -259°F so it forms a liquid at approximately atmospheric pressure. As natural gas becomes liquid it reduces volume nearly 600 fold, thus making both storage and long distance

transportation economically feasible. Natural gas in its liquid state must be regasified and introduced to the consumer at the same pressure as other natural gas. The cooling process does not alter the gas chemically and the regasified LNG is indistinguishable from other natural gases of the same composition. DD

**liquefied petroleum gas**
A gas containing certain specific hydrocarbons which are gaseous under normal atmospheric conditions, but can be liquefied under moderate pressure at normal temperatures. Propane and butane are the principal examples. Abbreviated LPG. DD

**liquid-dominated convective hydrothermal resources**
Thermally-driven convective systems of meteoric water in the upper part of the earth's crust which transfer heat from a deep igneous source to a depth sufficiently shallow to be tapped by drill holes. The exploration target is a reservoir located in the upflowing part of the convective system. The thermal energy is stored both in the solid rock and in the water and steam which fill the pores and fractures. Often referred to as "hot spots." In major zones of upflow, coexisting steam and water may extend to the surface and result in geysers and hot boiling springs. J

**liquid fuel**
Any liquid used as fuel which can be poured or pumped. Petroleum or crude oil is a natural liquid fuel, while distilled oil, coal tar, and residual oil are prepared liquid fuels. D

**liquid-liquid extraction (solvent extraction)**
A process in which one or more components are removed from a liquid mixture by intimate contact with a second liquid, which is itself nearly insoluble in the first liquid and dissolves the impurities and not the substance that is to be purified. D

**liquid measure**
Includes the following: 4 gills (gi) or 16 ounces equal 1 pint (pt); 2 pints equal 1 quart (qt); 4 quarts equal 1 gallon (gal); 42 gallons equal 1 barrel (bbl); and 2 barrels equal 1 hogshead (hhd). D

**Liquid Metal Fast Breeder Reactor (LMFBR)**
A nuclear breeder reactor cooled by molten sodium in which fission is caused by fast neutrons. G

**liter; litre**
A metric unit of capacity that equals the volume occupied by 1 kilogram of water at 4°C and at the standard atmospheric pressure of 760 millimeters; it equals 1.000028 cubic decimeters. D

**lithification**
A type of coalbed termination in which the disappearance takes place because of a lateral increase in impurities resulting in a gradual change into bituminous shale or other rock. D

**lithium**
The lightest metal; a silver-white alkali metal. Lithium 6 is of interest as a source of tritium for the generation of energy from a controlled fusion reaction. Molten lithium will also be the heat exchanger. V

**littoral**
(1) Of or pertaining to a shore. (2) A coastal region.

**live steam**
Steam direct from the boiler and under full pressure; to be distinguished from exhaust steam, which has been deprived of its available energy. D

**LLL**
Lawrence Livermore Laboratory.

**LMFBR**
Liquid metal fast breeder reactor.

**LMU**
Logical mining units.

**LNG**
Liquefied natural gas.

**load**
(1) The amount of gas delivered or required at any specified point or points on a system; load originates primarily at the gas consuming equipment of the customers. (2) To load a governor is to set the governor to maintain a given pressure as the rate of gas flow through the governor varies. The amount of electric power delivered or required at any specified point or points on a system. Load originates primarily at the power consuming equipment of the customers. See design load; see demand. DD

**loader**
A mechanical shovel or other machine for loading coal, ore, mineral, or rock. D

**load factor**
The ratio of the average load in kilowatts supplied during a designated period to the peak or maximum load in kilowatts occurring in that period. Load factor, in percent, also may be derived by multiplying the kilowatt hours in the period by 100 and dividing by the product of the maximum demand in kilowatts and the number of hours in the period. K

**load factor rate schedule**
A rate schedule in which the number of kilowatt-hours in any energy block is based on a demand characteristic. For example: 3 cents per kilowatt-hour for the first 30 kilowatt-hours per kilowatt of the maximum demand; 1 cents per kilowatt-hour for the energy in excess of 30 kilowatt-hours per kilowatt of the maximum demand. L

**load impedance**
The load impedance of an energy load is the impedance which would be measured at the terminals of that load if they were not connected to a source. D

**load power**
The load power of an energy load is the average rate of flow of energy through the terminals of that load when connected to a specified source. D

**loam**
A mixture of sand, silt, or clay, or a combination of any of these with organic matter, humus. It is sometimes called topsoil in contrast to the subsoil that contain little or no organic matter. D

**LOCA**
Loss-of-cooling-accident.

**local mean time**
Hour angle of mean sun. Local apparent noon is transit time at which upper limb of sun crosses local meridian. This must be corrected by use of the equation of time from the Nautical Almanac when making sun observations. D

**LOFT**
Loss-of-fluid test.

**log**
Logarithm (common).

**long-lived isotope**
A radioactive nuclide that decays at such a slow rate that a quantity of it will exist for an extended period, usually radionuclides whose half-life is greater than 3 years. P

**long ton**
A unit that equals 20 long hundredweight or 2,240 pounds. Used chiefly in England. D

**longwall**
A long face of coal. D

**longwall mining**
A method of working coal seams that originated in England in the 17th century. The seam is removed in one operation by means of a long working face, or wall. The workings advance (or retreat) in a continuous line. The space from which the coal has been removed is either allowed to collapse or is completely or partially filled or stowed with stone or debris. Longwall mining emphasizes economy of extracting the maximum amount of scarce reserves. In contrast the conventional American pillar and block approach emphasizes the bountiful nature of U.S. coal reserves. G

**loss**
The general term applied to energy (kilowatt hours) and power (kilowatt) lost in the operation of an electric system. Losses occur principally as energy transformations from kilowatt hours to waste heat in electrical conductors and apparatus. K

**loss-of-coolant accident (LOCA)**
A reactor accident in which the primary coolant is lost from the reactor core. X

**loss of fluid test (LOFT)**
An experimental device, one-sixtieth the size of a commercial pressurized-water reactor used to simulate loss of coolant accidents. X

**loss of head**
(1) The decrease of energy head between two points resulting from friction, bends, obstructions, expansions, or any other cause. It does not include changes in the elevation of the hydraulic grade line unless the hydraulic and energy grade lines parallel each other. (2) The difference between the total heads at two points in a hydraulic system. E

**loss of vend**
Difference between weight of raw coal and that of salable products, expressed as a percentage. D

**low coal**
Coal occurring in a thin seam or bed. D

**lowest normal tides**
A plane of reference lower than mean sea level by half the maximum range. This does not take into account wind or barometric pressure fluctuations. D

**low explosive**
An explosive in which the change into the gaseous state is effected by burning and not by detonation as with high explosives. Blasting powder (black powder or gunpowder) is the only low explosive in common use. D

**low-grade**
(1) An arbitrary designation of dynamites of less strength than 40 percent. It has no bearing on the quality of the materials, as they are of as great purity and high quality as the ingredients in a so-called high-grade explosive. (2) Pertaining to ores that have a relatively low content of metal compared to other richer material from the same general area. (3) Designates coal high in impurities. D

**low-grade coal**
Combustible material which has only limited uses owing to undesirable characteristics. D

**low-level liquid waste**
Fluid materials that are contaminated by less than $5 \times 10^{-5}$ microcuries/milliliter of mixed fission products. P

**LOX**
Liquid oxygen. D

**LPA**
Liquid petroleum air.

**LPCI**
Low-pressure coolant inspection.

**LPCS**
Low-pressure core spray.

**lpg**
Liquefied petroleum gas. Also abbreviated LPG. D

**LP gas**
A gas consisting of volatile hydrocarbons, from propane to pentane, mixed with hydrogen and methane under pressure. It withstands pressure and hence may be transported in steel tanks under pressure in liquefied form. May be used as a fuel to operate combustion-type engines in lieu of gasoline. Sold under various trade names but more commonly known as propane, butane, LP gas, blau gas (or blue gas). D

**LRA**
Land resource area (for biomass).

**LRR**
Land resource region.

**lu**
Lumen. D

**lubricant**
A substance, especially oil, grease, or a solid such as graphite, which may be interposed between the moving parts of machinery to reduce friction and prevent contact between the bearing surfaces. See also specific lubricants under alphabetical listing. D

**lubricating oils**
Oils ranging from clear thin oil to thick, dark oil. All equipment with moving parts require lubrication. There are literally hundreds of these oils to fill exacting requirements—for industry, automobile engines, and for space lubricants. M

**lucimeter**
Instrument for measuring the mean intensity of solar global radiation (direct and diffuse) near the earth's surface in a specified time interval. N

**lumen**
The quantity of light required to illuminate 1 square foot to an average intensity of 1 foot-candle. By multiplying the mean candlepower (mcp) by 3.63, the quantity of light expressed in lumens for the solid cone with a spread of 130° is obtained. D

**luminance**
A measure of surface brightness that is expressed as luminous flux per unit solid angle per unit projected area. D

**luminescence**
An emission of light that is not ascribable directly to incandescence and therefore occurs at low temperatures, that is produced by physiological processes (as in the firefly), chemical action, friction, or electrical action (as the glow of gases in vacuum tubes when subjected to electric oscillations of high frequency or as the glow of certain bodies when subjected to cathode rays), by certain bodies while crystallizing, by suddenly and moderately heating certain bodies previously exposed to light or to cathode rays, or by exposure to light, or that occurs in radioactivity. D

**lunar day**
The time for one rotation of the earth with respect to the moon or the interval between two successive upper transits of the moon over a local meridian. The mean lunar day is approximately 28.84 solar hours or 1.035 times as great as the mean solar day. Also called tidal day. D

**Lurgi process**
A type of coal gasification (steam-coal-oxygen) which employs a bed of crushed coal traveling downward through the gasifier and operating at pressures up to 30 atmospheres. Steam and oxygen are admitted through a revolving steam-cooled grate which also removes the ash produced at the bottom of the gasifier. The gases are made to pass

upward through the coalbed, carbonizing and drying the coal. Large quantities of steam are used to prevent ash from clinkering and the grate from overheating, and a hydrogen-rich gas is produced.

**lv**
Liquid volume.

**lvb**
Low-volatile bituminous. D

**LWBR**
Light-water breeder reactor.

**LWR**
Light-water reactor.

# M

**m**
Meter, mile, or mass.

**MAC**
Maximum allowable concentration.

**machine mining**
Implies the use of power machines and equipment in the excavation and extraction of coal or ore. In coal mines, the term signifies the use of coal cutters and conveyors and perhaps some type of power loader working in conjunction with face conveyors. D

**machine wall**
The face at which a coal cutting machine works. D

**macro–**
A prefix meaning large, long, visibly large. D

**macroclastic coal**
A well-laminated coal with a high proportion of vitrinite bands and fragments together with some resins, spore exines, cuticles, and a little fusinite. D

**macromolecule**
Very large molecule, notably one polymerized to a size visible without need for magnification. D

**macroporosity**
Porosity visible without the aid of a microscope, such as pipes and blowholes in ingots. D

**macroscopic**
Visible at magnifications of from 1 to 10 diameters. Also visible without a microscope or in a hand specimen. D

**maf, MAF**
Moisture and ash-free. D

**magazine**
A building specially constructed and located for the storage of explosives. D

**magma**
A comprehensive term for the molten fluids generated within the earth from which igneous rocks are believed to have been derived by crystallization or by other processes of consolidation. A magma not only includes the material represented by all or part of an igneous rock but also any volatile fluxes and residual liquors that may have escaped during or after consolidation. D

**magma systems**
Magma geothermal systems are those systems where the thermal energy is contained in liquid or near-liquid rock at temperatures ranging from 600° to perhaps 15,000°C. Recovering usable energy from volcanic areas has great potential, but in many instances, very deep drilling may be required to reach these resources. J

**magmatic**
Of, pertaining to, or derived from magma. D

**magnetic anomaly**
Variation of the measured magnetic pattern from a theoretical or empirically smoothed magnetic field on the earth's surface. D

**magnetic bottle**
A magnetic field used to confine or contain a plasma in controlled fusion (thermonuclear) experiments. W

**magnetic confinement**
A confinement technique used in nuclear fusion in which electrons are stripped from the reacting nuclei (deuterium and tritium, for example) forming a "plasma" which can be controlled by a magnetic field. There are several different types of magnetic confinement systems under development. V

**magnetic field**
Space surrounding a magnet or current-carrying coil, in which appreciable magnetic force exists. D

**magnetic force**
The mechanical force exerted by a magnetic field upon a magnetic pole placed in it. D

**magnetic method**
A geophysical prospecting method which maps variations in the magnetic field of the earth which are attributable to changes of structure or magnetic susceptibility in certain near-surface rocks. D

**magnetic mirror**
A magnetic field used in controlled fusion experiments to reflect charged particles back into the central region of a magnetic bottle. W

**magnetic storage**
A futuristic concept in which energy can be stored in a magnetic field around a superconducting material. V

**magnetohydrodynamics**
A type of energy conversion involving magnetism, fluids, and motion. In theory, MHD devices can generate electric power with many fewer mechanical parts than can ordinary generators. Instead of a solid armature rotating in a magnetic field, a fluid is used which will conduct electricity, current induced in this being collected by electrodes and sent out to do work. MHD generators offer the possibility of converting heat energy from fuels into electric power more cheaply, and with higher efficiency. CC

**magnetometer**
A device to measure variations in the earth's magnetic field, used by geophysicists to provide a more detailed picture of the subsurface formations in pinpointing possible favorable areas for oil drilling. M

**main intake**
The trunk or principal intake airway of a mine. D

**main return**
The principal return airway of a mine.

**mains**
Pipes used to carry gas or oil from one point to another. As contrasted with service pipes, they generally carry gas in large volume for general or collective use. DD

**major diameter**
Formerly called outside diameter. It refers to the largest diameter of a thread on a screw or nut. D

**major marketers**
Distributors of gasoline for integrated companies. B

**major mine disaster**
Defined by the U. S. Bureau of Mines as any accident that results in the death of five or more persons. D

**make**
A formation or accumulation of profitable vein material; as a make (that is, a body) of ore in a vein. D

**malingerer**
In industrial accident insurance, one who feigns disability or prolongs his period of disability, in order to collect accident insurance or compensation. D

**maltha**
(1) Various natural tars resulting from the oxidation and drying of petroleum. (2) A black viscid substance intermediate between petroleum and asphalt. Also called malthite. D

**man cage**
A special cage for raising and lowering men in a mine shaft. D

**man car**
A kind of car for transporting miners up and down the steeply inclined shafts of some mines. D

**mandrel**
A miner's pick. D

**manless coal face**
A coal face manned by remotely controlled equipment that eliminates the need for men in dangerous places. D

**manlock**
An air lock through which men pass to a working chamber which is under air pressure. D

**manometer**
An instrument for measuring the expansion or the expansive power of gases or vapors; a pressure gage or vacuum gage. M

**manoscope**
A manometer. D

**man-rem**
A unit used in health physics to compare the effects of different amounts of radiation on groups of people. It is obtained by multiplying the average dose equivalent to a given organ or tissue by the number of persons in that population. See rem. P

**manufactured gas**
(1) All gases made artificially or as byproducts, as distinguished from natural gas. (2) Applied particularly to a utility sendout. M

**MAOP**
Maximum allowable operating pressure.

**march**
The border or limit of a mineral area leased to, or owned by, a mining company. D

**marginal**
Economic jargon for "incremental," and associated with a small future increment in a stock or activity." C

**marine band**
Roof shale overlying coal seams and being uncommonly high in content of freshwater or marine shells. D

**marine science**
Any of the single sciences which together comprise the field of oceanography, such as marine geology, chemical oceanography, or physical oceanography. D

**markovnikovite**
Variety of petroleum found in Russia. D

**marl**
A calcareous clay, or intimate mixture of clay and particles of calcite or dolomite, usually fragments of shells. Marl in America is chiefly applied to incoherent sands; but abroad, compact, impure limestones are also called marls. D

**marsh**
A tract of soft, wet or periodically inundated land, generally treeless and usually characterized by grasses and other low growth. D

**marsh gas**
Methane, $CH_4$. If the decaying matter at the bottom of a marsh or pond is

stirred, bubbles of methane rise to the surface, thus the name marsh gas. D

**martic**
A mixture of bituminous matter, such as asphalt, and some foreign material, such as sand. D

**maser**
Contracted version of "microwave amplification by simulation emission of radiation." A class of amplifier from which the optical laser was developed. D

**mask**
(1) A screen, usually made of tracing cloth, to subdue and diffuse the light behind a plumbing or other sighted object. (2) A face for a respirator. D

**mass**
Quantity of matter, not identical to weight but obtained by dividing the weight of a body by the acceleration due to gravity.

**mass-energy equation (mass-energy equivalence) (mass-energy)**
The statement developed by Einstein that "the mass of a body is a measure of its energy content," as an extension of his 1905 Special Theory of Relativity. The statement was subsequently verified experimentally by measurements of mass and energy in nuclear reactions. The equation, usually given as: $E=mc^2$, shows that when the energy of a body changes by an amount, E, (no matter what form the energy takes) the mass, m, of the body will change by an amount equal to $E/C^2$. The factor $C^2$, the square of the speed of light in a vacuum, may be regarded as the conversion factor relating units of mass and energy. This equation predicted the possibility of releasing enormous amounts of energy in the atomic bomb by the conversion of mass to energy. It is also called the Einstein equation. O

**mass number**
The sum of the neutrons and protons in a nucleus. The mass number of uranium 235 is 235. It is the nearest whole number to the actual atomic weight of the atom. D

**mass spectrometer**
A device for analyzing a substance in terms of the mass-to-charge ratios of its constituents. It is so designed that the beam constituents of a given mass-to-charge ratio are focused on an electrode and detected or measured electrically. The mass spectrum shows the distribution in mass, or in mass-to-charge ratio of ionized atoms, molecules, or molecular fragments. M

**mast**
A drill derrick or tripod mounted on a drill unit, which can be raised to operating position by mechanical means. D

**mastic**
A bitumen preparation employed as an adhesive or water-proofing agent. D

**matter**
Anything that takes up space and has weight.

**max.**
(Formulas and tables only)—maximum.

**maximum credible accident**
The most serious reactor accident that can reasonably be imagined from any adverse combination of equipment malfunction, operating errors and other foreseeable causes. The term is used to analyze the safety characteristics of a reactor. Reactors are designed to be safe even if a maximum credible accident should occur. W

**maximum permissible dose**
That dose of ionizing radiation established by competent authorities as an amount below which there is no reasonable expectation of risk to human health, and which at the same time is somewhat below the lowest level at which a definite hazard is believed to exist. See radiation protection guide. W

**MBA**
Material balance area.

**MBD**
Million U. S. barrels per day.

**MBDO, mbdo**
Million barrels per day, of oil.

**MBDOE, Mbdoe**
Million U.S. barrels per day of oil equivalent.

**MBG**
Medium Btu-gas.

**MBPP**
Missouri Basin Power Project.

**MCA**
Manufacturing Chemists' Association, Inc.

**MCF, mcf**
Thousand cubic feet. Also abbreviated Mcf.

**mean**
The middle position or value between two extremes. D

**mean life**
The average time during which an atom, an excited nucleus, a radionuclide, or a particle exists in a particular form. W

**mean radiant temperature (mrt)**
That single temperature of all enclosing surfaces which would result in the same heat emission as the same surface with various different temperatures. D

**mean size**
The weight average particle size of any sample, batch or consignment of particulate material. D

**measured reserves**
Identified resources from which an energy commodity can be economically extracted with existing technology, and whose location, quality, and quantity are known from geologic evidence supported by engineering evidence. G

**mechanical equivalent of heat**
(1) Amount of mechanical energy which can be transformed into a single heat unit. (2) The equivalent of 778 foot-pounds per British thermal unit. D

**mechanical shovel**
A loader limited to level or only slightly graded drivages. The machine operates a shovel in front of it and pushes itself forward; when full, the shovel is swung over the machine and delivers into a mine car or tub behind. D

**mechanics**
The branch of physics that deals with the phenomena caused by the action of forces on material bodies. It is subdivided into statics, dynamics, or kinetics, or into the mechanics of rigid bodies and hydromechanics. D

**mechanization**
Essentially, the introduction of power machines to replace manual labor. D

**med**
Minimum energy dwelling.

**mega-**
A prefix that multiplies a basic unit by one million. O

**megaton energy**
The energy of a nuclear explosion which is equivalent to that of an explosion of one million tons (or 1000 kilotons) of TNT.

**megawatt-day per ton**
(1) A unit used for expressing the burnup of fuel in a reactor. (2) Specifically, the number of megawatt-days of heat output per metric ton of fuel in the reactor.

**mela-**
A prefix meaning dark-colored. D

**melano-**
A prefix meaning black or dark. D

**melaphyr**
A general term for altered and amygdaloidal rocks of basaltic or andesitic types. D

**melting point**
(1) Temperature at which a solid substance melts or fuses. (2) For asphalt, the melting point is defined as the temperature at which the asphalt is soft enough to permit a steel ball to drop through a disk of asphalt supported in a ring suspended in water (ring-and-ball method). (3) The melting point of grease is determined by placing a small amount of the grease on the bulb of a thermometer and heating in hot air until the grease begins to run off. M

**meniscus**
The curved top surface of a liquid column. It is concave upwards when the containing walls are wetted by the liquid (as water in a vertical glass tube) and convex upwards when wetted with liquid (as mercury in a vertical glass tube). D

**mephitic**
(1) Foul. (2) Noxious. D

**mephitic air**
Carbon dioxide. D

**mephitis**
(1) A noxious exhalation caused by the decomposition of organic remains. (2) Gases emanating from deep sources, as in mines, caves, and volcanic regions. D

**mer**
Maximum efficient rate. Also abbreviated MER. D

**mercantile system**
A theory in political economy that wealth consists not in labor and its products, but in the quantity of silver and gold in a country, and hence that mining, the exportation of goods, and the importation of gold should be encouraged by the State. D

**mercaptan**
Any of a class of compounds with the general formula RSH that are analogous to the alcohols and phenols but which contain sulphur in place of oxygen. D

**mercurialism**
Chronic poisoning with mercury, as from excessive medication or industrial contacts with the metal or its fumes. D

**MERES**
Matrix of environmental residuals for energy systems (BNL).

**mero-**
A prefix meaning part, or fraction, from the Greek meros. D

**meroleims**
Coalified remains of parts of plants. D

**mesh number (grit number)**
The designation of size of an abrasive grain, derived from the openings per linear inch in the control sieving screen. D

**meso-**
(1) A prefix meaning to denote rocks belonging to the middle zone of metamorphism. (2) Produced by high temperature, hydrostatic pressure, and intense stress. D

**meson**
General term for the short-lived elementary particles with masses between that of the electron and that of the proton; for example, mu-mesons (muons), pi-mesons (pions), and K-particles. D

**mesothermal deposit**
A mineral deposit formed at moderate temperatures and moderate pressures, in and along fissures or other openings in rocks, by deposition at intermediate depths, chiefly from hydrothermal fluids derived from consolidating intruding rocks. A mesothermal deposit differs from a mineral deposit formed in the deep veins and from one formed at shallow depths in its mineral composition and in the character of the alteration of the wall rock accompanying its formation. D

**met**
A convenient and approximate unit of human heat production equivalent to the average metabolic heat produced by resting man, about 18.50 British thermal units per square foot per hour. D

**meta-**
A prefix which, when added to the name of a rock, signifies that the rock has undergone a degree of change in mineral or in chemical composition through metamorphism. D

**metabituminous coal**
Bituminous coal containing from 89 to 91.2 percent carbon (ash-free, dry basis). D

**metabitumite**
Hard black lustrous variety of hydrocarbon found in proximity of igneous intrusions. D

**metabolism**
The utilization of oxygen by all cells of the body for the production of energy and heat. In this process carbon dioxide is produced. D

**metacannel coal**
High-rank cannel coal. D

**metalignitous coal**
Coal containing form 80 to 84 percent carbon. D

**metallic elements**
Elements that are generally distinguishable from other elements (nonmetallic elements) by their luster, malleability, electrical conductivity, and their usual ability to form positive ions. D

**metallurgical coal**
Coal with strong or moderately strong coking properties that contains no more than 8.0 percent ash and 1.25 percent sulfur, as mined or after conventional cleaning. G

**metamerism**
Isomerism produced by the attachment of different radicals to the same atom or group, the same general chemical properties being retained. D

**metamorphic**
Characteristic of, pertaining to, produced by, or occurring during the metamorphism of certain rocks. D

**metamorphic grade; metamorphic rank**
The grade or rank of metamorphism depends upon the extent to which the metamorphic rock differs from the original rock from which it was derived. If a shale is converted to a slate or phyllite, the metamorphism is low grade; if it is converted to a mica schist containing garnet and sillimanite, the metamorphism is high grade. D

**metamorphism**
Any process by which consolidated rocks are altered in composition, texture, or internal structure by conditions and forces not resulting simply from burial and the weight of the subsequently accumulated overburden. Pressure, heat, and the introduction of new chemical substances are the principal causes of metamorphism, and the resulting changes, which generally include the development of new minerals, are a thermodynamic response to a greatly altered environment. D

**metashale**
Shale altered by incipient metamorphic reconstitution but not recrystallized and without the development of partings or preferred mineral orientation. D

**metasomasis**
A process of ore formation by the partial or complete replacement of a preexisting rock by the ore body. Limestone is usually the rock replaced, and the degree of replacement is greatest along shatter belts, joints and porous bands in the rock. D

**metasome**
A mineral developed within another mineral. D

**metastable phase**
The existence of a substance as a solid, liquid, or vapor under conditions in which it is normally unstable in that state. D

**meteor**
Originally, any atmospheric phenomenon. Sometimes still used in this sense, as in such terms as hydrometeor, optical meteor, etc. Now more commonly restricted to astronomical meteors, called shooting stars or falling stars, which are relatively small bodies of matter travelling through interplanetary space and which are heated to incandescence by friction when they enter the atmosphere of the earth. D

**meter**
(1) An instrument, apparatus, or machine for measuring fluids, gases, electric currents, etc., and for recording the results obtained, e.g., a gas meter, a water meter, or an air meter. See also specific meters under alphabetical listing. (2) The fundamental unit of length in the metric system. It is equal to 39.37079 in. or 3.2808 ft. M

**metering pump**
A portable, high-precision pump developed by the U. S. Bureau of Mines engineers to aid studies of rock and ground pressure changes. Small enough to be carried in a coat pocket, the hand-operated device preloads high-pressure hydraulic cells that, embedded in rock or concrete, measure the variations in load or pressure that accompany nearby excavations. The pump also meters, without leakage, the fluid, usually mercury, glycerin, or oil, that must be added to or withdrawn from a cell to obtain a desired pressure. D

**meter oil**
An oil of low cold-test, like the light lubricating oils from Texas crude oil. D

**methanation**
A process for production of methane by passing carbon monoxide and hydrogen mixtures over nickel catalysts. Medium-Btu gas can be upgraded to high-Btu gas by methanation. The reaction is $CO + 3H_2 \rightarrow CH_4 + H_2O$. Y

**methane**
$CH_4$, carbureted hydrogen or marsh gas or firedamp, formed by the decomposition of organic matter. The most common gas found in coal mines. It is a tasteless, colorless, nonpoisonous, and odorless gas; in mines the presence of impurities may give it a peculiar smell. Its weight relative to air is 0.555 and may therefore form layers along the roof and occupy roof cavities. Methane will not support life or combustion; with air, however, it forms an explosive mixture. It is a clean-burning gaseous hydrocarbon fuel, and is a principle component of natural gas. D

**methane drainage**
Three main systems of methane drainage have been developed: (1) the cross-measure borehole method which

consists of boring holes from 2-1/4 to 3-1/4 inches in diameter and 150 to 300 feet in length, into the strata above or below the seam, generally close to the working face. (2) The superjacent roadway system in which boreholes are drilled from a roadway situated above the seam being worked. (3) The pack cavity system in which corridors are left and supported in the goaf as the face advances, and from these firedamp is drawn off. D

**methane indicator**
A portable instrument to determine accurately the methane content in a given area. D

**methanol**
A light, volatile, flammable, poisonous liquid alcohol, $CH_3OH$, formed in the destructive distillation of wood or made synthetically and used especially as a solvent, antifreeze, or denaturant for ethyl alcohol and in the synthesis of other chemicals. It can be used for fuel, either pure, or as a mixture with other alcohols or gasoline.

**methanometer; methane tester**
An instrument for determining the methane content in mine air. D

**methyl alcohol**
(1) A poisonous liquid, $CH_3OH$, also known as methanol, which is the lowest member of the alcohol series. (2) Also known as wood alcohol, since its principal source is the destructive distillation of wood. D

**metra**
A pocket implement combining the uses of many instruments, such as thermometer, level, plummet, and lens. D

**metric system**
A system of weights and measures derived from the meter. The system includes: (1) measures of length, wherein the meter is the unit; (2) measures of surface, wherein the square meter is the unit; (3) measures of capacity, wherein the liter is the unit; and (4) measures of weights, wherein the gram is the unit. M

**metric ton**
One thousand kilograms, equal to 2,204.6 lbs. avoirdupois. M

**Mev**
One million or ($10^6$) electron volts. (Also written as MeV.)

**mezo; meso**
A term sometimes prefixed to the names of igneous rocks of Mesozoic age. D

**mf**
Moisture free.

**mg**
Milligram.

**mG**
Milligauss. D

**mgd**
Million gallons per day. E

**mg/l**
Milligrams per liter. E

**MHD**
Magnetohydrodynamics.

**mho**
The practical unit of conductance equal to the reciprocal of the ohm.

**mica**
A group of phyllosilicate minerals having similar chemical compositions and highly perfect basal cleavage; monoclinic. Mica is one of the best electrical insulators. D

**micrinoid**
A coal constituent similar to material derived from finely macerated vegetation. D

**micro-**
A prefix that divides a basic unit by one million.

**microcrystalline wax**
Composed of finer, less noticeable particles and does not crystallize like paraffin. M

**microcurie**
The millionth of a curie (37,000 disintegrations per second) or $10^{-6}$ curie. E

**microfragmental coal**
Coal composed of compact macerated mass of vegetable debris, such as durain, cannel, algal cannel, and boghead. D

**microgas survey**
A prospecting method which seeks to locate oil by the detection, in soil samples, of gases such as ethane, propane, and butane as evidence of leakage in the vicinity of oil pools. Methane is not significant as it is also formed by the decomposition of vegetable matter. D

**microlog**
A resistivity log in borehole surveying obtained with a device consisting of closely spaced electrodes, the arrangement of which is basically the same but in miniature, as the normal and lateral devices in the regular electric survey. It is designed to measure the resistivity of a small volume of rock next to the borehole. D

**micrometer**
An instrument for measuring very small dimensions or angles. Used in connection with a microscope or a telescope. There are a great variety of forms, but in nearly all, the measurement is made by turning a very fine screw, which gives motion to a scale, a spider line, a lens, a prism, or a ruled glass plate. D

**micrometrics**
The study of very fine particles.

**micromicrofarad**
One-millionth of a microfarad; $10^{-12}$ farad; abbreviation $\mu\mu f$.

**micromicron**
One-millionth of a micron; $10^{-6}$ micron or $10^{-9}$ millimeter; abbreviation $\mu\mu$.

**micromillimeter**
One-millionth of a millimeter; abbreviation $\mu mm$.

**micromineralogy**
Mineralogy based on the use of the microscope.

**micron**
(1) A unit of length, equal to one-millionth of a meter. (2) Used in measuring the dimensions of dust particles, etc. (1 micron - about 1/25,000 inch).

**micro-organisms**
In geochemical prospecting, may be taken to include bacteria, algae, fungi, and any others of the relatively small forms of plant and animal life that inhabit soils and natural waters.

**microphysiography**
Same as petrography.

**microsecond**
One-millionth of a second; abbreviations $\mu sec$ and $\mu s$.

**microsection**
(1) A transparently thin section of some substance mounted for examination with the microscope. (2) A thin section of rock so mounted for petrographic examination.

**microseism**
A very slight tremor or vibration of the earth's crust.

**microseismic movement**
Rather permanent, faint vibrations of the earth's crust (usually not exceeding 25 microns) caused by breakers on the coast or by storms far out at sea (up to 3,000 kilometers from the coast). Synonym for microseism.

**microspectroscopy**
A method of identifying metallic constituents by drilling out the minute portion to be analyzed, flowing collodion over the resulting chips, and transferring the collodion together with the chips to a pure carbon electrode for analysis in a standard spectrographic arc. D

**microvolt**
One-millionth of a volt; $10^{-6}$ volt; abbreviation $\mu$v.

**mid-Atlantic ridge**
A great mountain range extending the entire length of the Atlantic Ocean. This ridge is now thought to curve around the southern tip of Africa, cross the Indian Ocean, curve around southern Australia and New Zealand, and extend well up into the Pacific Ocean.

**middle band**
A stratum of rock, or more usually soft dirt, near the middle of a coal seam.

**migma**
A mush of partly fluid and partly solid rock material from which migmatite arises by consolidation. If the amount of its liquid portion becomes great enough, it will acquire mobility and may intrude into its surroundings in typical eruptive fashion.

**migration**
The movement of oil, gas, or water through porous and permeable rock. Parallel (longitudinal) migration is movement parallel to the bedding plane. Transverse migration is movement across the bedding plane. D

**migration of ions**
Change of position. Migration velocity is that with which ions move through solutions during electolysis.

**migration of oil**
The movement of seepage of oil through the rocks wherever they are sufficiently permeable to allow such passage, which is of considerable importance in oil geology.

**mile hr**
Miles per hour.

**mile sec**
Miles per second.

**mil-foot**
A standard of resistance in wire. The resistance of 1 foot of wire that is 1 mil in diameter.

**mill coal**
A noncoking coal mined from strip pits and used for zinc smelting.

**milli-**
A prefix that divides a basic unit by one thousand.

**milliampere**
One-thousandth of an ampere, abbreviation ma.

**millibar**
A unit of atmospheric pressure equal to one-thousandth of a bar or 1,000 dynes per square centimeter; abbreviation mb.

**millimicron**
One-thousandth of a micron, usually symbolized as 1 m$\mu$. Formerly much used as a measure for the wave length of visible light. Easily translated into angstroms, merely by the shift of a decimal point, since 1 m$\mu$ equals 10 angstroms.

**millicurie-hour**
A measure of gamma-ray exposure expressed as the product of the source

of millicuries and the time of exposure in hours.

**millidarcy**
The customary unit of measurement of permeability equal to one-thousandth of a darcy.

**millidegree**
A unit of temperature equal to one-thousandth of a degree; abbreviation is mdeg.

**millifarad**
One-thousandth of a farad; abbreviation is mf.

**milligal**
(1) A unit employed in the gravitation method of geophysical prospecting. (2) About one millionth of the average value of the acceleration due to gravity at the earth's surface, that is, 1 gallon = 1 centimeter per second per second.

**milligram**
A unit of weight in the metric system which equals 0.001 or $10^{-3}$ gram, 0.05432 grain, 0.000643 pennyweight, or 0.00003215 troy ounce; abbreviation is mg.

**milligram hours**
A measure of gamma-ray exposure expressed as the product of the equivalent radium content of the source, in milligrams, and the time of exposure in hours.

**milliliter**
One-thousandth of a liter. One liter is equal to 1.06 quarts. G

**millimass unit**
One-thousandth of an atomic mass unit; abbreviation is mamu.

**millimeter**
A unit of length that equals one-thousandth of a meter; abbreviation is mm.

**milling**
A process in the uranium fuel cycle by which ore containing only a very small percentage of uranium oxide ($U_3O_8$) is converted into material containing a high percentage of $U_3O_8$, sometimes referred to as yellowcake. X

**million electron volts**
A common unit of energy in nuclear science, which equals $10^6$ electron volts; abbreviation is mev.

**millirem (mrem)**
A unit used to measure radiation dose, which is one-thousandth of a rem. X

**millisecond delay**
A type of delay cap with a definite but extremely short interval between passing of current and explosion.

**millivolt**
One-thousandth of a volt. G

**milltons**
Net tonnage of ore available for milling after eliminating waste and unpayable material.

**MIL Spec (Sometimes mil-spec)**
A shortened form of military specification.

**min**
(1) In formulas and tables only, minimum. (2) Shortened form of minute.

**mine**
(1) An opening or excavation in the earth for the purpose of extracting minerals. (2) A pit or excavation in the earth from which metallic ores or other mineral substances are taken by digging. G

**mine bank**
An area of ore deposits that can be worked by excavation above the water level.

**mine cooling load**
The total amount of heat, sensible and latent, in British thermal units per

hour, which must be removed by the air in the working places.

**mine development**
The term employed to designate the operations involved in preparing a mine for ore extraction. These operations involve tunneling, sinking, crosscutting, drifting and raising.

**mined strata**
In mine substance, the strata lying vertically over the excavated area.

**mine dust**
Dust from rock drills, blasting, or handling rock. In the quantity inhaled by workers, dust may be classified as dangerous, harmless, and borderline, though the classification is purely arbitrary. Silica is a dangerous dust; bituminous coal dust is relatively harmless, and aluminum hydroxide is borderline. D

**mine fan signal system**
A system which indicates by electric light or electric audible system, or both, the slowing down or stopping of a mine ventilating fan.

**mine fires**
These very dangerous occurrences may arise as the result of spontaneous combustion, the ignition of timbers by gob fires, electric cable defects, or the heating and ignition of conveyor belts due to friction. D

**mine heads**
In a mine ventilation system, the cumulative energy consumptions are called the mine heads. These heads are in reality pressure differences, determined in accordance with Bernoulti's principle. D

**mine-mouth generation**
Generation of electrical energy based on the premise that it is cheaper to ship electricity than coal.

**mine-mouth plant**
A steam-electric plant or coal gasification plant built close to a coal mine and usually associated with delivery of output via transmission lines or pipelines over long distance as contrasted with plants located nearer load centers and at some distance from sources of fuel supply.

**miner**
One who mines; as (1) one engaged in the business or occupation of getting ore, coal, precious substances, or other natural substances out of the earth; (2) a machine for automatic mining (as of coal); and (3) a worker on the construction of underground tunnels and shafts as for roads, railways, waterways. D

**mineral**
An inorganic substance occurring in nature, though not necessarily of inorganic orgin, which has (1) a definite chemical composition, or more commonly, a characteristic range of chemical composition, and (2) distinctive physical properties or molecular structure. With few exceptions, such as opal (amorphous) and mercury (liquid), minerals are crystalline solids. D

**mineral claim**
A mining claim.

**mineral deposit**
Any valuable mass of ore. Like ore deposit, it may be used with reference to any mode of occurrence of ore, whether having the characters of a true, segregated, or gash vein, or any other form. D

**mineral economics**
Study and application of the technical and administrative processes used in management, control, and finance connected with the discovery, development, exploitation, and marketing of minerals.

**mineral engineering**
Term covers a wide field in which many resources of modern science and engineering are used in discovery, development, exploitation, and use of natural mineral deposits. D

**mineral fuels**
Coal and petroleum.

**mineral interests**
Mineral interests in land means all the minerals beneath the surface. Such interests are a part of the realty, and the estate in them is subject to the ordinary rules of law governing the title to real property. D

**mineral land**
Land which is worth more for mining than for agriculture. The fact that the land contains some gold or silver would not constitute it mineral land if the gold and silver did not exist in sufficient quantities to pay to work. Land not mineral in character is subject to entry and patent as a homestead however limited its value for agricultural purposes. D

**mineral tar**
A viscid variety of petroleum. Tar derived from various bituminous minerals, such as coal, shale, peat, etc. Shale tar. D

**miner's inch**
The miner's inch of water does not represent a fixed and definite quantity, being measured generally by the arbitrary standard of the various ditch companies. Generally, however, it is accepted to mean the quantity of water that will escape from an aperture 1-inch square through a 2-inch plank, with a steady flow of water standing 6 inches above the top of the escape aperture, the quantity so discharged amounting to 2,274 cubic feet in 24 hours. D

**miner's weight**
The term used in a coal mining lease as the basis for the price per ton to be paid for mining. It is not a fixed, unvarying quantity of mine-run material, but is such a quantity of material as operators and miners may, from time to time, agree as being necessary or sufficient to produce a ton of prepared coal. D

**mine sample**
A small quantity of coal or mineral taken at underground exposures for analysis and laboratory tests. See also quartering. D

**mine static head**
The energy consumed in the ventilation system to overcome all flow head losses. It includes all the decreases in total head (supplied from static head) which occur between the entrance and the discharge of the system. D

**mine tons**
Gross tonnage of ore including waste and unpayable material.

**mine total head**
The sum of all energy losses in the ventilation system. Numerically, it is the total of the mine static and velocity heads. D

**mine velocity head**
The velocity head at the discharge of the system. Throughout the system, the velocity head changes with each change in duct area or number and is a function only of the velocity of air flow. It is not a head loss. The velocity head for the system must technically be counted a loss, because the kinetic energy of the air is discharged to the atmosphere and wasted. Therefore, it must be considered a loss to the system in determining overall energy loss. D

**minimum ignition energy**
The minimum ignition energy required for the ignition of a particular flammable mixture at a specified temperature and pressure. D

**minimum interfacial energy, law of**
The tendency of seed crystals to assume a position on previously formed crystals that is the most stable position as far as forces of crystallization are concerned. This law explains a parallel grouping of crystals and the symmetrical or the parallel arrangement of inclusions in large single host crystals. D

**mining**
(1) The science, technique, and business of mineral discovery and exploitation. (2) Strictly, the word connotes underground work directed to severance and treatment of ore or associated rock. (3) Practically, it includes opencast work, quarrying, alluvial dredging, and combined operations, including surface and underground attack and ore treatment. D

**mining claim**
That portion of the public mineral lands which a miner, for mining purposes, takes and holds in accordance with mining laws. D

**mining engineering**
That branch of engineering chiefly concerned with the discovery, development, and exploitation of coal, ores and minerals. The term also embraces the cleaning, sizing, and dressing of the product. D

**mining explosives**
High explosives used for mining and quarrying can be divided into four main classes; namely, (1) gelatins; (2) semi-gelatins; (3) nitrogylcerin powders; and (4) non-nitroglycerin explosives. D

**mining geology**
The study of geologic structures and particularly the modes of formation and occurrence of mineral deposits and their discovery. D

**mining hazards**
The dangers peculiar to the winning and working of coal and minerals. These include collapse of ground, explosion of released gas, inundation by water, spontaneous combustion, inhalation of dust and poisonous gases, etc. D

**mining methods**
The systems employed in the exploitation of coal seams and ore bodies. The method adopted depends on a large number of factors, mainly, the quality, shape, size, and depth of the deposit, accessibility and capital available. D

**minor mine disaster**
A minor mine disaster is defined by the U. S. Bureau of Mines as any accident coming within one of the following categories: (1) a mine accident (not an explosion or fire) causing the death of less than five persons and considerable property damage; (2) a mine explosion or ignition causing injury to one or more persons or considerable property damage but no loss of life; (3) a mine explosion or ignition, resulting in the death of less than five persons; (4) a mine fire causing injury to one or more persons or considerable property damage but no loss of life; and (5) a mine fire resulting in the death of less than five persons. D

**MIS**
Modified in-situ process.

**miscible**
Capable of being mixed (stability and uniformity throughout the mixture are usually inferred). M

**miscible flooding**
A secondary recovery process in which propane is injected into a reservoir followed by dry gas. Since propane and dry gas are miscible (they are readily mixed), a liquid wall or bank is formed behind the crude oil. Pumping

alternate slugs of dry gas or water into the reservoir behind miscible bank forces it, and the oil ahead of it, toward the producing well. M

**mistral**
Cold, dry wind blowing in a southerly direction across the normally warm coastal region (Riviera) in southern France, the opposite of the sirocco.

**MIT**
Massachusetts Institute of Technology.

**MIUS**
Modular integrated utility system.

**mixed base crudes**
Crude oils containing quantities of both paraffin wax and asphalt. M

**mixed explosion**
One in which each ingredient, fire-damp and coal dust, are present below their lower limits, but in combination produce sufficient heat of combustion to propagate an explosion. D

**mixed gas**
A gas in which manufactured gas is co-mingled with natural or liquefied petroleum gas (except where the natural or liquefied petroleum gas is used only for "enriching" or "reforming") in such a manner that the resulting product has a Btu value higher than that previously produced by the utility prior to the time of the introduction of natural or liquefied petroleum gas. DD

**ml**
Milliliter.

**mm**
Millimeter.

**MMBtu**
Million Btu in gas.

**mmf, m**
Moisture, mineral matter free.

**mobile equilibrium**
Van Hoff's law (1884) shows that a temperature rise induces an endothermic change in the equilibrium constant, and fall in temperature an exothermic change. D

**mobile source**
A moving source of air pollution such as an automobile. E

**mobility**
The tendency of an element to move in a given geochemical environment. D

**moderator**
A material, such as ordinary water, heavy water, or graphite, used in a reactor to slow down high-velocity neutrons, thus increasing the likelihood of further fission. O

**module**
A common unit particularly specified for dimensional coordination. D

**modulus**
(1) Factor used in conversion of units from one system to another. (2) Formula or constant which defines properties (for example, density, elasticity) of materials. D

**modulus of resilience**
The strain energy per unit volume absorbed up to the elastic limit under the condition of uniform uniaxial stress. D

**moil**
A tool for breaking and wedging out rock or coal. D

**moisture bed**
The total moisture (percent) in a seam of coal before working.

**molded coal**
An artificial fuel made of charcoal refuse and coal tar, molded into cylinders, dried and carbonized.

**mole**
A large diameter drill mounted on a movable framework and capable of tunneling holes 5-30 feet in diameter.

**mole.**
Weight in grams of a compound in terms of its molecular weight.

**molecular weight**
The sum of the atomic weights of the atoms composing a molecule. M

**molecule**
(1) Unit of matter. (2) The smallest portion of an element or a compound which retains chemical identity with the same particular substance en masse. M

**molten**
Reduce to the fluid state by heat; melted; fused; as, molten metal. D

**molten-iron process**
A process to gasify coal without causing a sulfur oxide pollution problem. It uses a molten-iron bath with air or oxygen. The product gases are essentially methane, carbon monoxide, and hydrogen which with methanation can be made into pipeline-quality gas. It is said to be the only process suitable for gasifying any coal, including anthracite and lignite. G

**molten salt reactor**
A fused-salt reactor. O

**momentum**
The product of mass times velocity. D

**MON**
Motor octane number.

**monetary policy**
Management of the money supply, credit conditions, and interest rates, to the degree they can be influenced by the Federal Reserve through the banking system, to stimulate or restrain the economy.

**money supply**
The amount of money in an economy, narrowly defined as the sum of paper currency and coin plus demand (or checking-account) deposits.

**monitor**
An instrument that measures the level of ionizing radiation in an area. O

**monitoring**
Periodic or continuous determination of the amount of pollutants or radioactive contamination present in the environment. E

**monochromator**
A device for isolating monochromatic or narrow bands of radiant energy from the source, often used in chemical analysis. E

**monophology**
The observation of the form of lands.

**monorail in coal mining**
A relatively new underground transport system in which the carriages, or buckets, are suspended from, and run along, a single continuous overhead rail or taut wire rope. The monorail is used in coal mines to transport supplies to the workings. It may be installed alongside the gate conveyor and worked by endless or main rope.

**mother lode**
The principal lode or vein passing through a district or particular section of country.

**motile**
Exhibiting or capable of spontaneous movement. E

**motor mix**
The antiknock fluid for gasoline used primarily in automotive engines. M

**motor oil**
(1) An oil suitable for use in an engine crankcase. (2) Applies to oils used to lubricate electric motors. M

**moving annual total**
In study of process costs (in large or in detail), a series of costs-per-unit observed and recorded at regular intervals (usually in monthly financial summaries cross reference to analyzed detail cost). Twelve months are covered and each month the new month's figures are added and those for the corresponding month of the previous year are removed. Therefore, like periods are always compared and seasonal fluctuations are smoothed out. Abbreviation M.A.T. D

**movement of force**
The turning effect on a body about a point called the pivot or fulcrum. In practice, turning effect is commonly called leverage.

**moving-bed catalytic cracking**
A cracking process in which the catalyst, consisting of granules 4- to 12-mesh, is continuously cycled between the reactor and the regenerator. Examples: the Airlife Thermofor and the Houdriflow processes. D

**mp**
Melting point. D

**MPD**
Magnetoplasmadynamics.

**mpg**
Miles per gallon.

**mph**
Miles per hour.

**MRDE**
Mining research and development establishment.

**mrem(s)**
Millirem(s).

**MRG**
Methane rich gas.

**MRI**
Midwest Research Institute.

**MS**
(1) Machine steel. (2) Margin of safety. (3) Medium steel. D

**M.S.A. all-service gas mask**
A filter-type box respirator produced by the Mine Safety Appliance Company and which is approved by the U.S. Bureau of Mines for two hours of continuous or intermittent use. D

**M.S.A. carbon monoxide detector**
An apparatus for estimating the amount of carbon monoxide in mine atmospheres. D

**M.S.A. methanometer**
This methane indicator is one in which the sample is made to flow continuously over the filaments while the determination is being made. D

**MSBR**
Molten salt breeder reactor.

**msf**
Magnetostrictive force. D

**MSHA**
Mine Safety and Health Act.

**MSL**
Mean sea level. D

**MST**
(1) Mean solar time. (2) Mountain standard time. D

**MSW**
Municipal solid waste.

**MT**
Metric tons. D

**mta**
Million tons per annum.

**MTCH**
Mining Technologies Clearinghouse.

**mtn**
Mountain. D

**Mton**
Megaton. D

**muck**
(1) Unconsolidated soils, sands, clays, loams encountered in surface mining. (2) Generally, earth which can be severed and moved without preliminary blasting. D

**mud**
Any soil containing enough water to make it soft. D

**mudcap**
A charge of dynamite, or other high explosive, fired in contact with the surface of a rock after being covered with a quantity of wet mud, wet earth, or sand, no borehole being used. D

**muddling off**
Commonly thought of a reduced productivity caused by the penetrating, sealing, or plastering effect of a drilling fluid. Actually there is little penetration into the capillaries of an ordinary producing formation. D

**mud drilling**
Drilling operations in which a mud-laden circulation fluid is used. D

**mud flush**
To clear fragmented materials from a borehole by circulating a mud-laden fluid. D

**mud geyser**
(1) A geyser that erupts sulfurous mud. (2) A type of mud volcano. D

**mud hog**
(1) A synonym for mud pump. (2) Pressure tunnel worker. D

**mud logging**
A method of determining the presence or absence of oil, gas, and salt water in the various formations penetrated by the drill bit. The drilling fluid and the cuttings are continuously tested on their return to the surface, and the results of these tests are correlated with the depth of origin. D

**mud pit**
A pit in which drilling mud is mixed, prepared, stored, or caught as it overflows from the drill-hole collar. D

**mud ring**
The section of a boiler where scale, alkalies and sediment collect. D

**mud swivel**
A modification of a water swivel specially designed for use when a mud-laden drill fluid is circulated in borehole-drilling operations. Also called mud pot. D

**mud system**
The technique and use of drilling mud as a circulating medium in drilling operations. D

**mud up**
The act or process of filling, choking, or clogging the waterways of a bit with consolidated drill cuttings. Also called sludging; sludging up. D

**mule**
A small car, or truck, attached to a rope and used to push cars up a slope or inclined plane. D

**mullock**
The accumulated waste or refuse rock about a mine. D

**multideck screen**
A screen with two or more superimposed screening surfaces mounted rigidly within a common frame. D

**multigrade oil**
One of the multi-viscosity number oils in which one oil combines three SAE viscosity number grades. For example, multigrade SAE 10W-30 grade may be used where SAE 10W, SAE 20-20W, or SAE 30 grades are specified. Multigrade oils are usually made to

meet the requirements of API Service MS, DG, and DM. They have been made possible by improved refining processes and the use of improved additives. M

**multiple completion wells**
Sometimes in drilling an oil well, more than one formation is located which would be commercially productive. Generally, in such a case, a separate tubing string is run inside the casing for each productive formation. Production from the separate formations is directed through the proper tubing strings and is isolated from the others by "packers" which seal annular space between the tubing strings and casing. M

**multiple detectors**
Two, or more, seismic detectors whose combined energy output is fed into a single amplifier-recorder circuit. This technique is used to effect a cancellation of undesirable near-surfacing waves. Synonym for multiple geophones; multiple recording groups. D

**Multiple Mineral Development Act; Multiple Use Act**
The Act of August 13, 1954, permitting oil and gas leases and mining claims on the same land, with the oil and gas lessee obtaining his rights under his lease and the mining claimant being accorded his rights to go to patent subject, however, to the oil and gas lessee's interest. D

**multiple-seam mining**
Mining two or more seams of coal, frequently close together, that can be mined profitably where mining one alone would not be profitable. D

**multishift working**
The working of two or three shifts per day on production. Machines represent a heavy capital outlay and the aim is to make them productive as long as possible in the 24 hours. D

**mungle shale**
An oil shale, in the West Calder district, Scotland. D

**muon**
(1) Contraction of mu-meson. (2) An elementary particle with 207 times the mass of an electron. It may have a single positive or negative charge. D

**municipal solid waste (MSW)**
The materials collected from urban areas in the form of organic matter, glass, plastic, waste paper, etc., not including human wastes. Y

**mush**
(1) Soft and damp small coal. (2) A coal which has been so crushed that it is unprofitable to mine. D

**mutabilite**
A soft "corklike" bitumen of porous or resinous consistency, partly soluble in organic solvents. D

**mutation**
A permanent transmissible change in the characteristics of an offspring from those of its parents. W

**mv**
Millivolt. D

**mvb**
Medium-volatile bituminous. D

**mw**
Milliwatt (one thousandth watt). D

**MW**
Megawatt.

**MWd**
Megawatt day(s).

**MWe**
Megawatt-electrical.

**MWt**
Megawatt-thermal.

# N

**n**
(1) Nano. (2) Used as a prefix to units of measure, indicating multiplication of the basic unit by $10^{-9}$ or division by 1 billion. (3) Symbol for index of refraction or refractive index. D

**na**
Nonattainment.

**NAAQS**
National Ambient Air Quality Standards.

**NAB**
National Alliance of Businessmen.

**nadir**
In aerial photography, the point on the ground plane that is vertically beneath the lens of an aerial mapping camera. Sometimes called the plumb point or ground nadir point. D

**NAE**
National Academy of Engineering.

**naked light**
Any light which is not so enclosed and protected as to preclude the ignition of an ambient firedamp-air mixture. Also called open light. D

**naked light mine**
A nongassy coal mine where naked lights may be used by the miners. D

**nano-**
A prefix that divides a basic unit by one billion ($10^9$).

**Nansen bottle**
An oceanographic water sampling bottle made of a metal alloy which reacts little with seawater, equipped with a rotary valve at each end so that when it is rotated at depth the valves close and lock shut, entrapping a water sample and setting the reversing thermometers. D

**naptha**
In modern use, an artificial, volatile, colorless liquid obtained from petroleum, a distillation product between gasoline and refined oil. D

**naptha gas**
Illuminating gas charged with the decomposed vapor of naptha. D

**napthode**
Concretions of bituminous limestone rich in carbonaceous matter. D

**NARUC**
National Association of Regulatory Utility Commissioners.

**NAS**
National Academy of Science.

**NASA**
National Aeronautics and Space Administration.

**nascent state**
The state or condition of a radical or element at the very instant it is freed from a previous combination, as in the case of hydrogen or oxygen. In a nascent condition, the chemical activity is greater than in the ordinary condition. M

**National Association of Regulatory Utility Commissioners**
An advisory council composed of Federal and state regulatory commissioners having jurisdiction over transportation agencies and public utilities. K

**native**
Occurring in nature, either pure or uncombined with other substances, usually applied to metals, such as native mercury, native copper. D

**native gas**
The total volume of gas indigenous to the storage reservoir at the time gas storage started. DD

**natural circulation reactor**
A reactor in which the coolant (usually water) is made to circulate without pumping, that is, by natural convection resulting from the different densities of its cold and reactor-heated portions. O

**natural frequency**
The frequency of free oscillation of a system. D

**natural gas**
Naturally occurring mixtures of hydrocarbon gases and vapors, the more important of which are methane, ethane, propane, butane, pentane, and hexane. M

**natural gas indicator**
This indicator consists of a naptha-burning safety lamp with a mirror attached to one side so that the action of the flame may be observed from above. This type of lamp should not be used if there is any possibility that manufactured gas or acetylene is present. D

**natural gasoline**
One of the liquids produced out of natural gas. M

**natural liquid gas**
Those liquid hydrocarbon mixtures which are gaseous at reservoir temperatures and pressures, but are recoverable by condensation or absorption. Natural gasoline and liquefied petroleum gases fall in this category. DD

**natural radiation, natural radioactivity**
Background radiation. O

**natural uranium**
Uranium as found in nature, containing 0.7% of U-235, 99.3% of U-238, and a trace of U-234. It is also called normal uranium. O

**nautical measure**
One nautical mile or knot equals 6,080 feet; 3 nautical miles equal 1 league; and 60 nautical miles equal 1 degree (at the equator). D

**nautical mile**
Any of various units of distance, used for sea and air navigation, based on the length of a minute of arc of a great circle of the earth and differing because the earth is not a perfect sphere: (1) a British unit that equals 6,080 feet or 1,853.3 meters; also called Admiralty mile; (2) a U. S. unit, no longer in official use, that equals 6,080.20 feet or 1,853.248 meters; and (3) an international unit that equals 6,076.1033 feet or 1,852 meters, used officially in the United States since July 1954. D

**NBS**
National Bureau of Standards.

**NCA**
National Coal Association.

**NCB**
The National Coal Board (United Kingdom). D

**NCRPM**
National Council on Radiation Protection and Measurement.

**NDIR**
Non-dispersive infrared analyzers.

**NEA**
National Energy Act.

**NEB**
National Energy Board.

**neck**
(1) The narrow entrance to a room next to the entry, or a place where the room has been narrowed because of poor roof. (2) A narrow stretch of land, such as an isthmus or a cape. D

**needle coal**
Variety of lignite, composed of fibrous needlelike mass of vascular bundles of palm stems. D

**needling**
To cut holes, notches, or ledges in a coal or rock surface to receive the ends of timber supports.

**neel point**
The temperature at which ferromagnetic and antiferromagnetic materials become paramagnetic. E

**negative gradient**
Describes conditions in a layer where the temperature decreases with increasing depth. D

**NEIC**
National Energy Information Center.

**NEMA**
National Electrical Manufacturers Association.

**NEP**
National Energy Plan.

**NEPA**
National Environmental Policy Act.

**NEPCO**
New England Petroleum Corporation.

**nephelometer**
An instrument which measures the scattering of light by determining the amount of light emitted at right angles to the original beam direction. Such devices are useful in studies of particles (size and amount) suspended in water. D

**NER**
Net energy analysis.

**NERA**
National Energy Resources Association.

**NERC**
National Electric Reliability Council.

**NEST**
Nuclear Emergency Search Team.

**net calorific value**
Net heat of combustion. In the case of solid fuels and liquid fuels of low volatility, a lower value calculated from gross calorific values as the heat produced by combustion of unit quantity, at constant atmospheric pressure, under conditions, such that all water in the products remains in the form of vapor. D

**net for distribution**
(1) On an electric system or company basis this means the kilowatt-hours available for total system or company load. (2) Specifically, it is the sum of net generation by the system's own plants, purchased energy, and net interchange (in less out). (3) On a national basis, it is the sum of the net generation of the total electric utility industry, plus or minus net interchange

with Canada and Mexico, plus purchases from industrial sources. K

**net generating station capability**
The capability of a generating station as demonstrated by test or as determined by actual operating experience less power generated and used for auxiliaries and other station uses. Capability may vary with the character of the load, time of year (due to circulating water temperatures in thermal stations or availability of water in hydro stations), and other characteristic causes. Capability is sometimes referred to as effective rating. K

**net generation**
Gross generation less kilowatt hours consumed out of gross generation for station use equals net generation. K

**nether**
The lower part of, as in nether roof, and opposed to the term upper. D

**net radiation**
The difference between the total incoming and total outgoing radiation. Incoming radiation is made up of (a) short wave from sun and sky, and (b) thermal, mainly from atmospheric layers close to ground. Outgoing radiation from a surface comprises (a) reflected short wave component depending on incident irradiance and reflectance of the surface, and (b) long wave thermal radiation component depending on ground surface temperature and long wave emittance. Q

**net system capability**
The net generating station capability of a system at a stated period of time (usually at the time of the system's maximum load), plus capability available at such time from other sources through firm power contracts less firm power obligations at such time to other companies or systems. K

**net unit value**
The difference between the gross unit recoverable value and the cost of mining, treating, and marketing the ore; in other words, the net operating profit. D

**network**
A system of transmission or distribution line so cross-connected and operated as to permit multiple power supply to any principal point on it. K

**neutral flame**
A gas flame in which there is no excess of either fuel or oxygen. D

**neutralization**
Making neutral or inert, as by the addition of an alkali to an acid solution. D

**neutralization number**
The weight, in milligrams, of potassium hydroxide needed to neutralize the acid in 1 g of oil. The neutralization number of an oil is an indication of its acidity. D

**neutrino**
An electrically neutral elementary particle with a mass so small that it is extremely difficult to detect. It is produced in many nuclear reactions; for example, in beta decay, and has high penetrating power. Neutrinos from the sun usually pass right through the earth. D

**neutron [Symbol n]**
An uncharged elementary particle with a mass slightly greater than that of the proton, and found in the nucleus of every atom heavier than hydrogen. A free neutron is unstable and decays with a half-life of about 13 minutes into an electron, proton, and neutrino. Neutrons sustain the fission chain reaction in a nuclear reactor. O

**neutron capture**
The reaction that occurs when an atomic nucleus absorbs or captures a neutron. The probability that a given

material will absorb neutrons is proportional to its neutron-capture cross section and depends on the energy of the neutrons and the nature of the material. D

**neutron economy**
The degree to which neutrons in a nuclear reactor are used for desired ends instead of being lost by leakage or useless absorption. Desired ends may include propagation of the chain reaction, converting fertile material to fissionable material, producing desired isotopes, and experimental use. D

**neutron log**
Strip recording of the secondary radioactivity arising from the bombardment of the rocks around a borehole. Used, generally in conjunction with other types of logs, for the identification of the fluidbearing zones of rocks. D

**new field wildcats**
Wells drilled in an area where neither oil nor natural gas has ever been found. M

**nfe**
Nitrogen-free extract.

**NFPA**
National Fire Protection Association.

**NFS**
Nuclear fuel services.

**NGL**
Natural gas liquids.

**ngor**
Net gas-oil ratio. D

**NGPA**
(1) Natural Gas Processors Association. (2) Natural Gas Policy Act.

**nicking**
The cutting of a vertical groove in the seam to liberate the coal after it has been holed or undercut. D

**NIER**
National Industrial Equipment Reserve.

**nigrite**
A product of the coalification of fix bitumens rich in carbon, insoluble or only slightly soluble in organic solvents.

**nil**
Nothing; zero. D

**nip out**
The disappearance of a coal seam by the thickening of the adjoining strata, which takes its place. D

**nitric oxide (NO)**
A gas formed in great part from atmospheric nitrogen and oxygen when combustion takes place under high temperature and high pressure, as in internal combustion engines. NO is not itself a pollutant; however, in the ambient air, it converts to nitrogen dioxide, $NO_2$, a major contributor to photochemical smog. E

**nitrogen dioxide ($NO_2$)**
(1) A compound produced by the oxidation of nitric oxide in the atmosphere. (2) This gas is considered the most dangerous of mine gases and is produced by the incomplete detonation of some explosives. It is very irritating to the air passages.

**nitrogen oxides**
A product of combustion of fossil fuels whose production increases with the temperature of the process. It can become an air pollutant if concentrations are excessive. G

**nitrous oxide**
A gas with the chemical formula $N_2O$. This gas is produced by the blasting of certain nitroglycerine explosives, especially if there is incomplete detonation. It is also produced in the exhaust of diesel locomotives. It is also used as an

anesthetic in dentistry, known as laughing gas. D

**nivenite**
A variety of uraninite high in uranium and carrying 10 percent or more of the yttrium earths and 6.7 to 7.6 percent thoria. D

**NLGI**
National Lubricating Grease Institute.

**NLGI number**
One of a series of numbers classifying the consistency range of lubricating greases, based on the ASTM cone penetration number. The National Lubricating Grease Institute grades are in order of increasing consistency (hardness). D

**NMRS**
Nuclear materials and reports system.

**NMSS**
Nuclear materials safety and safeguards.

**NOAA**
National Oceanic & Atmospheric Administration.

**noble gas**
An inert gas such as helium, neon, argon, krypton, xenon and radon, which do not normally combine chemically with other elements. D

**noble metals**
A metal with marked resistance to chemical reaction, particularly to oxidation and to solution by inorganic acids. The term as often used is synonymous with precious metal. D

**nocturnal radiation**
The radiation or loss of heat at night from a warm object such as a house to the cold sky and surroundings. H

**node**
A point, line or surface in a standing wave system where some characteristic of the wave field has essentially zero amplitude. D

**nodular**
Having the shape of or composed of nodules, which is said of certain ores. D

**nodule**
Rounded masses of pyrite found deposited in coalbeds or in the roof or floor which range from a fraction of an inch to several feet in diameter. D

**nog**
A block of wood wedged tightly into the cut in a coal seam after the coal cutter has passed. It forms a temporary support for the coal and the roof above. D

**noise**
(1) Any undesired audible signal. (2) In acoustics, noise is any undesired sound. E

**nominal capacity**
A national figure in general descriptions of the plant, applying to the plant as a whole and to the specific project under consideration. It may be taken as representing the approximate tonnage expected to be supplied to the plant during the hour of greatest load. D

**non-associated natural gas**
Free natural gas not in contact with, nor dissolved in, crude oil in the reservoir. DD

**nonbanded coal**
Coal that does not display a striated or banded appearance on the vertical face. It contains essentially no vitrain and consists of clarain or durain, or of material intermediate between the two. D

**noncaking coal**
Coal which does not form cake, namely hard, splint, cherry and durain coal. D

**noncoking coal**
A bituminous coal that burns freely without softening or any appearance of incipient fusion. The percentage of volatile matter may be the same as for coking coal, but the residue is not a true coke. D

**noncombustible**
Any material that will neither ignite not actively support combustion in air at 1,200° F when exposed to fire. D

**nondestructive testing**
Methods of examination, usually for soundness, which do not involve destroying or damaging the part being tested. It includes radiological examinations, magnetic inspections, etc. D

**non-electrolytes**
Materials which when placed in solution do not make the solution conductive to electrical currents. E

**nonionic**
Refers to an uncharged or electrically neutral particle. E

**nonselective mining**
The object of nonselective mining is to secure a low cost, generally by using a cheap stoping method combined with large scale operations. This method can be used in deposits where the individual stringers, bands, or lenses of high-grade ore are so numerous and so irregular in occurrence and separated by such thin lenses of waste that a selective method cannot be employed. Nonselective methods of stoping include caving, top slicing, some forms of opening stoping, and shrinkage stoping under most conditions. D

**nontabular deposits**
Mineral deposits of irregular shape. D

**nontidal current**
A current brought about or caused by forces that are independent of the tides. It includes currents in the general circulatory system of the sea as well as currents arising from meteorologic conditions such as winds, fresh-water, runoff, and differences in density and temperature of the water. E

**non-utility generation**
Generation by producers having generating plants for the purpose of supplying electric power required in the conduct of their industrial and commercial operations. Generation by mining, manufacturing, and commercial establishments and by stationary plants of railroads and railways for active power is included. K

**nonweathering coal**
Coal having a weathering index, as defined by U. S. Bureau of Mines standards, of less than 5 percent. D

**normal air**
A mixture of dry air and water vapor, varying from 0.1 percent to 3 percent by volume. D

**normal pressure; standard pressure**
Usually equal to the weight of a column of mercury 760 millimeters in height. D

**normal solution**
A solution made by dissolving 1 gram-equivalent weight of a substance in sufficient distilled water to make 1 liter of solution. D

**normal temperature**
Normal temperature and pressure are taken as 0° C. (273° absolute) and 30 inches (760 millimeters) of mercury pressure. Also called standard temperature. D

**normative**
(1) In petrology, characteristic of, pertaining to, agreeing with, or occurring in the norm. (2) Used in the quantitative or norm system of classification of igneous rocks, a normative

mode being one which is essentially the same as the norm. D

**norm system**
A system of classification and nomenclature for igneous rocks based on the norm of each rock. Only undecomposed rocks of which accurate chemical analyses are available are classifiable in this system, which consequently is more used in detailed petrologic studies than in ordinary geologic or mining work. D

**northeaster**
Strong wind or storm coming from the northeast, but blowing in a southwesterly direction.

**notching**
A method of excavating in a series of steps.

**Nottingham system; Barring mining**
A long-wall method of working coal seams in which the trams run on a rail tract along the face and handloaded at the sides. It follows that the system can only be adopted in relatively thick seams where the trams can travel along the face without any roof ripping. The method is now largely replaced by face conveyors. D

**noxious**
(1) Causing or tending to cause injury, especially to health; (2) Hurtful; pernicious; for example, noxious gases. D

**NPA**
National Petroleum Association.

**NPC**
National Petroleum Council.

**NPDES**
National pollutant discharge elimination system.

**NPRA**
National Petroleum Refiners Association.

**NPT**
Non-Proliferation Treaty.

**NRC**
(1) National Research Council. (2) The National Regulatory Committee. (3) The Nuclear Regulatory Commission.

**NRDC**
National Resource Defense Council.

**NRECA**
National Rural Electric Cooperative Association, an organization representing the cooperatively owned electric utilities.

**NSC**
National Security Council.

**NSF**
National Science Foundation.

**NSPAC**
National Standard Policy Advisory Committee.

**NSPS**
New source performance standards.

**NSSS**
Nuclear steam supply system.

**NTIS**
National Technical Information Service of the U.S. Department of Commerce.

**N.T.P.**
Normal temperature and pressure; a temperature of 0° C and a pressure of 760 millimeters of mercury.

**nuclear breeder reactor**
A reactor which produces more nuclear fuel than it consumes. See breeder reactor. Z

**nuclear converter reactor**
A reactor in which the major process is the conversion of fissionable fuel into energy as distinguished from a "breeder reactor" which produces

more fuel than it uses. A converter reactor also "converts" some fertile material into fissionable fuel but produces less fissionable fuel than it consumes. V

**nuclear energy**
The energy liberated by a nuclear reaction (fission or fusion) or by radioactive decay.

**nuclear fusion**
The process of joining under controlled pressure-temperature conditions of certain light atoms, such as of hydrogen in the deuterium-tritium or D-T fuel cycle, to form heavier atoms, resulting in a smaller total mass of fusion products and thus in a release of energy from converted matter. Also referred to as thermonuclear fusion.

**nuclear magnetic resonance**
The resonance phenomenon encountered in energy transfers between a radio-frequency alternating magnetic field and a nucleus placed in a constant magnetic field H which is sufficiently strong to decouple the nuclear spin from the influence of the atomic electrons. M

**nuclear power**
Power released in exothermic nuclear reactions and useful in ways comparable with, for example, electric power.

**nuclear power plant**
Any device, machine, or assembly that converts nuclear energy into some form of useful power, such as mechanical or electrical power. In a nuclear electric power plant heat produced by a reactor is generally used to make steam to drive a turbine that in turn drives an electric generator. O

**nuclear reaction**
A reaction involving a change in an atomic nucleus, such as fission, fusion, neutron capture, or radioactive decay, as distinct from a chemical reaction, which is limited to changes in the electron structure surrounding the nucleus. O

**nuclear reactor**
A device in which a fission chain reaction can be initiated, maintained, and controlled. Its essential component is a core with fissionable fuel. It usually has a moderator, a reflector, shielding, coolant, and control mechanisms. Sometimes called an atomic "furnace," it is the basic machine of nuclear energy. O

**nuclear superheating**
Superheating the steam produced in a reactor by using heat from a reactor. Two methods: recirculating the steam through the same core in which it is first produced (integral superheating) or passing the steam through a second, separate reactor.

**nuclear weapons**
A collective term for atomic bombs and hydrogen bombs.

**nucleon**
A constituent of an atomic nucleus, that is, a proton or a neutron. W

**nucleonics**
The science and technology of an atomic nucleus, that is, a proton or a neutron. O

**nucleus**
The small, positively charged core of an atom. All nuclei contain both protons and neutrons, except the nucleus of ordinary hydrogen, which consists of a single proton. O

**nuclide**
A species of atom having a specific mass, atomic number, and nuclear energy state. These factors determine the other properties of the element, including its radioactivity. P

**NURE**
(1) Nuclear uranium resources estimate. (2) Nuclear uranium resource evaluation program.

**nut coal**
Chestnut coal. Also called nuts.

**nuts**
A commercial term for sized coal (irrespective of size).

**nutty slack**
Mixture of small coals, sized from 2 inches downward and probably of high ash content.

# O

**O**
Chemical symbol for oxygen.

**OAPEC**
Organization of Arab Petroleum Exporting Counties. It was founded in 1968 for cooperation in economic and petroleum affairs. Original members were Saudi Arabia, Kuwait, and Libya. In 1970, Abu Dhabi, Algeria, Bahrain, Dubai, and Qatar joined. See also OPEC. G

**OBR**
Oxygen blown reactor.

**observation error**
An error caused by misreading of a signal or measuring device, or to faulty recording. D

**observer**
In seismic prospecting, the man in charge of the recording crew, including the shooters and linemen. D

**OBV**
Octane blending values.

**occlude**
To take in and retain (a substance) in the interior rather than on an external surface; to sorb. Used especially of metals sorbing gases. D

**occluded gases**
Gases which enter the mine atmosphere from pores, as feeders and blowers, and also from blasting operations. These gases pollute the mine air chiefly by the absorption of oxygen by the coal, and in addition by chemical combination of oxygen with carbonaceous matter. D

**ocean basin**
That part of the floor of the ocean that is more than about 600 feet below sea level. D

**ocean coal**
Coal seams lying beneath the sea. D

**oceanic stratosphere**
The cold, deep layers of the ocean consisting of waters of polar or subpolar origin. D

**oceanography**
The broad field of science which includes all fields of study which pertain to the sea. D

**ocean temperature**
The mean surface temperature of both the Pacific and Atlantic Oceans is 17°C; that of the Indian Ocean is 18°C. Maximum temperatures are respectively, 32°C, 30°C, and 35°C. D

**ocean thermal energy conversion**
A process of generating electrical energy by harnessing the temperature differences between surface waters and ocean depths.

**ocean thermal gradients**
The temperature differences between deep and surface water, possibly the largest natural resource on earth. Heat engines have been built which can work across these gradients, but they are of relatively low thermal efficiency. The process would probably be limited to tropical waters with high surface temperatures. Few environmental adverse effects are anticipated. B

**ocean tidal power**
Energy source producing a small impact on total world energy needs producing electricity from the back-and-forth water power harnessed, for example, by adjustable-blade turbine units in a barrier separating the inland tidal basin from the sea. The average tidal height of most beaches, however, is only three feet and insufficient as use for potential power generation. While tidal power may be aesthetically and ecologically speculative, it consumes no exhaustible resources and produces no pollutant wastes. B

**ochers**
A name given to various native earthy materials used as pigments. They consist essentially of hydrated ferric oxide and admixed with clay and sand in varying amounts and in impalpable subdivision. D

**OCR**
Office of Coal Research.

**OCS**
Outer continental shelf.

**octane**
(1) Any of the group of isomeric hydrocarbons, $C_8H_{18}$, of the paraffin series: specially, normal octane, a colorless liquid boiling at 124.6°C (256.1°F), found in petroleum. (2) A rating scale used to grade gasoline as to its antiknock properties. M

**octane number**
A term numerically indicating the relative antiknock value of gasoline. For octane numbers 100 or below, it is based upon a comparison with the reference fuels iso-octane (100 octane number) and n-heptane (0 octane number). The octane number of an unknown fuel is the percent by volume of isooctane with nheptane which matches the unknown fuel in knocking tendencies under a specified set of conditions. Above 100, the octane number of a fuel is based on the engine rating, in terms of milliters of tetraethyllead in isooctane, which matches that of the unknown fuel. M

**od**
Outside diameter.

**OD (optical density) color**
A term given to color values of oils calculated from the depth of oil (or depth of oil dissolved in a colorless solvent) having an optical density equal to that of the "neutral filter working standard" placed in the base of one of the metal cells of the Duboscq colorimeter. Neutral filters are pieces of gray-colored glass (disks) with an optical density of approximately 0.3. OD color is 1.57 times true color. M

**oe**
Oersted. D

**OEB/D**
Oil equivalent barrels per day.

**OECD**
Organization for Economic Cooperation and Development.

**OEO**
Office of Economic Opportunity, an independent agency of the Federal government.

**OEP**
Office of Emergency Planning.

**oersted**
The practical, centimeter-gram-second electromagnetic unit of magnetic intensity. A unit magnetic pole, placed in a vacuum in which the magnetic intensity is 1 oersted, is acted upon by a force of 1 dyne in the direction of the intensity vector. D

**off-peak energy**
Energy supplied during periods of relatively low system demands as specified by the supplier. K

**offset line**
In surveying, a line established parallel to the main survey line, and usually not far from it. D

**off shore windpower system (OWPS)**
A proposed system to generate electricity by wind turbines mounted on off-shore platforms advocated by Professor W. E. Heronomus of the University of Massachusetts. G

**offshore winds**
Land breeze. Winds blowing seaward from the coast. D

**ohm**
(1) The unit of measurement of electrical resistance. (2) That resistance through which a difference of potential or electromotive force of one volt will produce a current of one ampere. K

**Ohm's law**
The electric current flowing in a circuit is proportional to the electromotive force and inversely proportional to the resistance. D

**OIE**
Office of Inspection and Enforcement.

**oil-base mud**
A mud-laden drill-circulation medium in which an oil is used as the laden liquid instead of water. D

**oil-bearing shale**
Shale inpregnated with petroleum. Not to be confused with oil shale. D

**oil burner**
Any device wherein oil fuel is vaporized or atomized and mixed with air in vapor proportion for combustion. Oil burners are classed in many ways according to methods of operation, ignition, gasifying, oil feed, etc. BB

**oil burning principles**
To ignite properly, oil spray must be mixed with air so that it will vaporize and gasify. The higher the temperature and the Baumé of the oil at the burner, the easier it is to spray and the air pressure may be correspondingly lower. The temperature at which it begins to vaporize is the limit to which any oil may be heated. The lower the pressure at which the oil and air can be used, the less will be the cost for power. BB

**oildag**
Colloidal dispersion of graphite in oil. D

**oil derrick**
A towerlike frame used in boring oil wells to support and operate the various tools. D

**oilfield**
A region rich in petroleum deposits; especially, a region containing numbers of producing oil wells. D

**oilfield rotary**
The type and size of drilling machines used to rotary-drill boreholes in search of petroleum. D

**oil gage**
An instrument of the hydrometer type arranged for testing the specific gravity of oils. D

**oil gas**
A gas resulting from the thermal decomposition of petroleum oils,

composed mainly of volatile hydrocarbons and hydrogen. The true heating value of oil gas may vary between 800 and 1600 Btu per cubic foot depending on the operating conditions and feedstock properties. DD

**oil gas tar**
A tar produced by cracking oil vapors in the manufacture of oil gas. D

**oil, heavy**
Heavy, thick and viscous, usually refinery residuals commonly specified as grades 5, 6, and Bunker C. DD

**oil lease**
The permission granted by a landowner to a company to prospect and drill for oil and gas under his land and to produce them when found. D

**oil, light**
Generally, all oils lighter than residual fuel oils No. 5 and No. 6. Oils that have a low specific gravity, usually products of controlled distillation of crude oil but also including the byproducts benzol and toluol. DD

**oil of vitriol**
Concentrated sulfuric acid. D

**oil pocket**
A cavity in the rocks containing oil. D

**oil pool**
An accumulation of oil in sedimentary rock that yields petroleum on drilling. The oil occurs in the pores of the rock and is not a pool or pond in the ordinary sense of these words. D

**oil refinery**
A plant where petroleum is distilled and otherwise refined. D

**oils**
A group of neutral liquids, comprising three main classes: (1) fixed (fatty) oils, from animal, vegetable, and marine sources, consisting chiefly of glycerides and esters of fatty acids; (2) mineral oils, derived from petroleum, coal, shale, etc., consisting of hydrocarbons; and (3) essential oils, volatile products, mainly hydrocarbons with characteristic odors, derived from certain plants. D

**oil sands**
Sands which have been bonded with oil, for example, linseed oil. Such sands are particularly suited for the production of large cores where high strength and considerable permeability are required; also referred to as tar sands or bituminous sands. D

**oil seepage**
The slow leakage of petroleum oil from its underground accumulations. D

**oil shale**
Not actually a shale but a type of finely-grained sedimentary rock which has been compressed into thin layers. The organic material it contains is properly referred to as kerogen, although in laymen's terms it is frequently called oil. Kerogen is a high-molecular-weight composition formed from algae and sea organisms or from waxy spores and pollen grain. For each barrel produced, an estimated 1-1/2 tons of rock must be processed, and in addition, large quantities of water are needed in the production process. M

**oil shale lands**
Lands on or under which oil shale is present. D

**oil shale retorting, in situ**
This process of extracting oil from shale consists of injecting intense heat, whether gaseous, scalding steam or nuclear energy, into a deep well and then forcing it through the shale. When it reaches 900°F, the heat decomposes the organic matter into three parts: oil (approximately 2/3), gas (1/10) and coke (1/4). Once the oil

has been separated from the shale, it will then flow into an underground pool from which it can be pumped out for retorting and refining through another well. It can then be pumped from the surface to refineries for further processing.

**oil spill**
The accidental discharge of oil into oceans, bays or inland waterways. Methods of oil spill control include chemical dispersion, combustion, mechanical containment and absorption. E

**oil trap**
A geologist's term for a place where oil collects underground.

**oil well**
A dug or bored well, from which petroleum is obtained by pumping or by natural flow. D

**oil-well casing**
Ordinary outside-coupled pipe used as borehole casing or drivepipe. Also called oilfield casing. D

**oil-well cement**
A hydraulic cement which sets at a slower rate than portland cement. D

**oil zone**
A formation that contains capillary or supercapillary voids, or both, that are full of petroleum and will move under ordinary hydrostatic pressure. D

**OIP**
Oil in place.

**OMB**
Office of Management and Budget in the Executive Office of the President.

**on-peak energy**
Energy supplied during periods of relatively high system demands as specified by the supplier. K

**ONR**
Office of Naval Research of the Department of the Navy.

**onshore winds**
Sea breezes, winds blowing shoreward from the sea.

**onstream time**
The length of time a unit is in actual production. M

**OOG**
Office of Oil and Gas in the Department of the Interior.

**opacity**
Degree of obscuration of light. For example, a window has zero opacity; a wall is 100 percent opaque. The Ringelmann system of evaluating smoke density is based on opacity. E

**OPEC**
Organization of Petroleum Exporting Countries. Founded in 1960 to unify and coordinate petroleum policies of the members. The members and the date of membership are: Abu Dhabi (1967); Algeria (1969); Indonesia (1962); Iran (1960); Iraq (1960); Kuwait (1960); Libya (1962); Nigeria (1971); Qatar (1961); Saudi Arabia (1960); and Venezuela (1960). OPEC headquarters are in Vienna, Austria. See also OAPEC. G

**opencast method**
The method consists in removing the overlying strata or overburden, extracting the coal, and then replacing the overburden. D

**open-cycle reactor system**
A reactor system in which the coolant passes through the reactor core only once and is then discarded. O

**open hole**
(1) Coal or other mine workings at the surface or outcrop. Also called opencast; opencut; open pit. (2) A borehole which is drilled without cores. D

**open-loop circuits**
In an open-loop, one medium serves as the collector and storage, or storage and distribution heat transfer medium. S

**open-pit mining; opencut mining**
A form of operation designed to extract minerals that lie near the surface. Waste, or overburden, is first removed, and the mineral is broken and loaded, as in a stone quarry. Important chiefly in the mining of ores of iron and copper, distinguished from the "strip mining" of coal and the "quarrying" of other nonmetallic materials such as limestone, building stone, etc. D

**operative temperature**
Operative temperature is that temperature of an imaginary environment in which, with equal wall (enclosing areas) and ambient air temperatures and some standard rate of air motion, the human body would lose the same amount of heat by radiation and convection as it would in some actual environment at unequal wall and air temperatures and for some other rate of air motion. D

**opr**
Oil production rate. D

**OPS**
(1) Offshore power systems. (2) Office of Pipeline Safety.

**option**
A privilege secured by the payment of a certain consideration for the purchase, or lease, of mining or other property, within a specified time, or upon the fulfillment of certain conditions set forth in the contract. D

**OR**
Operations research.

**orange heat**
A division of the color scale, generally given as about 900°C. D

**ORAU**
Oak Ridge Associated University.

**orbicular**
Containing spheroidal aggregates of megascopic crystals, generally in concentric shells composed of two or more of the constituent minerals; said of the structure of some granular igneous rocks, such as corsite. D

**organic**
Being, containing, or relating to carbon compounds, especially in which hydrogen is attached to carbon whether derived from living organisms or not. Usually distinguished from inorganic or mineral. D

**organic-cooled reactor**
A reactor that uses organic chemicals, such as mixtures of polyphenyls (diphenyls and terphenyls), as coolant. O

**organic petrochemicals**
Petrochemicals which contain atoms of both hydrogen and carbon, tracing their origin to living matter—the remains of plants and animals from which all hydrocarbons are derived. They are classified into two types according to the molecular arrangement of their carbon atoms, the aliphatics and the aromatics. M

**"Organic Theory"**
This theory of petroleum's origin holds that crude oil is an organic mineral formed by the decay and alteration of remains of prehistoric marine animals and plants. M

**oriental**
Frequently used in the same sense as precious when applied to minerals; from an old idea that gems came principally from the East, for example, oriental amethyst, oriental chrysolite, oriental emerald. D

**original oil-in-place**
The estimated number of barrels of crude oil in known reservoirs prior to

any production, usually expressed as "stock tank" barrels or the volume that goes into a stock tank after the shrinkage that results when dissolved gas is separated from the oil. G

**ORMAK**
Oak Ridge Tokomak.

**ORNL**
Oak Ridge National Laboratory.

**orogenic**
Pertaining to the processes by which great elongate chains and ranges of mountains are formed. D

**OSAHRC**
Occupational Safety and Health Review Committee.

**oscillograph**
An instrument which renders visible, or automatically traces, a curve representing the time variations of electric phenomena. The recorded trace is an oscillogram. D

**oscilloscope**
An instrument for showing visually graphical representations of the waveforms encountered in electrical circuits. D

**OSHA**
Occupational Safety and Health Act.

**OSM**
Office of Surface Mining.

**osmosis**
The passage of a solvent through a membrane from a dilute solution into a more concentrated one, the membrane being permeable to molecules of solvent but not to molecules of solute. D

**OTA**
Office of Technology Assessment.

**OTEC**
Ocean thermal energy conversion.

**outage**
(1) The period during which a generating unit, transmission line, or other facility, is out of service. It can be a "forced outage," due to emergency reasons, or a "scheduled outage," for inspection or maintenance. (2) The loss of a volatile liquid such as gasoline ascribed to evaporation or pilferage. Also called shrinkage. L

**outpost well**
A well drilled to extend a known oil or gas pool. If successful in this objective, it is an extension well. If unsuccessful, it is a dry outpost. D

**output**
(1) The quantity of coal or mineral raised from a mine and expressed as being so many tons per shift, per week, or per year. (2) The power or product from a plant or prime mover in the specific form and for the specific purpose required. D

**oven coke**
Coke produced by the carbonization of coal for the primary purpose of manufacturing coke. D

**overburden**
Material of any nature, consolidated or unconsolidated, that overlies a deposit of useful materials, ores, or coal, especially those deposits that are mined from the surface by open cuts. G

**overheating**
Heating a metal or alloy to such a high temperature that its properties are impaired. When the original properties cannot be restored by further heat treating, by mechanical working, or by a combination of working and heat treating, the overheating is known as burning. D

**overload**
(1) In general, a load or weight in excess of the designed capacity. (2)

The term may be applied to mechanical and electrical engineering plants, to loads on buildings and structures, and to excess loads on haulage ropes and engines. D

**overload capability**
The maximum load that a machine, apparatus, or device can carry for a specified period of time under specified conditions when operating beyond its normal rating but within the limits of the manufacturer's guarantee, or within safe limits as determined by the owner. L

**overloader**
A loading machine of the power-shovel type for quarry and opencast operations. The bucket is filled, the machine retracted, and the bucket swung over to the discharge points. D

**over-the-road hauling**
Hauling over public highways employing a haul unit commonly termed a truck. Various restrictions, such as weight, width of vehicle, safety features, guard against spillage, etc., must be considered in the type of equipment used. D

**overwind**
In hoisting through a mine shaft, failure to bring cage or skip smoothly to rest at proper unloading point at surface. D

**OWPS**
Offshore wind-power system.

**OWRR**
Office of Water Resources Research in the Department of Interior.

**oxidant**
Any oxygen-containing substance that reacts chemically in the air to produce new substances. Oxidants are the primary contributors to photochemical smog. E

**oxidation**
(1) Combination with oxygen. (2) Increase in oxygen content of molecular compound. (3) Increase in valency of electropositive part of compound, or decrease in valency of electronegative part. D

**oxidation of coal**
The absorption of oxygen from the air by coal, particularly in the crushed state; this engenders heat which can result in fire. Ventilation, while dispersing the heat generated, supports oxidation which increases rapidly with a rise in temperature. Fresh air should not gain access to the coal. D

**oz**
Ounce. D

**ozone**
An allotropic form of oxygen corresponding to the formula $O_3$. An unstable blue gas with a pungent odor and powerful bleaching action. Commercial mixtures containing 1 to 2 percent of ozone are produced by electrical irradiation of air or oxygen. Used as an oxidizing agent; in purification of drinking water; in treating of industrial wastes; in food preservation in cold storage; for bleaching waxes, oils, and textiles; in promoting production of peroxides; and as a bactericide. M

# P

**P**
Symbol for pressure. Also abbreviation for page and abbreviation for pico used as a prefix to units of measure, ondicating multiplication of the basic unit by $10^{-12}$ or division by 1 trillion. Also chemical symbol for phosphorous. D

**packaged power reactor**
A small nuclear power plant designed to be crated in packages small enough to be conveniently transported to remote locations. O

**PAD**
Petroleum Administration for Defense.

**paired electrons**
Said of two valence electrons when they form a nonpolar bond between two atoms. D

**pair production**
The transformation of a high-energy gamma ray into a pair of particles (an electron and a positron) during its passage through matter. D

**paleontology**
A science that deals with the life of past geological periods, is based on the study of fossil remains of plants and animals, and gives information especially about the phylogeny and relationships of modern animals and plants and about the chronology of the history of the earth. D

**palingenesis**
The process of formation of a new magma by the melting or fusion of country rocks with heat from another magma, with or without the addition of granitic material. D

**PAN**
Pexoxyacyl nitrates (for air pollution).

**panemone**
A vertical-axis wind collector capable of reacting to horizontal winds from any direction. U

**parabolic mirror**
A solar energy collecting device, its shiny, large, curved surface focuses solar heat and light on boilers, which then provide steam to operate engines. CC

**parabolic reflector**
A system to obtain solar energy by concentrating sunlight into air-filled heat pipes, each attached to small water storage tanks, to produce steam without the need for a central storage tank and the associated pumping costs. B

**paraffin**
A white, tasteless, odorless, and chemically inert waxy substance composed of saturated hydrocarbons with a boiling point greater than 572°F. It is obtained from petroleum oil, principally Eastern crudes. D

**paraffin base crude oil**
A variation of crude oil containing a high degree of paraffin wax and little or no asphalt. Besides wax, they also yield large amounts of high-grade lubricating oil. M

**paraffin coal**
A light-colored bituminous coal used for the production of oil and paraffin. D

**paraffin wax**
A colorless, more or less clear crystalline mass, without odor or taste and slightly greasy to the touch. M

**paragenesis**
A general term for the order of formation of associated minerals in time succession, one after another. To study the paragenesis is to trace out in a rock or vein the succession in which the minerals have developed. D

**parameter**
An arbitrary constant which characterizes a mathematical expression. E

**parent**
A radionuclide that upon radioactive decay or disintegration yields a specific nuclide (the daughter), either directly or as a later member of a radioactive series. W

**Parr's classification of coal**
This is based on the proximate analysis and calorific value of the ash-free, dry coal. The heating value of the raw coal is obtained, and from these data a table is drawn up at one end of which are the celluloses and woods of about 7,000 Btu's per pound. Parr then plots these figures against the percentage volatile matter in unit coal. D

**particle**
A minute constituent of matter, generally one with a measurable mass. The primary particles involved in radioactivity are alpha particles, beta particles, neutrons, and protons. O

**particle accelerator**
Modern machines which accelerate subatomic particles to such great velocities that as these particles strike atoms, the nucleus of the atom may be altered or split. Among those now in use are the cyclotron, the linear accelerator, the Van de Graaf generator, Proton synchroton, and the Bevatron. D

**particulate loading**
The introduction of particulates into the ambient air. E

**particulate matter**
Fine solid particles, such as ash, that remain individually dispersed in emissions from fossil-fuel plants. X

**partitioning**
The process of separating liquid waste into two or more fractions. P

**Pascal's law**
The component of the pressure in a fluid in equilibrium that is due to forces externally applied is uniform throughout the body of fluid. D

**PASEM**
Plan of assistance to solar energy manufacturers.

**passive solar system**
One in which solar energy alone is used for the transfer of thermal energy. This system requires no pumps, blowers, or other heat transfer medium moving devices which use energy other than solar. Collection, storage and distribution are achieved by natural heat transfer phenomena employing convection, radiation and conduction in conjunction with the use of thermal capacitance as a heat flow control mechanism. S

**patent**
(1) A document which conveys title to the ground, and no further assessment work need be done; however, taxes must be paid. (2) A government protection to an inventor, securing to him for a specific time the exclusive right of manufacturing, exploiting, using and selling an invention. D

**pathogenic**
Causing or capable of causing disease. E

**pcd**
Pounds per capita per day, waste.

**PCRV**
Prestressed concrete reactor vessel.

**PCS**
Primary coolant system.

**PDU**
Process development unit.

**pea coal**
In anthracite only, coal small enough to pass through a mesh three-fourths to one-half inch square, but too large to pass through a 3/8 inch mesh. Also known as No. 6 coal. D

**peak day**
The 24-hour day period of greatest total gas sendout. DD

**peaking capability**
(1) Generating capability normally designed for use during the maximum load period of a designated time interval. (2) Generating units or stations which are available to assist in meeting that portion of peak load which is above base load. D

**peak load**
The maximum load in a stated period of time. L

**peak load plant**
A power plant which is normally operated to provide power during maximum load periods. L

**peak power**
See demand power.

**peak responsibility**
The load of a customer, a group of customers, or part of a system at the time of occurrence of the system peak. L

**peak-shaving**
The process of supplying power from an extraneous source to help meet the peak demand on a system. U

**peat**
There are two types of peat, low moor (Flachmoor) and high moor (Hochmoor). Low moor peat is the most common starting material in coal genesis. The material is formed in swamps from accumulated plant organic matter. Stagnant ground water is necessary for peat formation to protect the residual plant material from decay. Peat has a yellowish brown to brownish black color, is usually of fibrous consistency, and can be either plastic or friable; in its natural state it can be cut; further, it has a high moisture content, 75% or higher in its natural state. Y

**peat coal**
(1) A natural product intermediate between peat and lignite. (2) An artificial fuel made by carbonizing peat. D

**pebble bed reactor**
A reactor in which the fissionable fuel (and sometimes also the moderator) is in the form of packed or randomly placed pellets, which are cooled by gas or liquid. O

**pelagic**
(1) Related to water of the sea as distinct from the sea bottom. (2) Related to sediment of the deep sea as distinct from that derived directly from the land. D

**penetrometer**
An instrument which automatically records the depth of drilling and the penetration rate. D

**Pennsylvania crude**
A type of oil produced in Pennsylvania, New York, West Virginia and parts of Ohio. It contains a high percentage of paraffin-base lubricating oil stock. M

**pennyweight**
One-twentieth troy ounce. Used in the United States and in England for the valuation of gold, silver and jewels. Abbreviation pwt. D

**percentage extraction**
The proportion of a coal seam which is removed from the mine. The remainder may represent coal in pillars or coal which is too thin or inferior to mine or lost in mining. D

**percolate**
(1) To pass through fine interstices. (2) To filter, as water percolates through sand. D

**performance number**
One of a series of numbers for converting fuel antiknock values in terms of a reference fuel into an index which is an approximate indication. M

**periodic law**
A law in chemistry discovered by Medeleev. The physical and chemical properties of the elements are periodic functions of their atomic weights. D

**permissible**
A machine or explosive is said to be permissible when it has been approved by the United States Bureau of Mines for use underground under prescribed conditions. All flameproof machinery is not permissible, but all permissible machinery is flameproof. D

**permissible hydraulic fluids**
Commercially available, fire-resistant fluids developed by the oil industry in cooperation with the U. S. Bureau of Mines. They are water-in-oil emulsions, and can be substituted for flammable hydraulic fluids by users of large machinery, whether the equipment is operated underground or on the surface. D

**personnel monitoring**
Determination by either physical or biological measurement of the amount of ionizing radiation to which an individual has been exposed, such as by measuring the darkening of a film badge or performing a radon breath analysis. O

**pesticide**
A chemical substance, such as DDT, which kills such pests as mice, insects, fungi, etc. Petroleum products are used as a base or carrier for many pesticides. M

**petrochemical**
A contraction of the words "petroleum" and "chemical," originally coined to designate chemicals of petroleum origin. At present, it is so loosely used and covers such a wide variety of products that it cannot be defined specifically. M

**petrochemistry**
The chemistry of rocks. O

**petrol**
Term commonly used in Great Britain for motor spirit or gasoline. M

**petroleum**
A material occurring naturally in the earth, predominantly composed of mixtures of chemical compounds of carbon and hydrogen with or without other nonmetallic elements such as sulfur, oxygen, nitrogen, etc. Petroleum may contain, or be composed of, such compounds in the gaseous, liquid, and/or solid state, depending on the

nature of these compounds and the existent conditions of temperature and pressure. M

**petroleum coke**
Primarily a fuel, it is almost pure carbon. It has useful properties: it burns with little or no ash, it conducts electricity, it is highly resistant to chemical action, it does not melt, and it has excellent abrasive qualities. The fact that it is almost pure carbon reduces the chances of contaminating the metals or chemicals being refined. M

**pF**
The numerical measure of energy with which water is held in the soil, expressed as the common logarithm of the head, in centimeters of water, necessary to produce the suction corresponding to the capillary potential. E

**PFBC**
Pressurized fluidized bed combustion.

**PGA**
Purchased gas adjustment.

**PGC**
Pyrolysis-gasification-combustion (turbine).

**pH**
A measure of the acidity or alkalinity of a material, liquid or solid. pH is represented on a scale of 0 to 14 with 7 representing a neutral state; 0, the most acid; and 14 the most alkaline. E

**phosphor**
(1) A luminescent substance. (2) A material capable of emitting light when stimulated by radiation. O

**photochemical conversion**
The method of breaking down water to produce hydrogen and oxygen involving the absorption of light by chemical dyes, a possible long-term method of producing hydrogen. Z

**photochemical oxidants**
Secondary pollutants formed by the action of sunlight on the oxides of nitrogen and hydrocarbons in the air; they are the primary contributors to photochemical smog. E

**photochemical smog**
Air pollution associated with oxidants rather than with sulfur oxides, particulates, etc. Produces necrosis, chlorosis and growth alterations in plants and is an eye and respiratory irritant in humans. E

**photochemistry**
A branch of chemistry dealing with the effect of radiant energy (such as light) in producing chemical changes. E

**photoelectric cell**
A device which registers changes in the amount of light falling on it by converting it to electricity. D

**photogrammetry**
The art of making surveys of measurements by the aid of photography.

**photolysis**
The decomposition of molecules by the absorption of light energy, e.g., the breakdown of water to hydrogen and oxygen. Q

**photometer**
An instrument for measuring the intensity of light, or for comparing intensities from two sources. D

**photon**
The carrier of a quantum of electromagnetic energy. Photons have an effective momentum but no mass or electrical charge. O

**photosynthesis**
(1) The formation of carbohydrates by the reaction of carbon dioxide and water in the presence of solar energy and catalysts in green plants. (2) A fundamental process of nature by

which green plants use the energy of light to produce various substances; a method of producing hydrogen. Y

**photovoltaic cell**
Solar energy gathering device which changes light rather than radiant heat into electrical energy. Also called "solar batteries." Newest of these cells are those made of very thin strips of silicon which has first been carefully purified and then with equal care delicately contaminated, giving about one hundred watts from each square yard when exposed to sunlight. Thus, they need 90 square feet per kilowatt. CC

**PHS**
Public Health Service of the U.S. Department of Health and Human Services.

**PHWR**
Pressurized heavy-water-moderator/coolant reactor.

**physics**
The science, or group of sciences that treats of the phenomena associated with matter in general, especially in its relation to energy, and of the laws governing those phenomena, excluding the special laws peculiar to living matter (biology) or to special kinds of matter (chemistry). Physics treats of the constitution and properties of matter, mechanics, acoustics, heat, optics, electricity, and magnetism. More generally, it includes all the physical sciences. D

**pick mines**
Mines in which coal is cut with picks. D

**pico-**
A prefix that divides a basic unit by one trillion ($10^{12}$). Same as micromicro.

**picocurie**
A unit of radioactive disintegration; 1 picocurie = $10^{-12}$ curies = 2.22 disintegrations per minute. E

**piezoelectric**
Having the ability to develop surface electric charges when subjected to elastic deformation, and conversely. D

**pig**
A heavily shielded container (usually lead) used to ship or store radioactive materials. O

**pile**
Old term for nuclear reactor. This name was used because the first reactor was built by piling up graphite blocks and natural uranium. W

**pillar**
An area of coal or ore left to support the overlying strata of hanging-wall in a mine. Pillars are sometimes left permanently to support surface works or against old workings containing water. Coal pillars, such as those in pillar-and-stall mining, are extracted at a later period. D

**pilot plant**
A small scale industrial process unit operated to test the application of a chemical or other manufacturing process under conditions that will yield information useful in the design and operation of full-scale manufacturing equipment. The pilot unit serves to disclose the special problems to be solved in adapting a successful laboratory method to commercial-size units. It is usually larger than a laboratory scale model. M

**pinch effect**
In controlled fusion experiments, the effect obtained when an electric current, flowing through a column of plasma, produces a magnetic field that confines and compresses the plasma. O

**pinging (pinking)**
A description of the sound of detonation. M

**pipeline**
A line of pipe with pumping machinery and apparatus for conveying a liquid, or gas. D

**piston**
In engines and pumps, a reciprocating device in a cylinder or tube which receives pressure from, or delivers to, a fluid. M

**pit**
Loosely speaking, a coal mine. Not commonly used by coal men, except in reference to surface mining where the workings may be known as a strip pit. D

**pitch**
(1) The angle at which a coal seam inclines below a horizontal line. (2) The solid or semisolid residue from the partial evaporation of tar. (3) Strictly, pitch is a bitumen with extraneous matter such as free carbon, residual coke, etc. D

**pitchblende**
A massive variety of uraninite or uranium oxide, found in metallic veins. Contains 55 to 75 percent $UO_2$; up to 30 percent $UO_3$; usually a little water and varying amounts of other elements. D

**pitch coal**
A brittle, lustrous bituminous coal or lignite. D

**pitch coke**
Coke made by the destructive distillation of coal-tar pitch. D

**pithead**
The top of a mine shaft including the buildings, roads, tracks, plant, and machines around it. D

**placer**
(1) A place where gold is obtained by washing. (2) An alluvial or glacial deposit, as of sand or gravel, containing particles of gold or other valuable mineral. In the United States mining law, mineral deposits, not veins in place, are treated as placers, so far as locating, holding, and patenting are concerned. Various minerals, besides metallic ores, have been held to fall under this provision, but not coal, oil or salt. D

**plant factor**
The ratio of the average power load of an electric power plant to its rated capacity. Sometimes called capacity factor. O

**plasma**
An electrically neutral gaseous mixture of positive and negative ions. Sometimes called the "fourth state of matter," since it behaves differently from solids, liquids and gases. High temperature plasmas are used in controlled fusion experiments. O

**plasma fuel**
This is the white-hot substance that is the principal ingredient of suns and stars. Neither solid, liquid, nor gaseous, it could become the great energy-thrust provider of the future. The heavy hydrogen of the oceans may be the best source for this fusion-type power. O

**plate tectonics**
A theory proposing that land masses or blocks of rigid materials in the upper mantle are in motion relative to one another and through collision explain the occurrence of volcanic, seismic and orogenic activities at plate margins. Z

**PLBR**
Prototype large breeder reactor.

**Plowshare**
The Atomic Energy Commission program of research and development on peaceful uses of nuclear explosives. The possible uses include large-scale excavation, such as for canals and harbors, crushing ore bodies and producing heavy transuranic isotopes. The term is based on a Biblical reference, Isaiah 2:4. W

**plumb**
(1) Synonym for vertical. (2) A plumb bob or a plummet.

**plutonium (Symbol Pu)**
A heavy radioactive, man-made, metallic element with atomic number 94. Its most important isotope is fissionable plutonium-239, produced by neutron irradiation of uranium-238. It is used for reactor fuel and in weapons. O

**pluvial**
Of a geologic change, resulting from the action of rain or sometimes from the fluvial action of rainwater flowing in stream channels. D

**pm**
(1) Particulate matter. (2) Permanent magnet.

**pneumatic**
Set in motion or operated by compressed air. D

**poc**
Pool operating center.

**point source**
In air pollution terminology, a stationary source of a large individual emission, generally of an industrial nature. This is a general definition; point source (or area source) is legally and precisely defined in Federal regulations. E

**polarization**
The process by which ordinary light is changed into polarized light. The plane at right angles to the plane of transverse vibration is called the plane of polarization. D

**polar molecule**
One in which the electrical charges of its constituent atoms or groups produce a dipole. D

**pollution**
The presence of matter or energy whose nature, location or quantity produces undesired environmental effects. E

**polychromatic**
Showing a variety or a change of colors.

**polyelaterite**
Soft elastic bitumen which differs from elaterite by a nontacky consistency, more intensive fluorescence and complete insolubility in organic solvent. It is believed that these properties are due to a higher state of polymerization.

**polyethylene**
One of the polymers $(C_2H_4)_n$, ranging from colorless liquid to white solid, depending on molecule size. Made by polymerization of ethylene. The low-molecular-weight polymers are used as lubricants; the high-molecular-weight polymers as plastics. M

**polymer**
A substance produced from another by polymerization. M

**polymer gasoline**
A product of polymerization of normally gaseous hydrocarbons to hydrocarbons boiling in the gasoline range. M

**polymerization**
A reaction combining two or more molecules to form a single molecule having the same elements in the same proportions as in the original molecules; the union of light olefins to

form hydrocarbons of higher molecular weight. The process may be thermal or catalytic. Of importance in the production of motor fuel and aviation fuel components from cracked gases. If unlike olefin molecules are combined, the process is referred to as "copolymerization" and produces "copolymers." High polymers are used as synthetic lubricating oils, plasticizers, etc. M

**pool reactor**
A reactor in which the fuel elements are suspended in a pool of water that serves as the reflector, moderator, and coolant. Popularly called a swimming pool reactor, it is usually used for research and training. O

**porosity**
The amount of void space in a reservoir, regardless of whether or not that space is accessible to fluid penetration. Effective porosity refers to the amount of connected pore spaces, that is, the space available to fluid penetration. D

**port**
An opening in a research reactor through which objects are inserted for irradiation or from which beams of radiation emerge for experimental use. O

**positive drive**
A driving connection in two or more wheels or shafts that will turn them at approximately the same relative speeds under any conditions. D

**positive gradient**
Layer of water where temperature increases with depth.

**positron**
Positive electron of same mass as negative electron, which has only transitory existence. D

**potential**
The words potential and voltage are synonymous and mean electrical pressure. The potential or voltage of a circuit, machine, or any piece of electrical apparatus means the potential normally existing between the conductors of such circuit or the terminals of such machine or apparatus. D

**potential difference**
The difference in pressure between any two equipotential lines. D

**potential energy**
Energy due to position. BB

**potential gradient**
Potential gradient means an ascending or descending value of voltage related to a linear measurement, as a distance along the earth surface or ground. D

**pound-calorie**
A hybrid term between the English and metric units and defined as the amount of heat required to raise 1 pound of water 1°C. D

**pound force**
A force which, when acting on a body of mass one pound, gives it an acceleration equal to that of standard gravity. D

**pour point**
The lowest temperature at which oil will pour or flow when it is chilled without disturbance under definite conditions. In the United States these conditions are prescribed in ASTM Method D 97. M

**power**
(1) Any form of energy available for doing any kind of work; for example, steampower and waterpower. (2) Specifically, mechanical energy, as distinguished from work done by hand. D

**power, apparent**
Apparent power is proportional to the mathematical product of the volts and amperes of a circuit. This product generally is divided by 1,000 and

designated in kilovoltamperes (kVA). It is comprised of both real and reactive power. K

**power coefficient**
The ratio of the power extracted to the power available. U

**power density**
The rate of heat generated per unit volume of a reactor core. O

**power distillate**
The untreated kerosene condensates and still heavier distillates down to 28°Bé from Mid-Continent petroleum. Used as fuel in internal combustion engines. D

**power (electric)**
The time rate of generating, transferring, or using electric energy, usually expressed in kilowatts. K

**power factor**
The ratio of the mean actual power in an alternating-current circuit measured in watts to the apparent power measured in volt-amperes, being equal to the cosine of the phase difference between electromotive force and current. D

**power, firm**
Power or power-producing capacity intended to be available at all times during the period covered by a commitment, even under adverse conditions. K

**power gas**
A low Btu gas (as Mond gas) made for producing power, especially for driving gas engines. D

**power, interruptible**
Power made available under agreements which permit curtailment or cessation of delivery by the supplier. K

**power, non-firm**
Power or power-producing capacity supplied or available under an arrangement which does not have the guaranteed continuous availability feature of firm power. K

**power per unit band**
The limit approached by the quotient obtained by dividing (1) the power of the energy being transmitted by a given system, at a given time and in a given frequency band, by (2) the width of this band as the width of the band approached zero. D

**power pool**
A power pool is two or more interconnected electric systems planned and operated to supply power in the most reliable and economical manner for their combined load requirements and maintenance program. K

**power, reactive**
The portion of "Apparent Power" that does no work. It is commercially measured in kilovars. Reactive power must be supplied to most types of magnetic equipment, such as motors. It is supplied by generators or by electrostatic equipment, such as capacitors. K

**power reactor**
A reactor designed to produce useful nuclear power, as distinguished from reactors used primarily for research or for producing radiation or fissionable materials. O

**power, real**
This is the energy or work-producing part of "Apparent Power." It is the rate of supply of energy, measured commercially in kilowatts. The product of real power and length of time is energy, measured by watt hour meters and expressed in kilowatt hours. K

**power shovel mining**
Power shovels are used for mining coal, iron ores, phosphate deposits and copper ores. The shovels may be used either for mining or for stripping and removing the overburden or for both types of work, although at some coal

mines the shovels used for stripping are considerably larger than those used for mining. D

**power upon the air**
In coal mine ventilation, the horsepower applied is often known as the power upon the air. This may be the power exerted by a motive column due to the natural causes, to a furnace, or it may be the power of a mechanical motor. The power upon the air is always measured in the foot-pounds per minute. D

**ppb**
Parts per billion.

**PPBR**
Program planning budgeting review.

**ppm**
Parts per million.

**precipitate**
To separate a solid form from a solution by chemical means. D

**precipitators**
In pollution control work, any of a number of air pollution control devices usually using mechanical/electrical means to collect particulates from an emission. E

**precision**
The closeness of approach of each of a number of similar measurements to the arithmetic mean, the sources of error not necessarily being considered critically. Accuracy demands precision, but precision does not require accuracy. D

**preheat**
To heat, previous to some treatment, as an oil to be subsequently distilled, or as a body of gas or oil to be used as fuel. M

**pressed distillate**
The oil left in petroleum refining after the paraffin has been separated from the paraffin distillate by cooling and pressing. D

**pressed fuel**
An artificial fuel prepared from coal dust, waste coal, etc., incorporated with other ingredients, as tar, and compressed in molds into blocks; briquettes. D

**pressure**
The force or thrust exerted on a surface, normally expressed as force per unit area. Pressure is exerted in all directions in a system. Common examples: air pressure in a tire, or water pressure at some depth in the ocean. M

**pressure, absolute (PSIA)**
Pressure above that of a perfect vacuum. It is the sum of gauge pressure and atmospheric pressure. DD

**pressure drop**
The decrease in pressure due to friction, which occurs when a liquid or gas passes through a pipe, vessel, or other piece of equipment. M

**pressure gauge—(PSIG)**
Pressure above atmospheric pressure. DD

**pressure maintenance**
When natural oil reservoir pressures are insufficient, gas or water in the formation are replaced by gas and water injection techniques. Both methods involve drilling carefully located auxiliary service wells to permit the most efficient injection pattern. Groups of producers in a field work together so that the pressure of the injected water or gas can be regulated according to conditions throughout the reservoir. This is essential for optimum recovery. M

**pressure vessel**
A strong-walled container housing the core of most types of power reactors;

it usually also contains moderator, reflector, thermal shield, and control rods. O

**pressurized**
Any structure, area, or zone fitted with an arrangement that maintains nearly normal atmospheric pressure inside while subjected to high pressure outside. D

**pressurized water reactor**
A power reactor in which heat is transferred from the core to a heat exchanger by water kept under high pressure to achieve high temperature without boiling in the primary system. Steam is generated in a secondary circuit. Many reactors producing electric power are pressurized water reactors. O

**pretreatment**
In waste water treatment, any process used to reduce pollution load before the waste water is introduced into a main sewer system or delivered to a treatment plant for more substantial reduction of the pollution load. E

**preventive maintenance**
(1) A system of plant and equipment inspections made at regular intervals and the condition of the items recorded and marked on a chart. (2) A system which enables breakdowns to be anticipated and arrangements made to perform the necessary overhauls and replacements in good time. D

**primary cell**
A cell which generates or makes its own electrical energy from the chemical action of its constituents. Examples of primary cells are the voltaic cell, Daniell cell, LeClanche cell, and the dry cell. D

**primary chemical**
A chemical obtained directly by extraction and purification from a natural raw material; for example, toluol from coal tar. D

**primary distribution feeder**
An electric line supplying power to a distribution circuit, usually considered to be that portion of the primary conductors between the substation or point of supply and the center of distribution. K

**primary energy**
Energy available from firm power. K

**primary fuel**
Fuel consumed in original production of energy as contrasted to a conversion of energy from one form to another. G

**primary treatment**
The first stage in waste water treatment in which substantially all floating or settleable solids are mechanically removed by screening and sedimentation. E

**primary voltage**
The voltage of the circuit supplying power to a transformer is called the primary voltage, as opposed to the output voltage or load-supply voltage which is called secondary voltage. In power supply practice the primary is almost always the high-voltage side and the secondary the low-voltage side of a transformer. K

**prime mover**
Any machine capable of producing power to do work. M

**prime rate**
The rate of interest charged by commercial banks for short-term loans to the most credit-worthy corporations.

**priming**
The act of adding water to displace air, thereby promoting suction, as in a suction line of a pump. D

**probability**
The ratio of the chances favoring an event to the total number of chances for and against it.

**probable reserves**
A realistic assessment of the reserves that will be recovered from known oil or gas fields based on the estimated ultimate size and reservoir characteristic of such fields. Probable reserves include those reserves shown in the proved category. G

**probo**
Product-oil-bulk-ore.

**process heat reactor**
(1) A reactor designed primarily for large scale production of plutonium-239 by neutron irradiation of uranium-238. (2) A reactor used primarily for the production of radioactive isotopes. O

**process weight**
The total weight of all materials, including fuels, introduced into a manufacturing process. The process weight is used to calculate the allowable rate of emission of pollutant matter from the process. E

**producer gas**
A gas manufactured by burning coal or coke with a regulated deficiency of air. The principal combustible component is carbon monoxide and the normal heat content is between 130 and 150 Btu per cubic foot. DD

**producing horizon**
Rock from which oil or gas is produced. D

**producing zone**
The part of a sandstone or other reservoir rocks that yields oil and/or gas. D

**production**
(1) The act or process of generating electric energy. (2) A functional classification relating to that portion of utility plant used for the purpose of generating electric energy, or to expenses relating to the operation or maintenance of production plant, or the purchase and interchange of electric energy. K

**production reactor**
(1) A reactor designed primarily for large scale production of plutonium by neutron irradiation of uranium 238. (2) A reactor used primarily for the production of isotopes. D

**profile**
A vertical section of a soil showing the nature and sequence of its various zones. D

**program**
Sequence of steps to be performed. Preparation is called programming, done by a programmer. D

**propane**
A gaseous paraffin hydrocarbon, $C_3H_8$, occurring in crude oil, natural gas, and refinery cracking gas. It is used as a fuel, a solvent, and a refrigerant. M

**prospecting**
The search for outcrops or surface exposure of mineral deposits. D

**protection**
Provisions to reduce exposure of persons to radiation. For example, protective barriers to reduce external radiation or measures to prevent inhalation of radioactive materials. W

**proton**
An elementary particle with a single positive electrical charge and a mass approximately 1837 times that of the electron. The nucleus of an ordinary or light hydrogen atom. Protons are constituents of all nuclei. The atomic number (Z) of an atom is equal to the number of protons in its nucleus. O

**proton synchrotron**
A type of particle accelerator for producing beams of very high energy protons (in the Bev range). O

**prototype**
The stage following the basic idea for a new machine. It is an experimental model which assists the inventor and manufacturer in solving the difficult details. The prototype may cover only the basic new principles or features of the machine but is adequate to conduct tests. D

**proved reserves**
The estimated quantity of crude oil, natural gas, natural gas liquids or sulfur which analysis or geological and engineering data demonstrates with reasonable certainty to be recoverable from known oil or gas fields under existing economic and operating conditions. G

**PSA**
Power supply area.

**PSC**
Public Service Commission.

**PSD**
Prevention of significant deterioration.

**P. S. detector tube**
A device to estimate the proportion of carbon monoxide in mine air. D

**psf**
Pounds per square foot.

**psi**
Pounds per square inch. G

**psia**
Pounds per square inch absolute. A measure of pressure that includes atmospheric pressure. G

**pspe**
Photosynthetic product enhancement.

**public domain**
All lands and waters in the possession and ownership of the United States, including lands owned by the several states, as distinguished from lands owned by individuals and corporations. D

**public land**
In the United States, the portion of the public domain to which title is still vested in the Federal Government. D

**PUC**
Public Utilities Commission.

**PUCA**
Public Utilities Control Authority.

**PUD**
Public Utilities District.

**pulsed reactor**
A type of research reactor with which repeated short, intense surges of power and radiation can be produced. The neutron flux during each surge is much higher than could be tolerated during a steady-state operation. O

**pulverized fuel**
Finely ground coal or other combustible material, which can be burned as it issues from a suitable nozzle through which it is blown by compressed air. D

**pump capacity**
The volume of fluid, at a specified pressure, that a pump can transfer or lift when powered by an engine or motor of any given horsepower. Pump capacity also depends on fluid viscosity, pump condition, line friction, etc. D

**pumped storage**
An arrangement whereby additional electric power may be generated during peak load periods by hydraulic means using water pumped into a storage reservoir during off peak periods. K

**pumped storage plant**
A power plant utilizing an arrangement whereby electric energy is

generated for peak load use by utilizing water pumped into a storage reservoir usually during off-peak periods. A pumped storage plant may also be used to provide reserve generating capacity. L

**pump pressure**
(1) The force per unit area or pressure against which the pump is acting to force a fluid to flow through a pipeline, drill string, etc. (2) The pressure imposed on the fluid ejected from a pump. D

**purchase capacity**
The amount of power available for purchase from a source outside the system to supply energy or capacity. K

**PUSH**
Purchase of solar heating.

**pushbutton coal mining**
A fully automatic and remotely controlled system of coal cutting, loading, and face conveying, including self-advancing roof support systems. D

**PVC**
Polyvinyl chloride. D

**PVDC**
Polyvinylidene chloride.

**PVS**
Photovaltic system.

**PVT**
Pressure-volume-temperature. D

**PWR**
Pressurized water reactor.

**pycnometer; pyknometer**
A device for weighing and thus determining the specific gravity of small quantities of oil or other liquids. D

**pyranometer**
An instrument for measuring sunlight intensity. It usually measures total (direct and diffuse) insolation over a broad wavelength range. H

**pyrheliometer**
An instrument that measures the intensity of the direct beam radiation (direct insolation) from the sun, when used in conjunction with a tube for rejecting non-parallel light rays. H

**pyrogen**
A substance produced by the action of heat. D

**pyrogenic**
Producing, or produced by, heat. D

**pyrolysis**
The chemical decomposition of waste in the absence of oxygen, where materials are heated at atmospheric pressure, and several products are obtained: low Btu gas, char (partially burned carbon residue), and a heavy tar-like oil. All three of these products can be used as fuels; however, the oil has a high viscosity, and a lower Btu value than ordinary fuel oil. N

**pyrometer**
An instrument for measuring temperatures beyond the range of thermometers usually by the increase of electric resistance in a metal when heated; either by the generation of electric current by a thermocouple when acted upon by direct heat or focused radiation; or by the increase in intensity of light radiated by an incandescent body as its temperature increases. D

**pyroscope**
A device that, by a change in shape or size, indicates the temperature or, more correctly, the combined effect of time and temperature (which has been called heat work). D

# Q

**Q**
Originally used as a symbol presenting a quintillion or $10^{18}$ Btu (1 billion billion). Now also used by some authors to represent a quadrillion or $10^{15}$ Btu. In either form, it is used to express very large energy figures.

**qt**
Quart.

**quad**
Short for quadrillion, the term used to express $10^{15}$ Btu.

**quadrant**
A quarter of a circle; an arc of 90°. D

**quadrillion**
Represents $10^{15}$ Btu; often abbreviated as quad.

**qualitative**
Describes a subject or object in terms of its qualities or characteristics; usually applied to non-measurable aspects. E

**quality control**
(1) Systematic setting, check and operation designed to maintain steady working conditions in continuous process such as mineral concentration. (2) To forestall trouble. (3) To check condition of ore, pulp or products at important transfer points. P

**quality factor**
The factor by which absorbed dose is to be multiplied to obtain a quantity that expresses, on a common scale for all ionizing radiations, the irradiation incurred by exposed persons. It is used because some types of radiation such as alpha particles are more biologically damaging than other types. W

**quantitative**
Describes a subject or object in terms of its measurable aspects or characteristics; implies the use of mathematics, especially statistics. E

**quantum**
Unit quality of energy according to the quantum theory. It is the equal of the product of the frequency of radiation of the energy and 6.6256 X $10^{-27}$ erg-sec. The photon carries a quantum of electromagnetic energy. O

**quantum theory**
The statement, according to Max Planck, German physicist, that energy is not emitted or absorbed continuously but in units or quanta. A corollary

of this theory is that the energy of radiation is directly proportional to its frequency. O

**quartz**
(1) A crystallized silicon dioxide. (2) The name of the mineral is prefixed to the names of many rocks that contain it, as quartz porphyry, quartz diorite. D

**quarry**
An open or surface working, usually for the extraction of building stone, as slate, limestone, etc. D

**quench**
To cool suddenly (as heated steel) by immersion, especially in water or oil. D

**quintal**
A metric unit that equals 100 kilograms. D

**quintillion**
Represents $10^{18}$ Btu, and abbreviated as quint.

**quitclaim**
(1) A release of a claim, a deed of release. (2) Specifically a legal instrument by which some right, title, interest, or claim by one person in or to an estate held by himself or another, is released to another, and which is sometimes used as a simple but effective conveyance for making a grant of lands whether by way of release or as an original conveyance. (3) In the United States, a document in which a mining company on occasion sells its surface rights but retains its mineral rights. D

# R

**r**
Roentgen; also ratio, rod and refraction.

**R**
Degree Rankine; also abbreviation for resistance, radius, gas constant and Reynold's number.

**rabbit**
A device used for the rapid insertion of irradiation samples into reactors and for their subsequent fast transfer to nearby laboratories.

**race**
(1) An aqueduct or channel for conducting water to or from the place where it performs work. (2) A groove along which some part of a machine moves, such as the annular ring in a ball bearing that guides and holds the balls in place.

**rad.**
Radius.

**Rad**
Radiation absorbed dose.

**radian**
A unit of plane angular measurement equal to the angle at the center of a circle subtended by an arc equal in length to the radius. One radian equals about 57.29°.

**radiant energy**
Energy sent out or emitted by rays or waves. M

**radiant heating**
Usually a system of heating by surfaces at higher than body temperatures whereby the rate of heat loss from human beings by radiation is controlled. D

**radiation**
(1) The emission and propagation of energy through matter or space by means of electromagnetic disturbance which display both wave-like and particle-like behavior; in this context the "particles" are known as photons. (2) The energy so propagated. The term has been extended to include streams of fast-moving particles (alpha and beta particles, free neutrons, cosmic radiation, etc.). Nuclear radiation is that emitted from atomic nuclei in various nuclear reactions, including alpha, beta and gamma radiation and neutrons. O

**radiation absorbed dose**
The basic unit of absorbed dose of one rad means the absorption of 100 ergs of radiation energy per gram of absorbing material. O

**radiation burn**
Radiation damage to the skin. W

**radiation chemistry**
That branch of chemistry that is concerned with the chemical effects (including decomposition) of high-energy radiation and particles on matter. D

**radiation damage**
A general term for the harmful effects of radiation on matter. W

**radiation dose**
Accumulated exposure to radiation during a specified period of time.

**radiation intensity**
In general, the quantity of radiant energy at a specified location passing perpendicularly through unit area in unit time. It may be given as number of particles or photons per square centimeter per second or in energy units, as ergs per square centimeter-second. D

**radiation protection guide**
The officially determined radiation doses which should not be exceeded without careful consideration of the reasons for doing so. These standards are equivalent to what was formerly called the maximum dose or maximum permissible exposure. O

**radiation pyrometer**
A device for ascertaining the temperature of a distant source of heat, such as a furnace.

**radiation shielding**
Reduction of radiation by interposing a shield of absorbing material between any radioactive source and a person, laboratory area, or radiation-sensitive device. O

**radiation source**
Usually a man-made, sealed source of radiactivity used in teletherapy, radiography, as a power source for batteries, or in various types of industrial gauges. O

**radiation sterilization**
(1) Use of radiation to cause a plant or animal to become sterile, that is, incapable of reproduction. (2) The use of radiation to kill all forms of life (especially bacteria) in food, surgical sutures, etc. W

**radiation survey**
Evaluation of an area or object with instruments in order to detect, identify and quantify radioactive materials and radiation fields present. P

**radioactive contamination**
(1) Deposition of radioactive material in any place where it may harm persons, spoil experiments, or make products or equipment unsuitable or unsafe for some specific use. (2) The presence of unwanted radioactive matter. P

**radioactive dating**
A technique for measuring the age of an object or sample of material by determining the ratios of various radioisotopes or products of radioactive decay it contains. For example, the ratio of carbon-14 to carbon-12 reveals the approximate age of bones, pieces of wood, or other archaeological specimens that contain carbon extracted from the air at the time of their origin. W

**radioactive decay**
The spontaneous transformation of an atomic nucleus during which it changes from one nuclear species to another with the emission of particles and energy. Also called "radioactive disintegration." V

**radioactive nuclides, natural**
Natural radioactivity is that exhibited by naturally occurring substances. Natural radioactive nuclides are classified as (i) primary, (ii) secondary, (iii) induced and (iv) extinct. Z

**radioactivity**
The spontaneous decay or disintegration of an unstable atomic nucleus,

usually accompanied by the emission of ionizing radiation. Often shortened to "activity." O

**radiobiology**
The study of the scientific principles, mechanisms, and effects of the interaction of ionizing radiation with living matter. D

**radioisotope—a radioactive isotope**
An unstable isotope of an element that decays or disintegrates spontaneously, emitting radiation. More than 1300 natural and artificical radioisotopes have been identified. O

**radioisotopic generator**
A small power generator that converts the heat released during radioactive decay directly into electricity. These generators generally produce only a few watts of electricity and use thermoelectric or thermionic converters. Some also function as electrostatic converters to produce a small voltage. Sometimes called an "atomic battery." O

**radiology**
(1) The science of radioactive substances, X-rays, and other high-energy radiations. (2) Specifically, the use of sources of radiant energy in the diagnosis and treatment of disease. D

**radiometer**
Essentially a heat-flow meter used to measure long-wall radiation as well as solar radiation. It can be used both for daytime and nighttime measurements and to measure the net heat transfer through a surface.

**radiomutation**
A permanent, transmissible change in form, quality or other characteristic of a cell or offspring from the characteristics of its parent, due to radiation exposure. See genetic effects of radiation, mutation. W

**radionuclide**
A synonym for radioactive nuclide. Z

**radioresistance**
A relative resistance of cells, tissues, organs, or organisms to the injurious action of radiation. Compare radiosensitivity. W

**radiosensitivity**
A relative susceptibility of cells, tissues, organs or organisms to the injurious action of radiation. Compare radioresistance. W

**radium (Symbol Ra)**
A radioactive metallic element with atomic number 88. As found in nature, the most common isotope has an atomic weight of 226. It occurs in minute quantities associated with uranium in pitchblende, carnotite and other minerals. W

**radon (Symbol Rn)**
A radioactive element, one of the heaviest gases known. Its atomic number is 86, and its atomic weight is 222. It is a daughter of radium in the uranium radioactive series. W

**rake**
(1) A train of mineral trucks. (2) A timber placed at an angle. (3) An angle, measured in degrees, formed by the leading face of a cutting tool and the surface behind the cutting edge.

**Ralston's classification of coal**
A classification based on the percentage of carbon, hydrogen, and oxygen in the ash-, moisture-, sulfur-, and nitrogen-free coal. These figures are plotted on trilinear coordinates giving well-defined zones of bituminous coals, lignites, peats, etc. D

**Ramsbottom coke test**
A carbon residue test which originated in England (IP Method 14). The corresponding ASTM Method is D 524, which is widely used in the United States. M

**random source of pollution**
Spillage of oil and hazardous chemicals on land or in the rivers, lakes, and

harbors due to transportation or any other accidents is a serious random source of pollution. E

**rank**
This term describes the stage of carbonification attained by a given coal. D

**Rankine scale**
A scale of absolute temperature based on the Fahrenheit scale. This temperature, in degrees Rankine, is equal to the temperature in degrees Fahrenheit plus 459.72. See degree Rankine. M

**RANN**
Research applied to national needs (NSF).

**RARE**
Roadless area review and evaluation.

**rash**
A substance grading about halfway between a coal and a shale and looks somewhat like an oil shale, which is characterized by a brown streak and leathery appearance with parting planes often smooth and polished, except that rash is more flexible. It usually occurs in very thin flakes or sheets at the bottom, the top, or within the seam. The color is usually dark but frequently grades into a lighter shade. O

**rate base**
The net plant investment or valuation base on which the utility is entitled to earn a fair return. L

**rated horsepower**
Theoretical horsepower of an engine based on dimensions and speed. D

**rated load**
This is the kilowatt power output which can be delivered continuously at the rated output voltage. It may also be designated as the 100 percent load or full-load rating of the unit. D

**rated output current**
The current derived from the rated load and the rated output voltage. D

**rated wind speed**
The lowest wind speed at which the rated output power of a wind machine is produced. U

**rating**
Limits placed on operating conditions of a machine, transmission line, apparatus, or device based on its design characteristics. Such limits as load, voltage, temperature, and frequency may be given in the rating. L

**raw coal**
Coal which has received no preparation other than possibly screening. D

**raw fuel**
A fuel which is used in the form in which it is mined or obtained, for example, coal, lignite, peat, wood, mineral oil, natural gas. D

**Rayleigh wave**
A surface wave associated with the free boundary of a solid. The wave is of maximum intensity at the surface and diminishes quite rapidly as one proceeds into the solid. Therefore, it has a tendency to hug the surface of the solid. D

**RBE**
Relative biological effectiveness.

**RCIC**
Reactor core isolation cooling.

**R & D**
Research and development.

**RD & D**
Research, development, and demonstration.

**RDF**
Refuse derived fuel.

**RDT & E**
Research, development, testing, and evaluating.

**Re**
Reynolds number.

**REA**
Rural Electrification Administration.

**reaction, physical**
A force opposing a given force in terms of Newton's laws. Chemically, either (1) change in pH value, or (2) molecular alteration. D

**reactivity**
A measure of the departure of a nuclear reactor from criticality. It is about equal to the effective multiplication factor minus one and is thus precisely zero at criticality. If there is excess reactivity (positive reactivity), the reactor is supercritical and its power will rise. Negative reactivity (subcriticality) will result in a decreasing power level. O

**reactor**
An atomic pile or confining equipment for control of a nuclear reaction intended to produce energy. Z

**reactor types**
There are three basic types of reactors, each characterized by the average energy of the neutrons which produce the fissions:
(1) Thermal reactors, where neutrons causing fission are in equilibrium with the moderator of the reactor and thus have energies corresponding to the temperature of the moderator.
(2) Intermediate reactors, in which the neutrons have energies in a broad band between those for thermal and fast reactors, i.e., 1 eV to 1000 eV.
(3) Fast reactors, in which the average energy of the neutrons is several hundred thousand electron volts. Z

**reagent**
Any substance which is used in detecting, examining, or measuring other substances because of its capacity for taking part in certain chemical reactions. M

**real calorific value**
The calorific value when determined by a calorimeter in the laboratory. D

**real wage, real income**
Purchasing power of earnings or income, adjusted for the inflation in consumer prices. Also real spendable earnings minus tax payments.

**receiving water**
Rivers, lakes, oceans or other bodies that receive treated or untreated waste waters. E

**receptor cells**
Elements within sense organs which convert various forms of energy in the environment into action potentials in the sensory nerves. Examples of energy conversion are: mechanical (pressure or touch), thermal (degrees of warmth), and electromagnetic (light). E

**recirculated air**
Air returned from a space to be heated, conditioned, or cleaned then redistributed to the space. D

**reclaimed oil**
A lubricating oil which, after undergoing a period of service, is collected, reprocessed, and sold for reuse. M

**reclamation (land)**
The operation or process of changing the condition or characteristics of land so that improved utilization can be achieved. This may be accomplished by various means such as irrigation of arid land, drainage of swamp or water-logged land or protection from flood menace of land constantly subject to overflow. E

**recoverable reserves**
Minerals expected to be recovered by present day techniques and under present economic conditions. G

**rectification**
The process by which electric energy is transferred from an alternating-current circuit to a direct-current circuit. D

**recuperator**
A continuous heat exchanger in which heat is conducted from the products of combustion to incoming air through flue walls. D

**recycling**
(1) A process for the recovery of useful substances from wastes; for example: (a) recycling of metal, glass, and paper from solid wastes for reuse, and (b) production of reusable water from waste-water. (2) An operation in which a substance is passed through the same series of processes, pipes, or vessels more than once. (3) In nuclear terminology, the reuse of fissionable material, after it has been recovered by chemical processing from spent or depleted reactor fuel, reenriched, and then refabricated into new fuel elements. E

**red oil**
Originally, the residual fraction obtained by fire-distilling pressed distillate. Such oils were acid-treated but not filtered. The term is now used to describe any oil of red color, regardless of refining process by which it is made. The bulk of the so-called red engine oils, bearing oils, and machinery oils are of this class. They may be considered as intermediate grades of lubricating oil for general purposes. Term sometimes used for oleic acid. M

**Redwood viscosimeter**
Standard British viscosimeter. The number of seconds required for 50 ml of an oil to flow out of a standard Redwood viscosimeter at a definite temperature is the Redwood viscosity prescribed by IP Method 70. Instrument is available in two sizes: Redwood No. 1 and Redwood No. II. When the flow time exceeds 2,000 sec, the No. II must be used. M

**refine**
To cleanse or purify by removing undesired components; to process a material to make it usable. See refining. M

**refinery**
A plant, with all its included equipment, for manufacturing finished or semifinished products from crude oil. M

**refining**
The separation of crude oil into its component parts, and the manufacture therefrom of products needed for the market. Important processes in refining are distillation, cracking, chemical treating, and solvent extraction. See also specific refining processes under alphabetical listing. M

**reflection**
(1) The bounding back of light rays or ether rays as they strike a solid surface. (2) In seismic prospecting, the returned energy (in wave form) from a shot which has been reflected from a velocity discontinuity back to a detector. D

**reflectivity**
The ratio of radiant energy reflected by a body to that falling upon it. D

**reflectometer**
A photometric or electronic device for measuring the reflectances of light or other radiant energy. E

**reflector**
A layer of material immediately surrounding a reactor core which scatters back or reflects into the core many neutrons that would otherwise escape. The returned neutrons can then cause more fissions and improve the neutron economy of the reactor. Common reflector materials are graphite, beryllium and natural uranium. W

**reflux**
In fractional distillation, that part of the distillate which may be returned to the column to assist in making a better separation into desired fraction. This operation is called refluxing. Reflux may be either circulating or induced. Circulating reflux is liquid which is withdrawn hot, cooled, and pumped back to the tower. Induced reflux is liquid formed within a fractionation tower by condensation of vapors by means of an internal cooling coil. D

**reformed gasoline**
Gasoline made by a reforming process. M

**reforming**
(1) The cracking of petroleum naptha or of straight-run gasoline of low octane number usually to form gasoline containing lighter constituents and having a higher octane number. (2) Generally, a chemical process using heat to break down a substance into desired components. D

**refraction**
A change of direction of a ray of light when it passes from one medium to another of different optical density. D

**refractive index**
(1) Ratio of speed of heat, light or sound traversing a medium to that in air. (2) Ratio of sine of angle or incidence to sine of angle of refraction of light refracted from vacuum into a medium, as measured in a refractometer. D

**refractometer**
(1) A firedamp detector; also interference methanometer. (2) An instrument designed for measuring the refractive indices of various substances. D

**refractory**
A material of a very high melting point with properties that make it suitable for such uses as furnace linings and kiln construction. D

**refrigeration**
The process of absorption of heat from one location and its transfer to the rejection at another place. Arbitrarily expressed in units of tons and is equal to the coil cooling load divided by 12,000 (a ton of ice in melting in 24 hours liberates heat at the rate of 200 British thermal units per minute, or 12,000 British thermal units per hour). D

**regeneration**
(1) The process of restoring a material to its original strength or properties. (2) In a catalytic process, the revivification or reactivating of the catalyst, sometimes done by burning off the coke deposits under carefully controlled conditions of temperature and oxygen content of the regeneration gas stream. M

**regulating rod**
A reactor control rod used for making frequent fine adjustment in reactivity. W

**regulator**
An opening in a wall or door in the return airway of a district to increase its resistance and reduce the volume of air flowing. This increased resistance induces a greater volume of air to flow through the other districts. A regulator consists of a sliding shutter which can be adjusted to any proportion of the maximum aperture. D

**reheater load**
The amount of sensible heat in British thermal units per hour, restored to the air in reheating. D

**Reid vapor pressure**
An important test for gasoline. It is a measure of the vapor pressure of a sample at 100°F and the test is commonly made in a bomb. The results are reported in pounds. This test is

usually carried out in the United States in accordance with ASTM Method D 323. M

**relative biological effectiveness (RBE)**
(1) A factor used to compare the biological effectiveness of different types of ionizing radiation. (2) The inverse ratio of the amount of absorbed radiation, required to produce a given effect, to a standard or reference radiation required to produce the same effect. W

**relative humidity**
The ratio of actual pressure of existing water vapor to maximum possible pressure of water vapor in the atmosphere at the same temperature, expressed as a percentage. M

**relief**
The elevations of inequalities of a land surface. D

**rem**
Roentgen equivalent man.

**remaining reserves**
Those quantities of crude oil, natural gas, natural gas liquids and sulfur, as estimated under proved or probable reserves, after deducting those quantities produced up to the respective date of the estimate. G

**rending**
The breaking of coal into lumps with a minimum of smalls. The relative slowness of low explosives makes them suitable for rending coal since they lack the greater shattering power of high explosives. D

**rep**
Roentgen equivalent physical.

**repressuring**
The injection of water, natural gas, or compressed air to increase the reservoir oil pressure. The oil flow in a well gradually diminishes as the gas pressure underground falls. The pressure may be increased by forcing air or gas into an output well and obtain flowing oil from output wells. D

**reprocessing**
Fuel reprocessing. Reprocessing as used in reactor engineering, the chemical and mechanical processes by which material removed after use in a reactor, such as plutonium 239 and the unused uranium 235 are recovered and prepared so that they may be reused. Z

**research**
Two broad meanings are: (1) reexamination of previously accepted data in the light of current expansion of basic knowledge; and (2) search in reality, specific to an entirely novel concept and calling for development of new approaches. D

**research method**
A test for determining the knock rating, in terms of ASTM Research octane numbers, of fuels for use in spark-ignition engines. The knocking tendency of the fuel is compared with those for blends of reference fuels of known octane number when run in the ASTM-CFR engine at 600 rpm under standard operating conditions, as prescribed in ASTM Methods D908 and D1656. D

**research reactor**
A reactor primarily designed to supply neutrons or other radiation for experimental purposes. It may also be used for training, materials, testing, and production of radioisotopes. O

**reserve**
To keep back; to keep in store for future or other use. D

**reserve generating capacity**
Extra generating capacity available to meet unanticipated demands for power, or to generate power in the event of loss of generation resulting

from scheduled or unscheduled outages of regularly used generating capacity. L

**reserve margin**
Extra generating capacity required to be standing by for addition to electrical grids, with no particular rapidity, in case demand in underestimated or normal supplies are unavailable owing to repairs, nuclear refueling, etc.; normally calculated as 10 to 20 percent of anticipated peak demand. C

**reserves, energy**
A term used to indicate the amount of fossil fuel remaining in the earth, such as measured reserves, based on present technology; proved reserves; indicated reserves; inferred reserves; demonstrated reserves; ultimate reserves; probable reserves; and possible reserves. Y

**reserve system capacity**
The difference between dependable capacity of the system, including net firm power purchases, and the actual or anticipated peak load for a specified period. L

**reservoir**
A natural underground container of liquids, such as oil or water, and gases. In general, such reservoirs were formed by local deformation of strata, by changes of porosity and by intrusions. D

**reservoir oil pressure**
The pressure within an oil pool. It may be sufficient to cause the oil to flow to the surface without pumping. Additional pressure may be created by injecting gas or compressed air as in repressuring. D

**reservoir storage**
The volume of liquid in a reservoir at a given time.

**residence time**
The average length of time a particle of reactant spends in contact with catalyst. M

**residual fuel oil**
The heavier, high-viscosity fuel oils which usually need to be heated before they can be pumped and handled conveniently. Industry is residual oil's major market—it fires open-hearth furnaces, steam boilers and kilns. Large apartment and commercial buildings rank second as residual oil consumers and gas and electric utilities third. M

**residue**
The solid materials remaining after completion of a chemical or physical process, such as burning, evaporation, distillation or filtration. E

**residue gas**
The natural gas remaining after the extraction of various liquid hydrocarbons. DD

**resilience**
The work which a body can do in springing back after a deforming force has been removed. If a body is stressed beyond its elastic limit, the resilience equals that proportion of the total work of deformation which the body can give back upon removal of the forces. D

**resin**
A polymer of unsaturated hydrocarbons from petroleum processing, e.g., in the cracking of petroleum oils, propane deasphalting, clay treatment of thermally cracked naphthas. Chief uses include: rubber and plastics; impregnants; surface coatings. See also synthetic resin. M

**resource recovery**
The process of obtaining materials or energy, particularly from solid waste. E

**respiration**
The production of carbon dioxide through slow utilization of carbon by an organism usually in the temperature range from 30° to 104°F. Y

**respirator**
A device (as a gas mask) for protecting the respiratory tract (as against irritating and poisonous gases, fumes, smoke, dusts) with or without equipment supplying oxygen or air. D

**resue**
To mine or strip sufficient barren rock to expose a narrow but rich vein, which is then extracted in a clean condition. D

**resuscitate**
To restore to animation or life; especially to restore from apparent death; revivify, as to resuscitate a drowned person. D

**retaining wall**
A thick wall designed to resist the lateral pressure of earth behind it. Retaining walls are often necessary at mine sites in valleys to gain space for sidings. D

**retort**
A vessel in which substances are subjected to heat for the purpose of distillation or decomposition. A retort is distinguished from a still in that it is more often used for the treatment of solid or semisolid substances. M

**retort carbon**
Retort carbon, or glance coal, is a very dense form of carbon, produced by the deposition of carbon in the upper part of the retorts in the manufacture of coal gas. It exhibits the luster of a metal, is sonorous when struck, and is a good conductor. D

**retort gas**
Gas resulting from the heating of coal in retorts, for example, in the byproduct process of coke manufacture. D

**retrofit**
(1) The retroactive modification of an existing building or machine. (2) In current usage, the most common application of the word "retrofit" is to the question of modification of existing jet aircraft engines for noise abatement purposes. E

**retrograde condensation and evaporation**
The phenomena associated with the behavior of a hydrocarbon mixture in the critical region, wherein, at constant temperature, the liquid phase in contact with the vapor may be vaporized by an increase in pressure, or the vapor condensed by a decrease in pressure. Likewise, in the same region, at constant pressure, the liquid phase may be vaporized by a reduction of temperature, or the vapor condensed by an increase in temperature. M

**return**
Any airway which carries the ventilating air out of the mine. D

**rev**
Revolution.

**RFF**
Resources for the future.

**rh**
Relative humidity.

**rheology**
The study of the deformation and flow of matter. Z

**rhm**
One roentgen per hour at 1 meter.

**rhn**
Rockwell hardness number.

**rhr**
Residual heat removal.

**rib**
(1) The solid coal on the side of a gallery or longwall face. (2) A pillar or barrier of coal left for support. D

**rich oil**
Absorption oil containing dissolved natural gasoline fractions. M

**riffle**
In mining, the lining of the bottom of a sluice, made of blocks or slats of wood or stones, arranged in such a manner that clinks are left between them. The whole arrangement at the bottom of the sluice is usually called the riffles. D

**rifling**
(1) Working coal which was left behind over the waste. (2) The spiral grooving in the walls of a drill hole and/or on the surface of a drill core. D

**rig**
A derrick complete with enginehouse and other equipment necessary for operation that is used for boring and afterwards pumping an oil well; an oil derrick. D

**rig time**
The hours, days, etc., a drill rig is actually in use in drilling and other related borehole drilling operations. D

**Ringelmann chart**
A series of illustrations ranging from light grey to black used to measure the opacity of smoke emitted from stacks and other sources. The shades of grey simulate various smoke densities and are assigned numbers ranging from 1 to 5. Ringelmann No. 1 is equivalent to 20 percent dense; No. 5 is 100 percent dense. Ringelmann charts are used in the setting and enforcement of emission standards. E

**riparian**
Of, pertaining to, situated, or dwelling on the bank of a river or other body of water. D

**riprap**
A foundation or sustaining wall of stones thrown together without order. D

**rising**
An excavation carried from below upward; a rise or riser. D

**Rittinger's law**
The energy required for reduction in particle size of a solid is directly proportional to the increase in surface area. D

**river mining**
Mining or excavating beds of existing rivers after deflecting their course, or by dredging without changing the flow of water. D

**rms**
Root mean square.

**road oil**
An asphaltic residual oil or a blend of such oil with distillates which do not volatize readily. It is used for dust laying or in the construction of various types of highways. D

**rocket fuel**
Propellant consisting of two components, oxidizer and fuel, which react to give gaseous products and release energy. Rocket fuels may be liquids or solids. In the latter case, the two components must be intimately premixed. In some instances, the liquid system may be a single liquid, in which case, it is called a monopropellant. M

**rock pressure**
In petroleum geology, the pressure under which fluids, such as water, oil and gas, are confined in rocks. No particular cause or origin of the pressure is implied. It is desirable to

substitute for the term rock pressure, as now used in oil, gas, and underground water technology, the more appropriate term, reservoir pressure. D

**Rockwell hardness test**
A method of determining the relative hardness of metals and case-hardened materials. The depth of penetration of a steel ball (for softer metals) or of a conical diamond point (for harder metals) is measured. D

**rod**
(1) A relatively long, slender body of material used in or in conjunction with a nuclear reactor. It may contain fuel, absorber, or material in which activation or transmutation is desired. (2) A bar, the end of which is slotted or screwed for the attachment of a drill bit. O

**rod dope**
Grease or other material used to protect or lubricate drill rods. Also called gunk; rod grease. D

**rod slap**
The impact of drill rods with the sides of a borehole, occurring when the rods are rotating. D

**roentgen**
A unit of exposure to ionizing radiation. It is that amount of gamma or x-rays required to produce ions carrying 1 electrostatic unit of electical charge (either positive or negative) in 1 cubic centimeter of dry air under standard conditions. Named after Wilheim Roentgen, German scientist who discovered x-rays in 1895. Abbreviation r. O

**roentgen equivalent man**
The unit dose of any ionizing radiation which produces the same biological effect as a unit of absorbed dose of ordinary x-rays. The RBE dose (in rems) = RBE x absorbed dose (in rads). Abbreviation rem. O

**roentgen equivalent physical**
An absolute unit of absorbed dose of any ionizing radiation with a magnitude of 93 ergs per gram. It has been superseded by the rad. O

**roily oil**
Crude oil more or less emulsified with water. M

**roll oil**
Oil used in rolling of brass, copper, and similar soft metals. M

**RON**
Research octane number.

**roof bolting**
A system of roof support in mines. Boreholes from 3 to 8 feet long are drilled upward in a roof and bolts of 1 inch diameter or more are inserted into the holes and anchored at the top by a split cone or similar device. The bolt end protrudes below roof level and is used to support roof bars, girders, or simple steel plates pulled tight up to the roof by a nut on the bolthead. The bolts are put up in a definite pattern. The idea is to clamp together several roof beds to form a composite beam with a strength considerably greater than the sum of the individual beds acting separately. D

**room and pillar technique**
A predominant technique in underground mining, it uses so-called conventional equipment to drill and undercut the coal face. The coal is then broken with explosives and loaded on machines for transport and the exposed roof stratum is supported by pillars which form rooms or entries.

**Röschen method**
A firedamp drainage method utilizing controlled drainage from the coal seams as they are being mined. This method, which is also known as the pack cavity method, was devised to extract gas from the mined-out areas

of advancing longwall mining systems by leaving corridors or cavities at regular intervals in the pack. D

**rosin oil**
Oil obtained by distillation of rosin; varies in color from almost colorless to dark brown; used in grease making and in the manufacture of printer's ink. M

**rotary boring**
A system of boring, using usually hollow rods, with or without the production of rock cores. Rock penetration is achieved by the rotation of the cutting tool. The method is used extensively in exploration, particularly when cores are required. It is the usual method in oil well boring with holes from 6 to 18 inches in diameter. D

**rotary burner**
Either vertical or horizontal, the basic principle of operation is centrifugal force, that is, the oil entering at the center of a rotary cup is whirled around very rapidly until the oil is thrown away from the cup. Being mixed with air, it will ignite. BB

**rotary drilling**
One of two basic types of drilling equipment used today, it is one where a bit is attached to the lower end of a string of pipe, called "drill pipe." Bits range from less than 4 to more than 22 inches in diameter. They are made of very hard steel, and some types have two or three rotating cones covered with sharp teeth for grinding through rock. The bit and pipe pass vertically through a rotary turntable on the derrick floor. As the pipe is turned and lowered into the earth, the bit bores a hole deeper and deeper. As the hole deepens, the drilling crew adds new lengths of drill pipe. M

**rotary percussion drill**
A new technique for improved drilling operations for oil using a circulating fluid to activate a hammer-like mechanism about the drill bit that creates a rapid series of percussion blows. In this way, rotary percussion drills bore and pound their way into the earth. M

**round coal**
Large coal, or large coal which has passed through the preparation plant and become more or less rounded. D

**ROW**
Right of way.

**royalty**
As used in an oil and gas lease, a share of the product or profit reserved by the owner for permitting another to use the property. D

**rpm**
Revolutions per minute.

**rule of thumb**
A statement or formula that is not exactly correct, but is accurate enough for use in rough figuring. D

**running coal**
(1) A term applied to bituminous coal because of its tendency to soften and cohere when burning. (2) It is also applied to a very friable coal which disintegrates and flows into the workings. D

**runoff**
The portion of rainfall, melted snow or irrigation water that flows across ground surface and eventually is returned to streams. Runoff can pick up pollutants from the air or the land and carry them to the receiving waters. E

**run of river plant**
A hydroelectric power plant using the flow of the stream as it occurs and having little or no reservoir capacity for storage of water. Sometimes called "stream flow" plants. K

**RWCU**
Reactor water clean-up.

# S

**s**
Symbol for surface area, specific surface. Also symbol for linear distance, and abbreviation for second. D

**S**
Chemical symbol for sulfur; abbreviation for south.

**saddle**
(1) A ridge connecting two higher elevations. (2) A low point in the crestline of a ridge. D

**SAE**
Society of Automotive Engineers.

**SAE horsepower**
An arbitrary rating, developed by the Society of Automotive Engineers, used to calculate horsepower of engines. M

**SAE viscosity number**
An arbitrary number; one of a system for classifying crankcase oils and automotive transmission and differential lubricants, according to their viscosities, established by the Society of Automotive Engineers. SAE numbers are used in connection with recommendations for crankcase oils to meet various design, service, and temperature requirements affecting viscosity only; they do not connote quality. M

**safety rod**
A standby control rod used to shut down a nuclear reactor rapidly in emergencies. O

**sag bolt**
Bolts installed at intersections to measure roof sag.

**salable output**
The total tonnage of clean coal produced at a mine as distinct from pithed output. It is the tonnage of coal as weighed after being cleaned and classified in the preparation plant. D

**salinity**
The total amount of solid material in grams contained in one kilogram of seawater when all the carbonate has been converted to oxide, the bromine and iodine replaced by chlorine, and all organic matter completely oxidized. Expressed as grams per kilogram of seawater or parts per thousand. D

**SALT**
Strategic Arms Limitation Treaty.

**salt dome**
A type of structure resulting from the upward thrust of a great mass of salt far below the earth's surface. When a salt dome rises through a layer of oil-

bearing sedimentary rock, oil may be trapped above the salt dome or in structures similar to faults along its flanks. Oil and gas are frequently found adjacent to, or on top of, an underground salt dome. M

**salt water intrusion**
The invasion of salt water into a body of fresh water, occurring in either surface or ground water bodies. When this invasion is caused by oceanic waters, it is called sea water intrusion. E

**sample**
Representative fraction of body of material, removed by approved methods, guarded against accidental or fraudulent adulteration, and tested or analyzed in order to determine the nature, composition, percentage of specified constituents, etc., and possibly their reactivity. D

**sand filter**
A filter for purifying domestic water, consisting of specially graded layers of aggregate and sand, through which water flows slowly downwards. A similar type of filter is used for treating sewage effluent, but has coarser sand. D

**sandhog**
A man who works in compressed air. D

**sanitary landfilling**
An engineered method of solid waste disposal on land in a manner that protects the environment; waste is spread in thin layers, compacted to the smallest practical volume and covered with soil at the end of each working day. E

**santa ana**
Hot, dry wind blowing from the Southern California desert often creating dangerously dry vegetation, increasing the incidence of brush fires.

**saponifier**
Any compound, as a caustic alkali, used in soapmaking to convert the fatty acids into soap. A term used in the flotation process. D

**sapropelic coal**
Coal of which the original plant material was more or less transformed by putrefaction. Complete seams of sapropelic coals are rare, but layers or bands of varying thickness within seams are more frequent. D

**saturant**
A bituminous or asphaltic substance, used to impregnate fibrous material and, particularly, to give weather-resistant properties to building materials. M

**saturation**
The extent or degree to which the voids in rock contain oil, gas, or water. Usually expressed in percent related to total void or pore space. D

**Saybolt color**
A color standard for petroleum products. The procedure for determining Saybolt color and a description of the Saybolt chromometer are given in ASTM Method D 156. M

**SBA**
Small Business Administration.

**SBLC**
Standby liquid control (system).

**scalar**
A quantity fully described by a number, such as speed. D

**scaler**
An electronic instrument for counting radiation-induced pulses from Geiger counters and other radiation detectors. D

**scattering**
A process that changes the trajectory of a particle. Scattering is caused by

collisions with atoms, nuclei, and other particles, or by interactions with fields of magnetic force. If the energy of the scattered particle is not changed by the collision, it is elastic scattering; if a change in energy occurs, it is inelastic scattering. D

**scavenger mining**
The taking out of coal so close to the surface as to undermine the topsoil, resulting in devastation above ground. D

**scf**
Standard cubic feet.

**schist**
A crystalline rock that can be readily split or cleaved because of having a foliated or parallel structure, generally secondary and developed by shearing and recrystallization under pressure.

**schlieren**
A method or apparatus for visualizing or photographing regions of varying density in a field of flow. D

**School's method**
A method for determining the uranium in any of its ores in which the uranium is extracted with dilute nitric acid. D

**scintillation**
A flash of light produced in a phosphor by an ionizing event. O

**SCOT**
Shell-claus off-gas treating.

**scouring**
A wet cleaning or drycleaning process involving mechanical operations. D

**scram**
The sudden shutdown of a nuclear reactor, usually by rapid insertion of the safety rods. O

**scraper**
(1) A rod for cleaning out shotholes prior to charging with explosives. (2) A steel tractor-driven surface vehicle, 6 to 12 cubic yard capacity, mounted on large rubber-tired wheels. (3) A tool for cleaning the dust out of the borehole. D

**scraper ripper**
A piece of strip-mine equipment that handles the jobs of breaking coal, loading and hauling. Features of the scraper ripper include ripping teeth on the lip for breaking the coal, and a flight conveyor for carrying the broken coal away from the lip. D

**scree**
(1) Long trails of loose rock collected on the slopes beneath steep mountain sides. (2) A sieve, or strainer. (3) A coal screen. D

**screened coal**
Coal that has passed over any kind of a screen and therefore consists of the marketable sizes. D

**screening**
Use of one or more screens (sieves) to separate particles of ore into defined sizes. Also called sizing. D

**scrubber**
An air pollution control device that uses a liquid spray to remove pollutants from a gas stream by absorption or chemical reaction. Scrubbers also reduce the temperature of the emission. E

**SCS**
Soil Conservation Service of the U.S. Department of Agriculture.

**SDWA**
Safe Drinking Water Act.

**sea coal**
(1) Old name for bituminous coal, either because it was exported by sea or because it was at first applied to coal washed ashore from deposits below sea level. (2) Pulverized bituminous coal used as a foundry facing. D

**sea level**
The level of the surface of the sea between high and low tide. Used as a standard in measuring heights and depths. E

**seam**
(1) A stratum or bed of coal or other mineral, generally applied to large deposits of coal. (2) Mechanical or welded joints. D

**seasonal power**
Power generated or made available to customers only during certain seasons of the year. L

**seasonal rate schedule**
A schedule containing an electric rate available only during certain specified seasons of the year. L

**SEA**
Science and Education Administration (USDA).

**sec**
Second.

**secondary cell**
A cell which receives its electrical energy from a charging operation, then stores this energy until it is required. Hence, the name storage battery or accumulator which is often given to secondary cells. D

**secondary recovery**
Similar to pressure maintenance techniques, secondary recovery methods are used at a point when natural pressure in an oil well has been almost exhausted. Secondary recovery repressurizes a reservoir, usually through injection of water or gas. Water flooding, miscible flooding, thermal recovery, "fireflooding," etc., are a few of the secondary recovery techniques. M

**secondary voltage**
The output or load-supply voltage of a transformer or substation is called the secondary voltage. K

**Second Law of Thermodynamics**
Second Law Efficiency is the ratio of the energy required to do a task in the best theoretically possible way (according to the thermodynamics of reversible processes, which proceed infinitely slowly) to the energy actually used to do the task. It thus measures the theoretical room for improvement in end use efficiency, since thermodynamics enables us to calculate how little energy can be used to do a job regardless of the method, known or unknown, by which it is done. C

**second mining**
The recovery of pillar coal after first-mining in chambers has been completed. D

**seconds A.P.I.**
A unit of viscosity as measured with a Marsh funnel according to American Petroleum Institute procedure. D

**section**
(1) A portion of the working area of a mine. (2) A very thin slice of anything, especially for microscopic examination. (3) A piece of land that is one square mile or 640 acres in area forming one of the 36 subdivisions of a township in a U.S. public-land survey. D

**sedentary**
Formed in place, without transportation, by the disintegration of the underlying rock or by the accumulation of organic material, said of some soils, etc. D

**sedimentation**
The settling of solid particles of soil, coal, or mineral from liquid, as a result of either gravity or centrifuging. The large particles sink at a faster rate than smaller particles of the same shape. The principle is applied to a number of coal and mineral washers, cyclones, and classifiers. D

**seepage pond**
An artificial body of surface water formed by discharge of liquid waste. P

**SEFOR**
Southwest experimental fast-oxide reactor.

**SEIA**
Solar Energy Industries Association.

**seismic**
Pertaining to, characteristic of, or produced by, earthquakes or earth vibration, as, seismic disturbances, or seismic records. D

**seismic analysis**
A quick, easy and inexpensive method of determining the consolidation of overburden. The process is based on the principle that sound or shock waves travel through different subsurfacing materials at varying speeds and along different paths. By this method the operator can determine whether overburden can be ripped or whether it will need to be drilled and blasted. D

**seismograph**
A device originally developed to record earthquakes, it is now widely used as an oil prospecting tool. An explosive is set off in a shallow, small-diameter hole. The waves from the explosion travel downward, striking successive rock formations below, from which they are reflected back to the surface. As these waves return, geophones (detectors) pick them up and the impulses are recorded. By correlating the intensity of the waves and the time intervals required for them to travel down and back, the geophysicist learns the general characteristics of the underground structure. M

**selective coating**
An optical coating for collectors that normally has a high absorptivity for incident sunlight and low emissivity for long wave radiation. In simple terms the absorber plate absorbs more heat and re-radiates less so the collector efficiency is improved and higher temperatures are achieved. A selective coating can also be used to increase reflectivity but the material chosen would need quite different characteristics. M

**selective mining**
A method of mining whereby ore of unwarranted high value is mined in such a manner as to make the low-grade ore left in the mine incapable of future profitable extraction. In other words, the best ore is selected in order to make good mill returns, leaving the low-grade ore in the mine. Frequently called robbing a mine. D

**self-contained cooling unit**
A combination of apparatus for room cooling complete in one package; usually consists of compressor, evaporator, condenser, fan motor and air filter. Requires connection to electric line. D

**semianthracite**
Coal intermediate between anthracite and semibituminous coal. D

**semibituminous coal**
It is harder and more brittle than ordinary bituminous coal. The term superbituminous was suggested instead of semibituminous as a more appropriate one. D

**semiconductor**
Crystal system in which, though the electrons are ionically bound, a slight rise of temperature frees the valence atoms so that the system becomes a conductor. Conduction of electricity proceeds in one direction only. D

**semidirunal**
Having a period or cycle of approximately one-half lunar day (12.42 solar hours). The tides and tidal currents are said to be semidiurnal when two flood periods and two ebb periods occur each lunar day. D

**semidull coal**
A variety of banded coal containing from 21 to 40 percent of pure, bright ingredients (vitrain, clarain, and fusain), the remainder consisting of clarodurain and durain. D

**semifluid**
(1) Having the attributes of both a liquid and a solid. (2) Similar to semisolid, but more closely related to a liquid than a solid. M

**semifusain**
A coal constituent transitional between vitrain and fusain. It displays gradual disappearance of cell structure, hardness, and yellowish color when observed in thin sections. Same as vitrifusain. D

**semifusinite**
A constituent intermediate between vitrinite and fusinite, showing a well-defined structure of wood and sclerenchyma. D

**semifusite**
Same as semifusain; vitrifusain. D

**semisolid**
(1) Having the attributes of both a solid and a liquid. (2) Similar to semifluid, but more closely related to a solid than a liquid. M

**semisplint coal**
Coal intermediate between durain coal and clarain coal (duroclarain), a banded coal containing 20 to 30 percent of opaque attritus and more than 5 percent anthraxylon. D

**sensible heat**
The heat added to, or taken from, a body when its temperature is changed. M

**separation processes**
Oil refining operation using solvent extraction, adsorption, crystallization and fractional distillation. M

**separative work units (SWU's)**
A measure of the work required to separate uranium isotopes in the enrichment process. It is used to measure the capacity of an enrichment plant independent of a particular product and tails. To put this unit of measurement in perspective, it takes about 100,000 SWU's per year to keep a 1,000 MWe LWR operating and 2,500 SWU's to make a nuclear weapon. X

**separator**
(1) A machine for separating, with the aid of water or air, materials of different specific gravity. (2) Strictly, a separator parts two or more ingredients, both valuable, while a concentrator saves but one and rejects the rest. D

**sequence control**
A method of control whereby, once action has been initiated, a number of electrical circuits will automatically function in a prescribed order. D

**SERI**
Solar Energy Research Institute.

**service area**
Territory in which a utility system is required or has the right to supply electric service to ultimate customers. K

**service (service line, service pipe)**
In the gas industry, the pipe which carries gas from the main to the customer's meter. DD

**service wells**
Those drilled for injecting liquids or gas into an underground formation in order to increase the pressure, and thereby force the oil toward the producing wells. Service wells also include wells used for the disposal underground of salt water produced with the oil and gas. M

**SESA**
Social and Economic Statistics Administration of the Federal Government.

**set**
(1) A timber frame for supporting the sides of an excavation shaft or tunnel. Also called sett. (2) A group of pumps that are used for lifting water from one level to another. These are two of the many definitions of this term. D

**settled production (Artificial Lift)**
A state which is reached when the initial pressure of a flush oil well extends itself. At this point, the well is usually put "on pump." Many wells never flow naturally and must be pumped from the start. Others drop in flow rate shortly after production begins, and become settled fairly early in their lives. Several varieties of pumps are used, such as: surface pumps, submersible electric pumps, and gas lifts. M

**sewage gas**
A gas which generates in, and can be collected from, sealed tanks where sewage sludge undergoes digestion, by anaerobic bacteria. It may also arise in badly ventilated sewers, where it becomes dangerous. It consists of two-thirds methane, and one-third carbon dioxide. D

**SGP**
Shell gasification process.

**shaft**
(1) An excavation of limited area compared with its depth, made for finding or mining ore or coal, raising water, ore, rock, or coal, hoisting and lowering men and material, or ventilating underground workings. (2) Specifically applied to approximately vertical shafts, as distinguished from an incline or inclined shaft. D

**shake**
In a coal mine, a vertical crack in the seam and roof. D

**shaker**
A mechanically vibrated screen through which a returning drill fluid is passed to screen out large chips, fragments, and drill cuttings before the drill fluid flows into the sump. Also called shale screen, or shale shaker. D

**shale**
The clay and mud-derived sedimentary rocks that are generally impermeable but constitute the bulk of sedimentary sequences.

**shale oil**
The organic oil obtained by pyrolysis of oil shale at 900° F. It can only be recovered by heat distillation rather than other conventional reservoir mechanics. It is 80-86% carbon and has a C:H ratio of 7:9. Y

**shale spirit**
The lower boiling fractions obtained in the refining of crude shale oil. D

**sharp gas**
Gas is sharp when at its most explosive point. D

**shear**
To make vertical cuts in a coal seam that has been undercut. D

**shield**
A body of material used to reduce the passage of radiation. In mining or tunneling, a framework or screen of wood or iron protecting the workers. O

**shift conversion**
The process for production of gas with a desired quantity of carbon monoxide from crude gases derived from gasification. A carbon monoxide enriched gas is saturated with steam and passed through a catalytic reactor where the carbon monoxide reacts with steam to produce hydrogen and carbon dioxide with the carbon dioxide removed in a washing or scrubbing operation. The hydrogen to carbon monoxide in the product can be shifted. Synthetic gas is purified, then the hydrogen to carbon monoxide ratio adjusted to 2 to

1 volumetric ratio, so that with heat and a catalyst, ethanol is produced. Y

**shipping measure**
(1) For measuring the entire internal capacity of a vessel: one register ton equals 100 cubic feet. (2) For measuring cargo: 1 U.S. shipping ton equals 40 cubic feet, or 32.143 U.S. bushels. D

**shock-proof**
As applied to the current-carrying parts of an electric system (excepting trolley wires) is taken to mean that contact with such parts is prevented by the use of grounded metallic coverings or sheaths. D

**shock wave**
The wave of air and dust which, in some cases, travels ahead of the flame of a coal dust explosion. It may occur when an ignition takes place near the closed end of a mine roadway, and the reaction products behind the flame cannot escape freely. D

**shoot**
To explode a charge in blasting operations. D

**shoring**
Timbers braced against a wall as a temporary support. D

**short circuit**
A short circuit is an abnormal connection of relatively low resistance, whether made accidentally or intentionally, between two points of different potential in a circuit. D

**short-lived isotope**
A radioactive nuclide that decays so rapidly that a given quantity is transformed into its daughter products within a short period (usually those with a half-life of days or less). P

**short ton**
A unit of weight that equals 20 short hundredweights or 2,000 avoirdupois pounds. Used chiefly in the United States, in Canada, and in the Republic of South Africa. D

**short wall**
(1) The reverse of longwall, frequently used to mean the face of a room. (2) A method of mining in which comparatively small areas are worked, as opposed to longwall. D

**shot**
A single explosive charge fired in coal, stone or ore. D

**shovel**
Any bucket-equipped machine used for digging and loading earthy or fragmented rock materials. D

**shp**
Shaft horsepower. D

**shrinkage (natural gas)**
The reduction in volume of wet natural gas due to the extraction of some of its constituents, such as hydrocarbon products, hydrogen sulfide, carbon dioxide, nitrogen, helium, and water vapor. DD

**shroud**
A structure used to concentrate or deflect a windstream. U

**shunt**
To shove or put aside or out of the way; sidetrack. D

**sial**
A layer of rocks underlying all continents, that ranges from franitic at the top to gabbroic at the base. The thickness is variously placed at 30 to 35 kilometers. The name derives from the principal ingredients, silica and alumina. D

**SIC**
Standard industrial classification.

**sieve**
Any screen that is used to separate particles according to size. D

**sieve analysis**
The determination of the percentage of particles which will pass through screens of various sizes. D

**silt**
Unconsolidated particles with a grain size between 0.002 and 0.02 millimeter.

**sima**
The basic outer shell of the earth; under the continents it underlies the sial, but under the Pacific Ocean it directly underlies the oceanic water. D

**simple engine**
A reciprocating engine from which steam or compressed air is exhausted to atmosphere after expansion in one cylinder only. D

**single-cycle reactor system**
A direct-cycle reactor system. O

**sinking**
A method of controlling oil spills that employs an agent to entrap oil droplets and sink them to the bottom of the body of water. The oil and sinking agent are eventually biologically degraded. E

**sintering**
The agglomerating of small particles to form larger particles, cakes, or masses; in case of ores and concentrates, it is accomplished by fusion of certain constituents. D

**SIP**
State implementation plan.

**sirocco**
Hot wind blowing across the Mediterranean Sea toward southern Europe from Africa. Noted for its ability to melt snow fields at a rapid rate.

**slab oil**
A white oil used by candymakers and bakers for oiling the marble slabs on which candy and pastry are worked. M

**slack**
Commonly used to describe the smaller sizes of coal passing through screen openings, approximately 1 inch or less in diameter. D

**slack barrel**
A barrel for nonliquid materials. The construction is lighter than that of the usual barrel. M

**slag**
A substance formed in any one of several ways by chemical action and fusion at furnace operating temperatures: (1) in smelting operations, through the combination of a flux, such as limestone, with the gangue or waste portion of the ore; (2) in the refining of metals, by substances such as lime added for the purpose of effecting or aiding the refining; or (3) by chemical reaction between refractories and fluxing agents such as coal ash, or between two different types of refractories. D

**slate**
Dark shale lying next to the coalbeds. It contains impressions of the plant life of distant ages, proving the vegetable origin of coal. D

**slim hole**
Oil driller's terms for diamond-drill borehole 5 inches or less in diameter. D

**sloam**
A layer of clay between seams of coal. Also spelled sloom. D

**slop**
A term loosely used to denote odds and ends of oil produced in the plant, which must be rerun or further processed in order to make them suitable for use. Also called slop oil. D

**SLS**
Saint Lawrence Seaway Development Corporation.

**sludge**
Any deposit, sediment, or mass, as the waste resulting from oil refining, the mud brought up by a mining drill, the precipitate in a sewage tank, the sediment in a stream boiler or crankcase, etc.

**sludge abatement**
The control of the discharge into watercourses (or on adjacent land), of mineralized or impure water, or sludge, or mining debris. D

**slug**
A short, usually cylindrical fuel element. O

**sluice**
(1) To mine an alluvial deposit by hydraulicking. (2) To drench, wash, or scour with gushes or floods, as of water. (3) Flush. D

**slump**
The downward slipping of a mass of rock or unconsolidated materials of any size, moving as a unit or as several subsidiary units, usually with backward rotation, on a more or less horizontal axis parallel to the cliff or slope from which it descends. D

**slurry**
(1) A free-flowing mixture of solids and liquid. (2) Specifically, a suspension of cracking catalyst in cycle stock. M

**slush**
To fill mine workings with sand, culm, etc., by hydraulic methods. Same as silt. To move ore or waste filling with a scraper (slusher) hoist. D

**slushing oil**
Oil used on metals to form a protective coating against rust and tarnish. M

**small coal**
(1) Thin seams of coal, also called low coal. (2) Coal with a top size less than 3 inches. D

**SMCRA**
Surface-Mining Control and Reclamation Act.

**SME**
Society of Manufacturing Engineers.

**smoke point**
A measure of the burning cleanliness of jet fuel and kerosene. The maximum flame height, in millimeters, at which a kerosene will burn without smoking when determined in the apparatus and under the conditions described in IP Method 57. The smoke point of jet fuels is determined by ASTM Method D 1322; this smoke test uses the same apparatus as that used by IP Method 57. M

**smut**
A thin band of soft, inferior coal. D

**SNG**
Synthetic natural gas, supplemental natural gas, or substitute natural gas.

**SNM**
Special nuclear material(s).

**snowbird mine**
A mine that produces or ships only small quantities of coal, and operates only when coal is costly by reasons of a scarcity or a shortage of cars for shipment. D

**snow load**
The live load which must be included when designing a flat or low-pitched roof in temperate and cold climates; in England it is assumed at 15 pounds per square foot. D

**SNUPPS**
Standard Nuclear Unit Power Plant System.

**SOC**
System operating center.

**social compact or contract**
Idea favored by some liberal economists that, in return for tax relief,

workers would agree to moderate their wage demands.

**sodium-graphite reactor**
A reactor that uses liquid sodium as coolant and graphite as moderator. O

**soft coal**
Bituminous coal as opposed to anthracite. D

**soft fire**
A flame with a deficiency of air. D

**soft radiation**
Ionizing radiation of long wavelength and low penetration. D

**soft rays**
Beta particles or gamma rays having little penetration. D

**soft rock**
Rock that can be removed by air-operated hammers, but cannot be handled economically by a pick. D

**soil**
Broadly and loosely, the regolith, or blanket of an unconsolidated rock material that lies on the bedrock. D

**solar array**
A number of individual solar collection devices arranged in a suitable pattern to collect solar energy effectively. Q

**solar collection efficiency**
The ratio of the amount of heat usefully collected to the total solar irradiation during the period under consideration. Instantaneous efficiencies during the middle of the day, when the incident angle is favorable, are generally higher than daylong efficiencies, which must take into account the high and unfavorable incident angles that prevail during the early morning and late afternoon hours. R

**solar compass**
Synonym for sun compass. D

**solar constant**
Intensity of the solar radiation on a surface normal to the sun's rays beyond the earth's atmosphere at the average earth-sun distance of 92,955,888 miles (one Astronomical unit). It is equal to 429 BTU/ft$^2$/hour. H

**solar cooling advantage**
Has largest supply of energy when the demand is highest. R

**solar declination**
An angle between the earth-sun line and the equatorial plane. The earth's tilted axis results in a day-by-day variation of this angle. This daily change is the primary reason for the variation in the distribution of solar radiation over the earth's surface, for the varying number of hours of daylight and darkness, and for the large summer-winter differences in solar radiation intensity at any given location.

**solar energy**
Power obtained from solar radiation through the use of collectors. Once collected, this form of energy must be converted to a useful form such as electricity. B

**solar energy utilization process**
There are three technological processes by which solar energy can be utilized: (1) heliochemical; (2) helioelectrical; and (3) heliothermal. R

**solar farm**
A proposed system of collecting solar energy through megawatt generating systems using special lenses to focus sunlight onto chemically-coated, nitrogen-filled pipes that would transfer the heat to a central storage unit using molten salts. B

**solar furnace**
An apparatus for production of a very high temperature in a small space by

focusing the radiation of the sun on a small focal point by the use of mirrors. FF

**solar house**
A term first applied to residences which utilized large southfacing expanses of glass to admit winter sunshine. More recently interest has been focused on the use of separate solar radiation collectors mounted on south-facing roofs or walls, with water or air as the heat transfer medium and with insulated tanks, rock beds, or heat-of-fusion materials for energy storage. R

**solar irradiation, maximum daily amount**
The amount which can be received at any given location and date which falls on a flat plate with its surface kept normal to the sun's rays so that it can receive both direct and diffuse radiation. R

**solarization**
An effect of strong sunlight (or artificial ultraviolet radiation) on some glasses, causing a change in their transparency. Glasses free from arsenic and of low soda and potash contents are less prone to this defect. D

**solar power**
The thermal energy of the sun. B

**solar radiaton intensity**
The tremendous amount of free energy outside the earth's atmosphere averaging 442 Btu/sq. ft-hr. That is 1.79 x $10^{14}$kw x $10^{14}$Hp/min. intercepted by the earth. Every two days, the sun gives the earth as much energy as there is locked in the known reserves of fossil fuels.

**solar salt**
Salt obtained by solar evaporation of seawater or salt lake water in shallow lagoons or ponds. D

**solar space heaters**
A heating system powered by solar energy using a black metal surface under glass on the roof to absorb sunlight and create a "greenhouse" effect to heat water or air which is circulated during the day and stored at night. While only considered to be supplemental space heating systems, they are able to heat all the water needed by most homes and buildings. B

**solar stills**
Distillation plants consisting of an airtight space in which evaporation of brackish water and condensation occur simultaneously. Solar energy penetrates this airtight space through a tilted transparent cover and is partially absorbed by brackish water in a basin at the bottom of the space.

**solar time**
Varying period, marked up successive crossings of the meridian by the sun. Hour angle of sun at observer's point (apparent time) corrected to true time by use of equation of time. Sundials show apparent solar time. D

**solar tower**
A tower placed so that the reflected direct radiation for heliostat mirrors can be focused at its top. Heat exchange takes place at the top of the tower and the heated fluid is used in a conventional steam power system at ground level. Q

**solid**
State of matter in which the constituent atoms, ions, or molecules are sufficiently restricted in their relative movement to result in a definite shape. D

**solid fuels**
Any fuel that is a solid, such as wood, peat, lignite, bituminous; and anthracite coals of the natural variety and the prepared varieties as, pulverized coal, briquettes, charcoal, and coke. D

**solidification**
The process of changing from a liquid or gas to a solid. D

**solidification range**
Temperature range over which solid mixtures (alloys, fluxes) melt. D

**solid smokeless fuel**
A solid fuel, such as coke, which produces comparatively no smoke when burnt in an open grate. The gas industry produces certain brands of smokeless fuel, such as Coalite, Rexco, Clean-Glow, and Phimax. D

**solubility**
The extent to which a substance will dissolve in a particular solvent. D

**soluble oil**
Oil which readily forms stable emulsions or colloidal suspensions in water. M

**solvent**
A substance used to dissolve another substance, for example, the water in a solution of salt in water. D

**solvent extraction**
A method of separating one or more substances from a mixture, by treating a solution of the mixture with a solvent that will dissolve the required substances, leaving the others. It is used in purifying certain fuels. D

**solvent-refined coal (SRC)**
A process of coal liquefaction in which coal is mixed with a liquid solvent, itself derived from the coal, then heated and passed (with additional hydrogen) to a high pressure reactor. Hydrogen and hydrogen sulfide are then separated from the mixture, it is then filtered, the solvent is distilled for reuse, and the final product is recovered either as liquid or solid.

**somatic effects of radiation**
Effects of radiation limited to the exposed individual, as distinguished from genetic effects, which also affect subsequent unexposed generations. Large radiation doses can be fatal. Smaller doses may make the individual noticeably ill, may merely produce temporary changes in blood-cell levels detectable only in the laboratory, or may produce no detectable effects whatever. Also called physiological effects of radiation. Compare genetic effects of radiation. W

**sonar**
The method or equipment for determining, by underwater sound, the presence, location, or nature of objects in the sea. The word sonar is an acronym derived from the expression "Sound Navigation and Ranging." D

**sone**
(1) A unit of loudness. (2) A simple tone with a frequency of 1000 Hz, 40 dB above a listener's threshold, generates a loudness of one sone (1 millison = 0.001 sone). E

**sonometer**
An instrument for measuring rock stress. Piano wire is tuned between two bolts cemented into drill holes in the rock, and change of pitch after distressing is observed and used to indicate stress. E

**soot**
A black substance, consisting essentially of carbon from the smoke of wood or coal, especially that which adheres to the inside of the chimney, also containing volatile products condensed from the combustion of the wood or coal, including certain ammonia salts. D

**sorption**
A term including both adsorption and absorption. Sorption is basic to many processes used to remove gaseous and particulate pollutants from an emission and to clean up oil spills. E

**sound absorption**
Sound absorption is the change of sound energy into some other form, usually heat, as it passes through a medium or strikes a surface. D

**sound-energy density**
At any point in a sound field, the sound energy contained in a given of the medium divided by the volume of that part of the medium. D

**sound intensity**
In a specified direction at any point, the average rate of sound energy transmitted in the specified direction through a unit area normal to this direction at the point considered. D

**source beds**
Rocks in which oil or gas has been generated. D

**source material**
In atomic energy law any material, except special nuclear material, which contains 0.58 percent or more of uranium, thorium, or any combination of the two. O

**sour crudes**
Crude oils containing relatively large amounts of sulfur and other mineral impurities. M

**sour natural gas**
Gas found in its natural state, containing such amounts of compounds of sulphur as to make it impractical to use, without purifying, because of its corrosive effect on piping and equipment. DD

**spad**
A means of marking an underground survey station that consists of a flat spike in which is drilled a hole for the threading of a plumbline. D

**sparse vitrain**
A field term to denote, in accordance with an arbitrary scale established for use in describing banded coal, a frequency of occurrence of vitrain bands comprising less than 15 percent of the total coal layer. D

**SPE**
Society of Petroleum Engineers.

**special batching sequence**
The sending of compatible crude oils or products, one behind the other, through pipelines. M

**special contract rate schedule**
An agreement for electric service between a utility and another party in addition to, or independent of, any standard rate schedule. L

**special nuclear material**
Plutonium, uranium-255, uranium-235, or uranium enriched to a higher percentage than normal of the 255 or 235 isotopes. P

**special purpose rate schedule**
A rate schedule limited in its application to some particular purpose or process within one, or more than one, type of industry or business. L

**specific energy**
In cutting or grinding, the energy expended or work done in removing a unit volume of work material, usually expressed as inch-pound per cubic inch or horsepower per minute per cubic inch. D

**specific gravity**
The weight of a substance compared with the weight of an equal volume of pure water at 4° C. D

**specific heat**
The ratio of the quantity of heat required to raise the temperature of a body one degree to that required to raise the temperature of an equal mass of water one degree. M

**specific power**
The power generated in a nuclear reactor per unit mass of fuel.

Expressed in kilowatts of heat per kilogram of fuel. The same as unit power. D

**specific speed**
A speed or velocity of revolution, expressed in revolutions per minute, at which the runner of a given type of turbine would operate if it were so reduced in size and proportion that it would develop one horsepower under one-foot head. The quantity is used in determining the proper type and character of turbine to install at a hydroelectric power plant under given conditions. E

**specific surface**
The surface area per unit of volume of soil particles. D

**spectral shift reactor**
A reactor design in which a mixture of light water and heavy water is used as the moderator and coolant. The ratio of light to heavy is varied to change (shift) the speed distribution (spectrum) of the neutrons in the reactor core. Since the probability of neutron capture varies with neutron velocity, a measure of reactor control is thus obtained. O

**spectroscope**
An instrument which disperses radiation into a spectrum for visual observation. M

**spectrum**
A visual display, a photographic record, or a plot of the distribution of the intensity of a given type of radiation as a function of its wave length, energy, frequency, momentum, mass, or any related quantity. O

**specular**
Mirrorlike, as specular iron ore, which is a hard variety of hematite. D

**specular coal**
Same as pitch coal. D

**spend**
(1) To exhaust of mining. (2) To dig out, as used in the phrase, to spend ground. D

**spent fuel**
Nuclear reactor fuel that has been irradiated (used) to the extent that it can no longer effectively sustain a chain reaction. Spent fuel contains radioactive waste materials, unburned uranium and plutonium. O

**sperm oil**
Oil obtained from the blubber of sperm whales and used in compounding lubricants. M

**sp gr**
Specific gravity.

**sp ht**
Specific heat.

**spill**
The accidental release of radioactive material. W

**spindle oil**
A well-refined oil suitable for the lubrication of light, highspeed machinery such as encountered in the textile industry. M

**spinning reserve**
(1) Generating capacity connected to the bus and ready to take load. (2) Capacity available in generating units which are operating at less than their capability. L

**splent**
A hard variety of bituminous coal that ignites with difficulty owing to its slaty structure, but makes a clear, hot fire. D

**split coal**
Coalbed separated by clay, shale, or sandstone parting which thickens so that both benches cannot be mined together. D

**split system**
(1) Historically, a combination of warm air heating and radiator heating. (2) Used for other combinations, such as hot water steam, steam warm air, etc. D

**spoil bank**
The piles of discarded earth remaining after strip mining. B

**spontaneous combustion**
The outbreak of fire in combustible material (such a oily rags or damp hay) that occurs without the direct application of a flame or a spark. It is usually caused by slow oxidation processes (such as atmospheric oxidation or bacterial fermentation) under conditions that do not permit the dissipation of heat. D

**spouter**
An oil or gas well, the flow of which has not been controlled by the engineers. D

**SPQ**
Synthetic pipeline quality gas.

**SPR**
Strategic petroleum reserve.

**sprayer**
A burner which breaks up the liquid fuel into very minute liquid particles, that is, it separates a jet of liquid into a finely divided spray, resembling "liquid dust." BB

**spray oil**
A petroleum product of low viscosity, usually similar to lubricating oil, used to combat pests which attack trees, etc. ASTM Method D447 gives some information on establishing the distillation range of such oils. M

**SPS**
Submerged production system.

**spudding**
As employed and understood by oil operators, denotes the first abrasion of the soil by the drill or that of first entrance of the drill into the ground. D

**sq**
Square.

**sq ft**
Square foot.

**sq km**
Square kilometer.

**sq mm**
Square millimeter.

**squeeze**
(1) A crushing of coal with the roof moving nearer to the floor. (2) The settling without breaking of a roof over a considerable area of working. Also called creep, crush, pinch, or nip. D

**squib**
A thin tube filled with black powder, forming a slow-burning fuse to explode a stemmed charge of black powder; a firing device that will burn with a flash which will ignite black powder. D

**SRC**
Solvent refined coal.

**SRTA**
Spherical reflector tracking absorber.

**SSNM**
Stragetic special nuclear materials.

**SSPP**
Solar-sea power plant.

**SSPS**
Satellite solar-power stations.

**stabilizer**
A fractionating tower for removing light hydrocarbons from an oil to reduce vapor pressure, particularly applied to gasolines. M

**stable**
(1) Incapable of spontaneous change. (2) Not radioactive. O

**stable air**
An air mass that remains in the same position rather than moving in its normal horizontal and vertical directions. Stable air does not disperse pollutants, and can lead to high build-ups of air pollution. E

**stable isotope**
An isotope that does not undergo radioactive decay. O

**stack**
(1) A smokestack. (2) A vertical pipe or flue designed to exhaust gases and suspended particulate matter. E

**stack effect**
The upward movement of hot gases in a stack due to the temperature difference between the gases and the atmosphere. E

**stadia**
(1) A temporary surveying station. (2) A method of surveying in which distances from an instrument to a rod are determined by observing the space on the rod scale intercepted by tow lines in the reticule of the telescope. D

**stagnation temperature**
The highest temperature a collector plate can reach. The surface coating on the plate must be selected to withstand this temperature. (Note: This definition is valid only in the context of solar collectors; not in the strict engineering sense.) H

**standard**
(1) Something that is set up and established as a rule for the measure of quantity, weight, extent, value, or quality. (2) Specifically, an original specimen measure or weight (as the international prototype meter and kilogram of the International Bureau of Weights and Measures) or an official copy of such a specimen used as the standard of comparison in testing other weights and measures. D

**standard air**
Air at 68°F dry-bulb temperature, 50-percent relative humidity, and 29.92 in. Hg., it has a density of 0.075 lb per cu ft. M

**standard pressure**
Pressure under which the mercury barometer stands at 760 mm, or 30 in. (equivalent to approximately 14.7 psi). M

**standards**
Agreed specific values, as defined by the organization concerned with such definition. These groups include: United States of America Standards Institute, formerly the American Standards Association; B.S.I., British Standards Institution; D.V.M., German Association for Testing Materials; A.FN.O.R., French Association for Standardizing Materials; I.S.A., International Standards Association. D

**standard temperature**
Commonly 60°F (15.56°C). M

**standby service**
Service that is not normally used but which is available through a permanent connection in lieu of, or as a supplement to, the usual source of supply. K

**standing gas**
A body of firedamp known to exist in a mine, but not in circulation. D

**stank**
To make watertight; to seal off; an airtight and watertight wall against old mine workings. D

**staple**
A shaft that is smaller and shorter than the principal one and joins different levels. D

**statement of performance**
A statement describing the scope and duty of a plant in terms, for example, of the tonnage of coal treated per hour, the processes used, the separations effected and sizes produced; sometimes also used to express the results of plant operation. D

**static pressure**
The difference, in consistent units, between the absolute pressure at a point, and the absolute pressure of the ambient atmosphere, being positive when the pressure at the point is above the ambient pressure, and negative when below. D

**stationary equipment**
Stationary equipment is installed in a given location and is not moved from that location in performing its function. This includes equipment such as substations, pumps, and storage-battery charging stations. D

**stationary source**
A pollution emitter that is fixed rather than moving (as an automobile). E

**station use (generating)**
The kilowatt hours used at an electric generating station for such purposes as excitation and operation of auxiliary and other facilities essential to the operation of the station. Station use includes electric energy supplied from house generators, main generators, the transmission system, and any other sources for this purpose. The quantity of energy used in the difference between the gross generation plus any supply from outside the station and the net output of the station. K

**statistics**
The collection, tabulation, and study of numerical facts and data. In industry, statistics will indicate trends which would be almost impossible to establish by other means. The statistical method is useful in: (1) estimating the real value of work done, goods, or machines in terms of useful service and maintenance costs; and (2) estimating the forecasting profits and markets. D

**STCRS**
Solar-thermal central receiver system.

**std**
Standard.

**STDS**
Solar-thermal distributed receiver system.

**steam**
Water in the form of a vapor which often is the medium or working substance by which some of the heat energy liberated from the fuel by combustion is transmitted to the engine and partly converted into mechanical work. BB

**steam accumulator**
A vessel to smooth out violent steam loads on a boiler. While the steam demand is relatively low the accumulator is charged and acts as a buffer when sudden steam demands occur. A similar effect is obtained by a recently developed type of shell boiler of the thermal storage type. D

**steam coal**
Coal suitable for use under steam boilers. D

**steam cracking**
High-temperature cracking of hydrocarbons in the presence of steam. M

**steam, dry**
Steam containing no moisture, and either saturated or superheated. BB

**steam-electric plant**
An electric power plant utilizing steam for the motive force of its prime movers. L

**steam gas**
Highly superheated steam. D

**steam injection**
A term applied to a secondary recovery method used in re-energizing oil reservoirs, steam injection is a variation of thermal recovery involving the injection of steam into the producing well over a period of about ten days to two weeks, shutting in the well for a week or so to allow the reservoir to become thoroughly heated, and then reopening the well. The heat contained in the steam becomes locked in the porous rock formation at the bottom of the well, thus reducing the viscosity of the crude oil and allowing it to flow far more readily into the well. This technique has been in wide use in relatively shallow reservoirs containing thick, viscous crudes. M

**steam jet system**
A process by which the sun's radiant energy can produce a useful cooling effect by the use of a concentrating collector, producing steam at a pressure high enough to make a steam-jet ejector function, and this in turn causes evaporation and chilling of water in a tank connected to the ejector. R

**steam power**
Any type of energy or power generated or developed through the use of a steam engine. It is usually applied to electrical energy or power generated by the use of steam engines or turbines to drive electric generators. E

**steam turbine**
A machine in which high pressure steam is made to do work by acting on, and rotating, blades in a cylinder. It is used for large stationary engines in power stations. Its comparative efficiency is high. D

**steam, wet**
Steam containing intermingled moisture, mist or spray. BB

**stent**
The amount of work expected from a coal miner in a day or week. D

**step rate schedule**
A rate schedule in which a single unit charge is applied to all kilowatt-hours consumed during the billing period, the unit charge per kilowatt-hour depending on the total amount of energy consumed. For example: 5 cents per kilowatt-hour when the monthly consumption is between 1 and 25 kilowatt-hours inclusive; 4 cents per kilowatt-hour for all kilowatt-hours used when the monthly consumption is between 26 and 50 kilowatt-hours inclusive; and 3 cents per kilowatt-hour for all kilowatt-hours used when the monthly consumption is over 50 kilowatt-hours. (Note: not now in common use.) L

**step up**
To increase the voltage (of a current) by means of a transformer. D

**sterilized coal**
That part of a coal seam which, for various reasons, is not mined. D

**still**
An apparatus in which a substance is changed by heat into vapor, with or without chemical decomposition. The vapor is then liquefied in a condenser and collected in another part of the apparatus. Used generally for separating the more volatile parts of liquids and obtaining them in pure form. D

**still gas**
(1) Any form or mixture of gas produced in refineries by cracking, reforming, and other processes, the principal constituents of which are methane, ethane, ethylene, butane, butylene, propane, propylene, etc. Used as a petrochemical feedstock or as a fuel at the refinery. (2) Refinery gas. M

**stock**
Petroleum, or products in storage, awaiting utilization, or transfer to the ultimate point of utilization, whether with or without change of ownership.

See also specific stocks under alphabetical listing. M

**stoichiometry**
(1) That branch of chemistry that deals with the numerical relationships between elements or compounds (atomic weights). (2) The determination of the proportions in which the elements combine (formulas); and the weight relations in reactions (equations). D

**stoker coal**
A screen size of coal specifically for use in automatic firing equipment. D

**stope**
(1) An excavation from which ore has been excavated in a series of steps. (2) A variation of step. D

**storage**
(1) In solar energy, a facility designed to receive energy in the form of heat (or cold) and preserve it for future use. (2) Fluid storage accommodates heat in proportion to its specific heat capacity and temperature increase. (3) Chemical storage usually involves a phase transformation of the storage material, and the heat storage capability is largely dependent upon the material's latent heat of fusion, or the energy released when chemical bonds are broken. (4) In heat storage, pebble beds, rock beds, or other solid materials are used to store heat and have an average specific heat coefficient of approximately 0.2. Their heat storage capability is dependent upon the large possible temperature differential between the storage bed and the circulating heat transfer media. H

**storage battery**
A combination of secondary cells or accumulators which, when once charged, may be used for a considerable time after as a source of electric current. D

**storage capacity**
The volume of a reservoir available to store water. L

**storage plant**
A hydroelectric plant associated with a reservoir having power storage. L

**storm oil**
Oil used to spread over water to prevent waves from breaking. It is applied with a canvas bag, called a sea drag, filled with oil and dragged over the surface of the water. M

**stove coal**
In anthracite only; two sizes of stove coal are made, large and small. Large stove, known as No. 3, passes through a 2-1/4 to 2 inch mesh and over a 1-7/8 to 1-1/2 inch mesh; small stove, known as No. 4, passes through a 1-7/8 to 1-3/8 inch mesh and over a 1-1/8 to 1 inch mesh. D

**stp**
Standard temperature and pressure. Also abbreviated STP. D

**STPS**
Solar thermal power systems.

**straight line rate schedule**
A rate schedule that provides for a constant charge per unit of energy regardless of the amount consumed. L

**straight-run gasoline**
(1) Gasoline produced from the primary distillation process. (2) Having insignificant antiknock qualities. M

**strain energy**
The work done in deforming a body. D

**strata**
(1) Sedimentary rock layers, plural of stratum. (2) Layers either artificial or natural. D

**strata gases**
These occur in the mineral deposit itself or in adjacent or nearby forma-

tion. Their origin may be in a particular formation in which they were laid down or formed subsequently by chemical action, or they may occasionally migrate into other formations, frequently because of release of pressure with mining. D

**stratified hot water tank**
A hot water tank in which the temperature at any level varies from a low temperature at the bottom of the tank to a maximum temperature at the top of the tank. H

**stratigraphic trap**
A formation in which the trap is formed as a result of thinning or permeability degradation in the reservoir rock or a break in its continuity. M

**stratigraphy**
That branch of geology which treats of the formation, composition, sequence, and correlation of the stratified rocks as parts of the earth's crust. D

**stratoscope**
An apparatus inserted in the drill hole which permits engineers to make a visual inspection of the strata. D

**straw oil**
Pale paraffin oil of straw color, used for many process applications. M

**stream day**
Denoting 24 hours of actual operation of a refinery unit; in contrast to a calendar day. M

**stringer**
A narrow vein or irregular filament of mineral traversing a rock mass of different material. D

**strip**
In coal mining, to remove the earth, rock and other material from a seam of coal, by power shovels, generally practiced only where the coal seam lies close to the earth's surface. D

**strip mining**
The mining of coal by surface mining methods as distinguished from the mining of metalliferous ores by surface mining methods which is commonly designated as open-pit mining. D

**stripped atom**
One from which one or more electrons has been removed, rendering it ionically charged. D

**stripper production**
Production in an oil well existing when a well reaches the point of producing below its "settled" rate. A stripper well is usually an older well which produces only a few barrels of oil per day, but is kept in production because its output is steady and the yield is good over a long period of time. They play a significant part in meeting our energy demands as they provide slow but sure production. M

**stripping**
An excavation with power shovels in which the coal seams are laid bare by the stripping of the surface soil and rock strata. Operators are resorting more to this form of mining as it greatly reduces cost of production, especially the item of labor. It is estimated that for every ton of anthracite that is stripped instead of mined, ten men are deprived of work. Stripping also eliminates the cost of timbering, hauling, pumping, ventilating, and complicated safety provisions. D

**stripping area**
In stripping operations, an area encompassing the pay material, its bottom depth, the thickness of the layer of waste, the slope of the natural ground surface, and the steepness of the safe slope of cuts. D

**stroke**
The maximum distance a piston moves within a cylinder before the direction of its travel is reversed. D

**Strontium 90 ($_{38}Sr^{90}$)**
A hazardous isotope produced in the process of nuclear fission. Strontium 90 has a "half-life" of 28 years. Thus it takes 28 years to reduce this material to half its original amount, 56 years to one quarter, 84 years to one-eigth, and so on. Strontium 90 typifies problems of radioactive waste storage which are faced in producing power by means of nuclear fission. V

**structural trap**
A reservoir, capable of holding oil or gas, formed from crustal movements in the earth that fold or fracture rock strata in such manner that oil or gas accumulating in the strata are sealed off and cannot escape. The most common structural traps are fault traps, anticlines, and salt domes. D

**stull**
A timber prop or platform. D

**subanthracite**
Coal intermediate between anthracite and superbituminous coal. D

**subaqueous mining**
Surface mining in which the material mined is removed from the bed of a natural body of water. D

**subatomic particle**
Any of the constituent particles of an atom, such as an electron, neutron, proton, etc. O

**subbituminous coal**
Black lignite or lignitic coal. D

**subcooling**
Cooling of a liquid refrigerant below the condensing temperature at constant pressure. D

**subcritical assembly**
A reactor consisting of a mass of fissionable material and moderator whose effective multiplication factor is less than one and that hence cannot sustain a chain reaction. Used primarily for educational purposes. O

**subcritical mass**
An amount of fissionable material insufficient in quantity or of improper geometry to sustain a fission chain reaction. D

**subhydrous coal**
Coal of hydrogen content below average for the rank of coal, for example, coals containing a high proportion of fusain. D

**sublime**
To cause to pass from the solid state to the vapor state by the action of heat and again to condense to solid form. D

**submersible pump**
A centrifugal pump, usually driven by electricity, which can be wholly submerged in water. D

**submetering**
The practice of remetering purchased energy beyond the customer's utility meter, generally for distribution to building tenants through privately owned or rented meters. L

**subsidence**
(1) A sinking down of a part of the earth's crust. (2) The lowering of the strata, including the surface, due to underground excavations, often coal mines. G

**subsoil**
Broadly and loosely, the part of the regolith (earth mantle) which lies beneath the true soil and which contains almost no organic matter. D

**substation**
An electrical installation containing generating or power-conversion equipment and associated electrical equipment and parts, such as switchboards, switches, wiring, fuses, circuit breakers, compensators, and transformers. D

**substitute natural gas (SNG)**
The conversion of other gases, liquids or other solid hydrocarbons to a gaseous fuel of calorific value, heat content, compatability, and quality equivalent in performance to that of domestic natural gas. DD

**suction**
Atmospheric pressure pushing against a partial vacuum. D

**suction anemometer**
An anemometer that measures wind velocity by the degree of exhaustion caused by the blowing of the wind through or across a tube. D

**sulfur minerals**
Occur naturally in association with volcanoes and hot springs, and in cap rocks in salt domes. Extensively produced from pyrites and pyritic minerals, either direct or as a byproduct, also as byproduct in gas stripping. Main uses are as sulfuric acid, sulfur dioxide (paper making), and in vulcanizing compounds, fungicides and insecticides. D

**sulfur oxides**
Compounds composed of sulfur and oxygen produced by the burning of sulfur and its compounds in coal, oil, and gas. Harmful to the health of man, plants and animals, and may cause damage to materials. W

**sulfur smog (classical smog)**
This smog is composed of smoke particles, sulfur oxides (SOx), and high humidity (fog). The sulfur oxide ($SO_3$) reacts with water to form sulfuric acid ($H_2SO_4$) droplets, the major cause of damage. V

**sulfur test ($CO_2$—$O_2$)**
Test described in ASTM Method D 1266 for determining the amount of sulfur in petroleum products, including liquefied petroleum gas, by lamp combustion. Combustion is controlled by varying the flow of carbon dioxide and oxygen ($CO_2$-$O_2$) to the burner. M

**summer oil**
A heavy, railway car and engine oil that has a flashpoint above 140°C and solidifies below -5°C. D

**summer peak**
The greatest load on an electric system during any prescribed demand interval in the summer (or cooling) season, usually between June 1 and September 30. K

**sun effect**
The quantity of heat from the sun tending to heat an enclosed space. D

**sun observations**
In surveying, fixation of longtitude and/or latitude of a station, or orientation of a survey line, by use of a theodolite to relate sun's position, sideral time and theodolite's location. D

**superanthracite**
Coal intermediate between anthracite and graphite. D

**superbituminous coal**
Same as semibituminous coal.

**supercharger**
A blower that increases the intake pressure of an engine. D

**superconductivity**
The abrupt and large increase in electrical conductivity exhibited by some metals as the temperature approaches absolute zero. D

**supercritical flow**
Flow with a mean velocity equal to or greater than Belanger's critical velocity. Also called shooting flow, sinuous flow, tortuous flow, and turbulent flow. E

**supercritical reactor**
A reactor in which the effective multiplication factor is greater than one; consequently a reactor that is

increasing its power level. If uncontrolled, a supercritical reactor would undergo an excursion. O

**superheat**
To heat a vapor not in contact with its own liquid so that it remains free from suspended liquid droplets. D

**superheater**
(1) One that superheats steam or other gases. (2) A coil or other device through which steam from a boiler passes to be superheated. D

**superheating**
The heating of a vapor, particularly steam, to a temperature much higher than the boiling point at the existing pressure. This is done in power plants to improve efficiency and to reduce condensation in the turbines. W

**supertanker**
A very large oil tanker. The definition changes with advancing marine technology. In the late 1940's, 45,000 dwt (dead weight ton) tankers were considered super tankers; now, common usage is 400,000 dwt, and still larger ships are planned. G

**surface combustion**
Combustion of injected, properly proportioned fuel and air on a surface, or within a definite zone; as when a mixture of air and gas or of air and oil vapor is forced through a porous wall, and ignited on the other side. D

**surface energy**
Product of surface tension (dynes per centimeter) and surface area ($cm^2$), expressed in ergs. Work required to increase surface area by unit area. D

**surface mining**
The mining in surface excavations. It includes placer mining, mining in open glory-hole or milling pits, mining and removing ore from open cuts by hand, or with mechanical excavating and transportation equipment, and the removal of capping or overburden to uncover the ores. D

**surplus**
Energy generated that is beyond the immediate needs of the producing system. This energy is frequently obtained from spinning reserve and sold on an interruptible basis. K

**surplus power**
Power other than primary power generated at a hydroelectric power plant. E

**survey**
To determine and delineate the form, extent and position (as a tract of land, a coast, or a harbor) by taking linear and angular measurements, and by applying the principles of geometry and trigonometry. P

**survey meter**
Any portable radiation detection instrument especially adapted for surveying or inspecting an area to establish the existence and amount of radioactive material present. P

**suspension**
The state of a material when its particulates are mixed with and buoyed in a fluid, but are not dissolved by it. M

**sustained yield**
The use of a renewable resource at a rate that permits resource regeneration for use continuing undiminished into the future (for example, timber cut so as to produce the same amount of wood each year). E

**sweat**
(1) To gather surface moisture in beads as a result of condensation. (2) The roof of a mine is said to sweat when drops of water are formed upon it by condensation of steam. D

**sweet crudes**
Crude oils having a fairly low sulfur content. M

**sweet natural gas**
Gas found in its natural state, containing such small amounts of compounds of sulfur that it can be used without purifying, with no deleterious effects on piping and equipment. DD

**switchgear**
This is a general term applying to switching, interrupting, controlling, metering, protective, and regulating devices, as well as assemblies of these devices with associated interconnections, accessories and supporting structures. The term is used primarily in connection with generation, transmission, distribution, and conversion of electric power. D

**switching station**
An assemblage of equipment for the sole purpose of tying together two or more electric circuits through switches, selectively arranged to permit a circuit to be disconnected, or to change the electric connections between the circuits. L

**SWU**
Separative work units, or selective work units.

**synchrocyclotron**
A cyclotron in which the frequency of the accelerating voltage is decreased with time so as to match exactly the slower revolutions of the accelerated particles. The decrease in rate of acceleration of the particles results from the increase of mass with energy as predicted by the Special Theory of Relativity. D

**synchronism**
The state when the phase difference between two or more periodic quantities is zero; they are then said to be in phase. D

**syncline**
A fold in rocks in which the strata dip inward from both sides toward the axis. Opposite of anticline. D

**syncrude**
A synthetic crude oil similar to petroleum oil produced by the hydrogenation of coal or coal extract. Y

**syndicate**
A group of persons or concerns who combine under a usually temporary agreement to carry out a particular transaction, such as underwriting a bond issue. D

**synergism**
Action of two agencies, usually two chemicals, to produce an end effect greater than or different from the sum of the effects of the two agencies acting separately. D

**synfuel**
Fuel produced artificially from hydrocarbons, peat or carbohydrates. Y

**syngas**
A synthetic gas produced by reactions of hydrocarbons or carbohydrates with oxygen and steam to produce a synthetic natural gas, consisting of carbon monoxide and hydrogen; may be abbreviated as SNG. Y

**syngenetic**
A term now generally applied to mineral or ore deposits formed contemporaneously with the enclosing rocks, as contrasted with epigenetic deposits, which are of later origin than the enclosing rocks. D

**synthane**
A coal gasification process being developed by the Bureau of Mines, to produce pipeline quality gas. G

**synthesis**
The act or process of making or building up a compound by the union of simpler compounds or of its elements. M

**synthetic**
A material produced by synthesis. M

**synthetic atmosphere**
A specific gaseous mass containing any number of constituents and in any proportion produced by man for a special purpose. E

**synthetic crude**
(1) The total liquid, multicomponent hydrocarbon mixture resulting from a process involving molecular rearrangement of charge stock. (2) Commonly applied to such product from cracking, reforming, viscosity breaking, etc. M

**synthetic oil**
Oil produced artificially as in the Berguis process or Fischer-Tropsch process. Synthetic oil or synthoil is produced from other hydrocarbons or from carbohydrates. D

**synthetic or substitute fuel**
A liquid, gaseous, or solid fuel produced by a man-made process; e.g., bitumen upgrading, coal gasification and oil-shale retorting. Z

**system, distribution**
Generally, mains, services, and equipment which carry or control the supply of fuel from the point of local supply to and including the sales meters. DD

**system, electric**
The physically connected generation, transmission, distribution, and other facilities operated as an integral unit under one control, management, or operating supervision. K

**system, field and gathering**
A network of pipelines (mains) transporting natural gas from individual wells to compressor station, processing point, or main trunk pipelines. DD

**system interconnection**
A connection between two electric systems permitting the transfer of electric energy in either direction. K

**system loss (in electrical industry)**
The difference between the system net energy or power input and output, resulting from characteristic losses and unaccounted for between the sources of supply and the metering points of delivery on a system. K

**system net input (in electrical industry)**
Net available energy that is put into a utility's system for sale within its own service area or otherwise used by the utility within its own service area. It is the net energy generated in a system's own plants, plus energy received from other systems, less energy delivered to other systems. K

**system output (in electrical industry)**
The net generation by the system's own plants plus purchased energy, plus or minus net interchange energy. K

**system pressure**
A gas system which operates at a pressure higher than the standard service pressure delivered to the customer is called a high pressure system. If the pressure is the same as that delivered to the customer without a pressure regulator, it is called a low pressure system. DD

**system required reserve**
The system reserve capacity needed as standby to insure an adequate standard of service. L

**system, transmission**
Pipelines (mains) installed for the purpose of transmitting gas from a source or sources of supply to one or more distribution centers, or to one or more large volume customers, or a pipeline installed to interconnect sources of supply. In typical cases transmission lines differ from gas mains in that they operate at higher pressures, are longer, and the distance between connections is greater. DD

# T

**t**
Temperature, usually ordinary temperature as expressed in degrees centigrade, or in degrees Fahrenheit. Also, abbreviation for ton. D

**T**
Temperature; absolute temperature, standard absolute temperature of the atmosphere. Also abbreviation for time, and ton. D

**table**
Any table-like surface for cleaning or sorting coal or ore, for example, rotary sorting table. D

**tachometer**
An instrument of the direct-reading type, indicating the speed of a shaft or machine in revolutions per minute. D

**tack coat**
A thin coat of hot road tar or bitumen emulsion applied over a road surface to improve adhesion with the subsequent single or two course bitumen final surfacing. D

**tactite**
A general term for rocks of complex mineralogy formed by the contact metamorphism of limestone, dolomite, or other carbonate rocks into which foreign matter from the intruding magma has been introduced by hot solutions. D

**tailings**
(1) The parts, or part, of any incoherent or fluid material separated as refuse, or separately treated as inferior in quality or value; leavings; remainders; dregs. (2) The sand, gravel, and cobbles which pass through the sluices in hydraulic mining were formerly generally designated as tailings, but of late years, especially in state and in United States legislative documents, they have been called mining debris or simply debris. D

**tall oil**
(1) A name derived from the Swedish word meaning pine. (2) A mixture of saponifiable materials derived from the waste liquor of sulfate paper manufacture. Used in compounding certain petroleum products. M

**tamp**
(1) To lightly pack a drilled hole with moist, loose material after the charge has been placed. (2) To pound or press soil to compact it. D

**tank reactor**
A reactor in which the core is suspended in a closed tank, as distinct from an open pool reactor. These are commonly used as research and test reactors. O

**TAO**
Technology assistance offices.

**taper**
A gradual and uniform decrease in size, as a tapered shank. D

**TAPPI**
Technical Association of the Pulp and Paper Industry.

**tar**
Any of various dark brown or black bituminous, usually odorous, viscous liquids or semiliquids that are obtained by the destructive distillation or carbonization of wood, coal, peat, shale, and other organic materials, and yield pitch on distillation. D

**tar distillate**
A fraction in petroleum refining containing heavy oils and paraffin. D

**TARGET**
Team to advance research for gas energy transformation.

**tariff**
A published volume of rate schedules and general terms and conditions. L

**tar sands**
Vast deposits of sand and clay which are heavily impregnated with oil. They are a rare natural phenomenon and have only been found in a few places in the world, and are likely to prove to be important additions to our oil reserves. M

**tax adjustment clause**
A clause in a rate schedule that provides for an adjustment in the customer's bill if the supplier experiences a change in tax rate from a specified base or the application of new taxes. L

**TCF**
Trillion cubic feet.

**Tcf/y**
Trillion cubic feet per year.

**T/D, TPD**
Tons per day.

**TDS**
Total dissolved liquids.

**tectonic**
(1) Pertaining to rock structures and topographic features resulting from deformation of the earth's crust. (2) Earthquakes not caused by volcanic action, landslides, or collapse of caverns. D

**TEEC**
Total energy and environmental conditioning.

**telain**
Anglicized from the German "telit." Greater fragments of plant tissues, which are completely soaked with vitrain, that is, the cell walls as well as the cell cavities. D

**telecontrolled power station**
A hydroelectric power station operated by remote control. D

**temp**
Temperature.

**temperature**
An arbitrary measurement of the amount of molecular energy of a body, or the degree of heat possessed by it. It should be distinguished from heat itself. Heat is a form of energy; temperature is a measurement of its intensity. M

**temperature standards**
For normal measurement, 0°C (regarding gas properties). For thermodynamics and physical properties either 18°C or 25°C. as defined in each stated case. D

**tensile strength**
Tenacity of resistance to pulling, measurable in dynes per square centimeter or, for metals and alloys, tons per square inch. Measured in latter case by testing a dimensioned bar to

breaking point on the tensile testing machine. D

**tensile stress**
Sometimes called modulus. The nominal stress developed by a material subjected to a specified stretching load, as in the tension test. Above the elastic limit, nominal tensile stress is considerably lower than the true stress because it does not reflect the decrease in cross-section area accompanying continued deformation. D

**tension**
In engineering, a pulling force or stress. In tension, metals are strong, while concrete and masonry are weak. D

**tephra**
A collective term for all clastic volcanic materials which, during an eruption, are ejected from a crater or from some other type of vent and transported through the air, including volcanic dust, ash, cinders, lapilli, scoria, pumice, bombs, and blocks. Synonym for volcanic ejecta. D

**TERA**
Total energy resource analysis.

**Tera (T)**
$10^{12}$. Y

**terminology**
The technical or special words or terms used in any science, art, industry, trade, etc. D

**terra**
The earth.

**terrain**
The tract or region of ground immediately under observation. D

**territorial waters**
The belt of sea adjoining a coast which is under the jurisdiction of the nation occupying the coast. E

**tertiary recovery**
Use of heat and other methods other than fluid injection to augment oil recovery (presumably occurring after secondary recovery). G

**TES**
Thermal energy storage.

**test well**
One that determines not only the presence of petroleum oil, but its commercial value, considering its abundance and accessibility. D

**textile oil**
Oil used to lubricate textile machinery. A wide assortment of nonlubricating oils used in the textile industry to render fabrics and yarns pliable to facilitate weaving and spinning operations. M

**TFTR**
Tokomak fusion test reactor (Princeton).

**therm**
Unit equal to 100,000 Btu. M

**thermal**
(1) Hot or warm. (2) Applied to springs which discharge water heated by natural agencies. (3) Identifies a type of electric generating station or power plant, or the capacity or capability thereof, in which the source of energy for the prime mover is heat. L

**thermal boring**
Use of high-temperature flame to fuse rock in drilling. Heat comes from ignition of kerosine with oxygen or other fuel system, at bottom of drill hole, and water with compressed air may be used to flush out the products. D

**thermal breeder reactor**
A breeder reactor in which the fission chain reactor is sustained by thermal neutrons. W

**thermal capacity**
The rating of a thermal electric generating unit or the sum of such ratings for all units in a station or stations. K

**thermal conductivity**
Rate of heat flow in a homogeneous material under steady conditions, per unit area, and per unit gradient in a direction perpendicular to the area. M

**thermal cracking**
A conversion process in which the less volatile heating oil fractions are subjected to higher temperatures under increasing pressure. The heat puts a strain on the bonds holding the larger, complex molecules together and causes them to break up into smaller ones, including those in the gasoline range. With this process, cracking not only increases gasoline quantity per barrel of crude oil, but gives substantial quality improvement. It is accomplished by the action of heat alone at the temperatures of 450° to 510°. M

**thermal efficiency**
(1) The ratio of the electric power produced by a power plant to the amount of heat produced by the fuel. (2) A measure of the efficiency with which the plant converts thermal to electric energy. (3) The ratio of the heat utilized to the total heat units in the fuel consumed. O

**thermal gradient**
In the earth, the rate at which temperature increases with depth below the surface. A general average seems to be around 30°C increase per kilometer of depth, or 150°F per mile. D

**thermal pollution**
Degradation of water quality by the introduction of a heated effluent. Primarily a result of the discharge of cooling waters from industrial processes, particularly electrical power generation. Even small deviations from normal water temperatures can affect aquatic life. Thermal pollution usually can be controlled by cooling towers. E

**thermal power**
(1) Any type of energy or power generated or developed through the use of heat energy. (2) In common practice, electrical energy or power generated by the use of steam engines or turbines to drive electric generators. E

**thermal process**
Any refining process which utilizes heat, without the aid of a catalyst. M

**thermal reactor**
A reactor in which the fission chain reaction is sustained primarily by thermal neutrons. Most reactors are thermal reactors. O

**thermal recovery**
A secondary recovery process to re-energize oil reservoirs by heat, rather than water or gas. M

**thermal shield**
A layer or layers of high density material located within a reactor pressure vessel or between the vessel and the biological shield to reduce radiation heating in the vessel and the biological shield. W

**thermal spring**
A stream of warm or hot water issuing from the ground, often after having been heated by buried lava and therefore, commonly occurring in volcanic regions where eruptions have ceased. E

**thermal storage**
A system which utilizes ceramic brick or other materials to store heat energy. V

**thermal stratification**
The formation of layers of different temperatures in bodies of water. E

**thermal structure**
The temperature variation with depth of sea water. D

**thermal unit**
A unit for the comparison or the calculation of quantities of heat. D

**thermal water**
The mineral-charged water that issues from a hot spring or geyser. D

**thermionic conversion**
The conversion of heat into electricity by evaporating electrons from a hot metal surface and condensing them on a cooler surface. No moving parts are required. O

**thermocouple**
A device consisting essentially of two conductors made of different metals, joined at both ends, producing a loop in which an electric current will flow when there is a difference in temperature between the two junctions. O

**thermodynamics**
Study of transformation of energy into other manifested forms and of their practical application. The three laws are: (1) Conservation of energy; energy may be transformed in an isolated system, but its total is constant. (2) Heat cannot be changed directly into work at constant temperature by any cyclic process. (3) Heat capacity and entropy of every crystalline solid becomes zero at absolute zero (0°K). D

**thermoelectric conversion**
The conversion of heat into electricity by use of thermocouples. D

**thermonuclear energy**
A term employed for energy derived from a nuclear fusion reaction; e.g. based on a deuterium-tritium fuel. Z

**thermonuclear reaction**
A reaction in which very high temperatures bring about the fusion of two light nuclei to form the nucleus of a heavier atom, releasing a large amount of energy. In a hydrogen bomb, the high temperature to initiate the thermonuclear reaction is produced by a preliminary fission reaction. O

**thermostat**
An automatic device for regulating temperature. M

**thermosyphon**
In a passive domestic hot water system, the cold water supply is pressure-fed to a storage tank located above the solar collectors. Exposure of the collectors to solar radiation causes the cold water to circulate by convection from bottom to top of the collectors, and once heated back to the storage tank. The heated water is then stored until demand is initiated and then gravity fed. S

**thick seam**
In general, a coal seam over 4 feet in thickness. D

**thin seam**
In general, a coal seam 2 feet and under in thickness. D

**thorium (Th)**
A naturally radioactive element with atomic number 90, and as found in nature, an atomic weight of aproximately 232. The fertile thorium 232 (${}_{90}Th^{232}$) isotope can be transmuted to fissionable uranium 233 (${}_{92}U^{233}$) by neutron irradiation. V

**three-part rate schedule**
A rate schedule that provides three components for determining the total bill: (1) customer charge: (2) demand charge; (3) energy charge. For example, 50 cents per month per meter; plus, a demand charge of $1.25 per month per kilowatt for the first 25 kilowatts of billing demand in the month; 90 cents per month per kilowatt for the excess of the billing

demand over 25 kilowatts; plus an energy charge of 1-1/2 cents per kilowatt-hour. (Note: not now in common use.) L

**three-phase circuit**
Usually a three-wire circuit using alternating current with three equal voltages. This should not be confused with the three-wire service supplying 110 volts and 220 volts. This latter is merely a two-voltage single-phase circuit arrangement and is used almost universally to provide power and lighting to homes and small business establishments. D

**three-phase current**
Alternating current (A.C.) in which three separate pulses are present, identical in frequency and voltage, but separated 120°. D

**threshold dose**
The minimum dose of radiation that will produce a detectable biological effect. O

**through coal**
(1) Coal as it is mined, that is, large and small mixed together; run-of-mine coal. (2) Coal after passage through a screen of stated size. D

**throughput**
Quantity of material (ore or selected fraction) passed through the mill or a section thereof in a given time or at a given rate. D

**THTR**
Thorium high-temperature reactor.

**tide**
Rise and fall of the surface of the sea due to the gravitational pull of the moon, generally taking place twice daily. In the open sea, this rise may not exceed 2 feet, whereas in the shallow seas bordering continents it may be more than 20 feet, and in narrow tidal estuaries from 40 to 50 feet. Since the moon travels in its own orbit in the same direction as the earth, a period of about 24 hours, 25 minutes will elapse between successive occasions when the moon is vertically above a given meridian. The interval between successive high tides will therefore be about 12-1/2 hours. D

**TIME**
Total industry marketing effort.

**tipper**
A vehicle, commonly a truck, with a body which can be raised at one end or sideways in order to discharge its contents. D

**tipple**
Originally the place where the mine cars were tipped and emptied of their coal, and still used in that sense, although now more generally applied to the surface structures of a mine including the preparation plant and loading tracks. D

**TIS**
Technical information center.

**TJ/y**
Terajoules per year.

**TLV**
Threshold level value.

**tn**
Ton.

**TOC**
Total organized carbon.

**TOE**
Metric tons of oil equivalent.

**tolerance**
A specified allowance (either plus or minus) of the given dimensions of a finished product to take care of inaccuracies in workmanship of parts to be fitted together. The amount allowed as tolerance is generally small as compared with the standard dimension of the part. D

**ton**
Any of various units of weight: (1) a unit that equals 20 long hundredweight or 2,240 pounds, used chiefly in England; also called a long ton. (2) a unit that equals 20 short hundredweight or 2,000 pounds, used chiefly in the United States, Canada, and the Republic of South Africa; also called a short ton. (3) a unit of internal capacity for ships that equals 100 cubic feet; also called a register ton. (4) a unit that approximately equals the volume of a long ton weight of sea water; used in reckoning the displacement of ships and equaling 35 cubic feet, also called a displacement ton. (5) A unit of volume for cargo freight usually considered to be 40 cubic feet; also called a freight ton. D

**top lease**
A lease granted by a landowner during existence of a recorded mineral lease which is to become effective if and when the existing lease expires or is terminated. D

**topman**
In mining, a workman who is employed at surface jobs around the mine plant. D

**topography**
The configuration of a surface area including its relief, or relative elevations, and the position of its natural and man-made features. E

**topping**
The distillation of crude oil to remove light fractions only. M

**topping cycle**
A means to increase thermal efficiency of a steam-electric power plant by increasing temperatures and interposing a device, such as a supercritical gas turbine, between the heat source and the conventional steam-turbine generator to convert some of the additional heat energy into electricity. G

**topsoil**
A general term applied to the surface portion of the soil including the average plow depth (surface soil) or the A-horizon, where this is deeper than plow depth. D

**torque**
(1) An engineering term defined as the product of a force times its lever arm. (2) A measure of the ability to produce rotation.

**torr**
The pressure exerted per square centimeter by a column of mercury 1 millimeter high at a temperature of 0°C where the acceleration of gravity is 980.665 centimeters per second. D

**torsion**
A body is under torsion when subjected to force couples acting in parallel planes about the same axis of rotation but in opposite senses. D

**total cooling effect**
The difference between the total heat content of the air-steam mixture entering a conditioner per hour and the total heat of the mixture leaving per hour. D

**total cooling load**
The sum of the sensible and latent heat component. D

**total energy**
The total energy at any section in a moving fluid consists of the sum of the internal static, velocity, and potential energies at that section. D

**total energy system**
A packaged energy system of high efficiency, utilizing gas fired turbines or engines which produce electrical energy and utilize exhaust heat in applications such as heating and cooling. V

**total pressure**
The pressure representing the sum of static pressure and velocity pressure at the point of measurement. E

**tower**
An apparatus for increasing the degree of separation obtained during the distillation of oil in a still. Towers may be divided into two general classes; those which secure separation by fractionation, and those which take advantage of partial condensation only. Towers of the first class are used when accurate work is necessary, as in the production of naphthas and gasoline to meet rigid distillation specifications. Towers operating by partial condensation are used to divide roughly the vapors from a still into several liquid portions. Commonly used in the former sense to mean bubble tower. M

**town's gas**
Gas manufactured from coal for use in cities for illumination and heating; usually a mixture of coal gas and carbureted water gas. D

**tracer gas**
A gas introduced in small quantities into the main body of the air to determine either the air current or the leakage paths in a ventilation system. D

**tracer, isotope**
(1) An isotope of an element, a small amount of which may be incorporated into a sample of material (the carrier) in order to follow (trace) the course of that element through a chemical, biological or physical process, and thus also follow the larger sample. The tracer may be radioactive, in which case observations are made by measuring the radioactivity. If the tracer is stable, mass spectrometers or neutrons activiation analysis may be employed to determine isotopic composition. (2) Tracers also are called labels or tags, and materials are said to be labeled or tagged when radioactive tracers are incorporated in them. W

**trade wind**
Wind blowing in one direction so reliably as to be used by yesterday's sailing ships in crossing the Atlantic.

**transducer**
(1) A device actuated by one transmission system and supplying related waves to another transmission system. (2) The input and output energies may be of different forms. Ultrasonic transducers accept electrical waves and deliver ultrasonic waves, the reverse also being true. D

**transducer loss**
The energy loss of a transducer connecting any energy source and energy load, measured by the ratio of the source power to the load power. D

**transfer admittance**
In a network made up of an energy source and energy load connected by a transducer, it is the quotient obtained by dividing the phasor representing the source current by the phasor representing the load voltage. D

**transformer**
An electromagnetic device for changing the voltage of alternating current electricity. L

**transformer oil**
A special type of oil of high dielectric strength, forming the cooling medium of electric power transformers. D

**transit**
A surveying instrument with the telescope mounted so that it can measure horizontal and vertical angles. Also called a transit theodolite. D

**translucent**
(1) Admitting and diffusing light so that objects beyond cannot be clearly distinguished. (2) Partly transparent. D

**transmission**
(1) The act or process of transporting electric energy in bulk from a source or sources of supply to other principal parts of the system or to other utility systems. (2) A functional classification relating to that portion of utility plant used for the purpose of transmitting electric energy in bulk to other principal parts of the system or to other utility systems, or to expenses relating to the operation and maintenance of transmission plant. K

**transmission level**
The transmission level of the energy at any point in an energy transmission system is the rate of flow of that energy as expressed in terms of (1) a specified reference rate of flow and (2) the transmission loss by which the actual rate of flow must be reduced to equal the reference rate. D

**transmissivity**
A coefficient relating the volumetric flow through a unit width of groundwater to the driving force (hydraulic potential), a function of both the porous medium, fluid properties, and saturated thickness of the aquifer. D

**transmutation**
A name given to the transformation of one element into another, taking place spontaneously with some, e.g., the radioactive elements, or induced by nuclear particles bombardment with others. Z

**trap**
Any geological barrier to oil migration such as a fault, dome or anticline (structural traps), or pinch-out of the reservoir (stratigraphic trap) that may localize the accumulation of an oil pool.

**treating**
The contacting of petroleum products with chemicals to improve the quality. M

**triangulation**
In surveying, the network of triangles into which any portion of the earth's surface is divided in a trigonometrical survey. D

**tributary**
Applied to any stream which directly or indirectly contributes water to another stream. D

**trillion**
$10^{12}$, often represented by T, which also means tera. Y

**tritium**
A radioactive isotope of hydrogen with a half life of 12.5 years. The nucleus contains one proton and two neutrons. It may be used as a fuel in the early fusion reactors. V

**tropic tide**
Tides occurring semimonthly when the effect of the moon's maximum declination is greatest. E

**troposhere**
The layer of the atmosphere extending seven to ten miles above the earth. Vital to life on earth, it contains clouds and moisture that reach earth as rain or snow. E

**trough collectors**
A type of solar collector which usually uses semicylindrical mirrored troughs to reflect solar heat on a longitudinal pipe and heat a circulating fluid. In a recent design the trough-type parabolic collector reflects solar energy onto steel absorber tubes plates with black chrome and placed at the parabolic focal lines of reflectors. The absorber tube is contained within an insulated housing, with an etched, tempered low-iron glass window. The reflector is an aluminum honeycomb sandwich with an adhesive-backed aluminized acrylic film. Z

**troy weight**
These are the weights used for precious metals. The equivalents are: 24

grains = 1 pennyweight; 20 pennyweights = 1 ounce; 12 ounces = 1 pound. The troy grain is the same as the avoirdupois grain, but the ounce is larger on the troy scale; 1 ounce troy = 31.103 grams; 1 ounce avoidupois = 28.35 grams. D

**TRU**
Transuranic wastes (contaminated).

**true bearing**
Azimuth angle of a survey line with respect to the north. D

**true depth**
The actual depth of a specific point in a borehole measured vertically from the surface in which the borehole was collared. Also called true vertical depth. D

**TSCA**
Toxic Substances Control Act.

**TSP**
Total suspended particles.

**TSS**
Total suspended solids.

**tunnel excavation**
Excavation carried out completely underground and is limited in width, and in depth or height. D

**turbidimeter**
A device used to measure the amount of suspended solids in a liquid. E

**turbidity**
A thick, hazy condition of air due to the presence of particulates or other pollutants, or the similar cloudy condition in water due to the suspension of silt or finely divided organic matter. E

**turbine**
A rotary engine actuated by the reaction or impulse, or both, of a current of fluid (as water, steam, gas, or mercury vapor) subject to pressure and usually made with a series of curved vanes on a central spindle arranged to rotate with the whole being enclosed by a casing provided with redirecting vanes and passageways which permit the inlet and outlet of the fluid in a desired manner. D

**turbojet engine**
An engine in which air is compressed by a rotating compressor, is heated by fuel combustion at compressor pressure, released through a gas turbine which drives the compressor, and finally ejected at high velocity through the rearward exhaust nozzle. M

**turboprop engine**
A gas turbine engine that uses most of its power to drive a propeller, instead of exhausting a very high-velocity gas jet to provide thrust. M

**turnaround efficiency**
The resulting efficiency when energy is converted from one form or state to another form or state, and then reconverted to the original form or state. V

**turn-over gas**
Total volume of stored gas available for delivery from a storage reservoir during one annual output cycle. DD

**TVA**
Tennessee Valley Authority.

**TW**
Terawatt.

**two-cycle engine**
An engine in which the piston completes two strokes between power impulses. There is a power impulse every revolution of the crankshaft. M

**two-part rate schedule**
An electric rate schedule of two components: (1) a charge for demand, and (2) a charge for the energy. L

**TWyr**
Terawatt year; an amount of energy equal to 1 trillion watt years.

# U-V

**u**
A symbol for internal energy, intrinsic energy per unit weight. D

**UCL**
University Coal Laboratories.

**UHF**
Ultra high frequency.

**UKAEA**
United Kingdom Atomic Energy Authority.

**UL**
Underwriters Laboratory.

**ullage**
(1) The amount which a tank, container, or vessel lacks being full (see outage). (2) A term generally used in connection with ship's tanks. M

**ultimate recoverable reserves**
The total quantity of crude oil, natural gas, natural gas liquids or sulfur estimated to be ultimately producible from an oil or gas field as determined by an analysis of current and engineering data. This includes any quantities already produced up to the respective date of the estimate. G

**ultimate recovery**
The quantity of oil or gas that a well, pool, field, or property will produce. It is the total obtained or to be obtained from the beginning to final abandonment. Also called ultimate production. D

**ultrasonics**
The acoustic field involving ultrasonic frequencies. D

**ultraviolet**
(1) Of radiation, beyond the visible spectrum at its violet end. (2) Having a wavelength shorter than those of visible light and longer than those of X-rays. D

**UMTA**
Urban Mass Transportation Administration of the U.S. Department of Transportation.

**UMWA**
United Mine Workers of America.

**unbranded gasoline**
Brands of smaller companies. B

**unconsolidated**
Uncemented and/or compacted. D

**underground fires**
There are two types of underground fires: (1) those which involve exposed surfaces and are known as open, freely burning fires; and (2) those which may be wholly or partly concealed, and are invariably caused by spontaneous heating of the coal itself, known as gob fires. D

**undulation**
A large, wavelike fold in the earth's crust. D

**unit coal**
(1) Applied to prepared coal as for analysis. (2) The pure coal substance considered altogether apart from extraneous or adventitious material (moisture and mineral impurities) which may, by accident or through natural causes, have become associated with the combustible organic substance of the coal. D

**unstable**
(1) Readily decomposed. (2) Liable to spontaneous combustion or oxidation. D

**universal coal cutter**
A coal cutter with a jib capable of cutting at any height or angle. D

**unproven area**
An area in which it has not been established by drilling operations whether oil and/or gas may be found in commercial quantities. D

**UNSCEAR**
United Nations Scientific Committee on the Effects of Atomic Radiation.

**unscreened coal**
Coal for which no size limits are specified. D

**unstable isotope**
A radioisotope. W

**upgrade**
To increase the commercial value of a coal by appropriate treatment. D

**upwind**
On the same side as the direction from which the wind is blowing. U

**uranium [ Symbol U ]**
A radioactive element with the atomic number 92 and, as found in natural ores, an average atomic weight of approximately 238. The two principle natural isotopes are uranium-235 (0.7% of natural uranium), which is fissionable, and uranium-238 (99.3% of natural uranium) which is fertile. Uranium is the basic raw material of nuclear energy. O

**USASI**
United States of America Standards Institute.

**USBM**
United States Bureau of Mines.

**USDA**
United States Department of Agriculture.

**USDC**
United States Department of Commerce.

**USDI**
United States Department of Interior.

**USDOE**
United States Department of Energy.

**USGS**
United States Geological Survey.

**USMB**
United States Metric Board.

**v**
Volume, velocity, volt, or voltage. D

**V**
Volume, velocity, linear velocity, average velocity, or volt(s). D

**va**
Volt-ampere; variance. D

**vacuum**
(1) A space entirely devoid of matter (called specifically "absolute vacuum," (2) A space, as the interior of a closed vessel, exhausted to some degree by an air pump or other artificial means. Any vacuum less than absolute is a partial vacuum. M

**vacuum distillation**
Distillation under reduced pressure. The boiling temperature is thereby reduced sufficiently to prevent decomposition or cracking of the material being distilled. M

**vacuum filtration**
The separation of solids from liquids by passing the mixture through a filter and where, on one side, a partial vacuum is created to increase the rate of filtration. D

**vadose**
In geology, pertaining to the circulation of liquids in the earth's crust below the surface and above the water table. E

**vake**
Soft, compact, mixed claylike material with a flat, even fracture found most often in volcanic terrains. D

**valency**
The number of hydrogen atoms or their equivalent, with which one atom of a given element will combine. D

**valley**
Any hollow or low-lying tract of ground between hills or mountains, usually traversed by streams or rivers, which receive the natural drainage from the surrounding high ground. Deep, narrow valleys are more appropriately termed glens, ravines, gorges, or canyons, according to their size and the steepness of the valley walls. Usually valleys are developed by the stream erosion but in special cases, faulting may also have contributed, as in rift valleys. D

**value**
(1) The desirability or worth of a thing as compared with the desirability of something else. (2) Worth, as the value of a mine. D

**valve**
Any contrivance inserted in a pipe or tube containing a lid, cover, ball or slide that can be opened or closed to control the flow or supply of liquids, gases, or other shifting material through a passage. D

**Van Allen radiation belts**
Several belts of ionizing radiation extending from a few hundred miles to a few thousand miles above the earth's surface. The radiation consists of protons and electrons which originate mostly in the sun and are trapped by the earth's magnetic field. D

**van der Waals' adsorption**
(1) Physical, as distinct from chemical cohesion. (2) Normal adhesive forces between molecules, characterized by relatively low heats of adsorption. D

**vapor**
The gaseous phase of substances that normally are either liquids or solids at atmospheric temperature and pressure, for example, steam and phenolic compounds. E

**vapor density**
The weight per unit volume of gas, e.g., grams per liter or pounds per cubic foot. M

**vapor-dominated convective hydrothermal resources**
Vapor-dominated geothermal systems produce superheated steam with minor amounts of other gases but little or no water. Thus, all of the fluid can be piped directly to the turbine. J

**vapor lock**
The displacement of liquid fuel in a feed line by vapors generated from the fuel, resulting in interruption of normal motor operation. It is caused by the vaporization of light ends in gasoline when, at some point in the fuel system, temperature exceeds the boiling points of these volatile fractions. M

**vapor plume**
The stack effluent consisting of flue gas made visible by condensed water droplets or mist. E

**vapor suppression**
A safety system that can be incorporated in the design of structures housing water reactors. In a vapor-suppression system, the space surrounding the reactor is vented into pools of water open to the outside air. If surges of hot vapors are released from the reactor in an accident, their energy is dissipated in the pools of water. Gases not condensed are scrubbed clean of radioactive particles by the bubbling. D

**vaporizing burner**
One in which the fuel oil is vaporized by heating in a retort. BB

**variable operating costs**
Costs associated with operation or utilization of a plant not including "fixed costs." L

**variance**
Sanction granted by a governing body for delay or exception in the application of a given law, ordinance or regulation. E

**variation**
(1) The angle by which the compass needle deviates from the true north. (2) Subject to annual, diurnal and secular changes. Called more properly declination of the needle. D

**varietal mineral**
A characterizing accessory mineral either present in considerable amounts in a rock or distinctive of a rock. D

**variometer**
A geophysical device for measuring or recording variations in terrestrial magnetism; a variable inductance provided with a scale. D

**vector**
An entity represented as a directed magnitude, such as velocity, which is defined as consisting of a speed and a direction. D

**vein**
A zone or belt of mineralized rock lying within boundaries clearly separating it from neighboring rock. It includes all deposits of mineral matter found through a mineralized zone or belt coming from the same source, impressed with the same forms and appearing to have been created by the same process. Sometimes used for a bed; for example, a coal seam or a bed of slate. D

**vein bitumen**
Synonym for asphaltite. D

**vent**
An opening or hole for the escape or passage of something, as of a gas or liquid, or for the relief of pressure within something, as a boiler. D

**ventilate**
To cause fresh air to circulate through (to replace foul air simultaneously removed), as in a room, mine, etc. D

**Venturi**
A contraction in a pipeline or duct to accelerate the fluid and lower its static pressure. Used for metering and other purposes. D

**Venturi meter**
A trademark for a form of the Venturi tube arranged to measure the flow of a liquid in pipes. Small tubes are attached to the Venturi tube at the throat and at the point where the liquid enters the converging entrance. The difference in pressure heads is shown on some form of manometer and from this difference and a knowledge of the diameters of the tubes, the quantity of flow is determined. D

**verifier**
In gas testing, an apparatus by which the amount of gas required to produce a flame of a given size is measured; a gas verifier. D

**vernier**
Small auxiliary scale in sliding contact with a main measuring scale on precision measuring instrument. It is calibrated so as to be slightly out of phase with the main scale. This gives a magnified reading of one main division and facilitates reading with accuracy proportional to the recurrence of the markings which are in phase. D

**vertical**
A term used to define a direction which is perpendicular to a horizontal, or level, plane. D

**vertical-axis rotors**
Windmills in which the axis of rotation is perpendicular to both the surface of the earth and the windstream. U

**VHF**
Very high frequency.

**vibrating conveyor**
A trough or tube flexibly supported and vibrated at relatively high frequency and small amplitude to convey bulk material or objects. D

**vibrating screen**
A screen which is vibrated to separate and move pieces resting on it. D

**virgin coal**
An area of coal which is in place (in situ), and unimpaired by mining activities. D

**virgin material**
Any basic material for industrial processes which has not previously been used. For example, wood pulp trees, iron ore, silica sand, crude oil, or bauxite. E

**virgin stock**
Oil derived directly from crude oil which contains no cracked material. Also called straight-run stock. M

**viscosimeter**
Apparatus used for determining the viscosity of a substance. M

**viscosity**
The measure of the internal friction or the resistivity to flow of a liquid. In measuring viscosities of petroleum products, the values of the viscosity are usually expressed as the number of seconds required for a certain volume of the oil to pass through a standard orifice under specified conditions. M

**vital capacity**
The term for the greatest volume of air that a man can expel from his lungs after a full inspiration. In other words, it is the greatest volume of air that can be moved in and out of the lungs in a single breath. The average man's vital capacity is between 4 and 5 liters. D

**vitiated air**
Air which has been rendered impure by the breath of men, or by being mixed with the various gases given off in mines. It is frequently called return air. D

**vitrain**
The term was introduced by M.C. Stopes in 1919 to designate the macroscopically recognizable, very bright bands of coals. Very bright bands or lenses, usually a few millimeters (3 to 5) in width; thick bands are rare. Clean to the touch. In many coals the vitrain is permeated with numerous fine cracks at right angles to stratification, and consequently breaks cubically—with conchoidal surfaces. In other coals the vitrain is crossed by only occasional perpendicular cracks. In the macroscopic description of seams only the bands of vitrain having a thickness of several millimeters are

usually noted. Examination with the microscope shows vitrain to consist of microlithotypes very rich in virtinite. After clarin, vitrain is the most widely distributed and common macroscopic constituent of humic coals.

**vitric**
An adjective designating (volcanic) ejecta consisting primarily of glassy material. D

**vitrify**
(1) To change into glass or into a glassy substance by heat and fusion. (2) To make vitreous; especially to produce (as in a ceramic ware) enough glassy phase or close crystallization by high-temperature firing to make nonporous. D

**vitrinite**
A group name comprising collinite and telinite. Differentiation between collinite and telinite depends in part on the method of observation. The distinction is more easily made in thin section or after etching a polished surface. Often there is uncertainty of distinction by reflected light and in such cases it is proper to use the general term vitrinite. D

**vitrinization**
The process of coalification that results in the formation of vitrain.

**vitro-**
A prefix meaning glassy and used before many rock names, as vitrophyre, in order to indicate a glassy texture. D

**VLCC**
Very large crude carrier.

**void**
A general term for pore space or other openings in rock. In addition to pore space, the term includes vesicles, solution cavities, or any openings either primary or secondary. Synonym for pore; interstice. D

**volatile matter**
Those products, exclusive of moisture, given off by a material as gas and vapor, determined by definite prescribed methods which may vary according to the nature of the material. D

**volatile solids**
The fraction or component of a solid that volatilizes at 650°C (for municipal solid waste). One estimate is that 200-300 million tons of volatile solids are produced per year from 1.7 billion tons of solids. (Anderson, 1972) Y

**volatization**
The process of vaporizing, usually by heating, of a solid or liquid. Y

**volt**
A unit of electrical force equal to that amount of electromotive force that will cause a steady current of one ampere to flow through a resistance of one ohm. G

**voltage**
The amount of electromotive force, measured in volts, that exists between two points. G

**voltage of a circuit**
The voltage of a circuit in an electric system in the electric pressure of that circuit measured in volts. It is generally a nominal rating based on the maximum normal effective difference of potential between any two conductors of the circuit. K

**voltaic cell**
A cell consisting of two electrodes and one or more electrolytes, which when connected in a closed circuit, will give out electrical energy. D

**volatility**
(1) The extent to which liquids vaporize. (2) The relative tendency to vaporize. M

**volcanism**
The action of volcanoes or volcanic forces in forming or changing the shape, composition, or characteristics of the lithosphere. E

**vulcanizing**
Process used to modify properties of rubber (strength, elasticity, stretch) by combination with sulfur, or a suitable sulfur-based compound, perhaps aided by heat and chemical accelerators. D

# W-Z

**w**
West; weight. D

**W**
Chemical symbol for tungsten; derived from wolfram, the German word for tungsten. Also abbreviation for west; weight; work; watt; water. Symbol for energy. D

**WAO**
Wet air oxidation.

**wacke**
Originally a German word used for more or less weathered basalt but later employed for various mixed and poorly sorted sediments. D

**wad**
A dark brown or black, impure mixture of manganese and other oxides. It contains 10 to 20 percent water, and is generally soft, soiling the hand. It is generally massive and of low specific gravity. A variety known as absolite carries as much as 32 percent cobalt. Also called black ocher, earthy manganese, bog manganese. D

**wadding**
Paper or cloth placed over an explosive in a hole. D

**wafers**
A name given to the rough slice obtained by sawing directly from a mother crystal or section. The process of manufacturing wafers is variously known as wafering, wafering from the crystal or slab, wafering from the mother crystal. D

**wage adjustment clause**
A clause in a rate schedule that provides for an adjustment of the customer's bill if the wage scale of the utility's employees varies from a specified standard. L

**wagon**
An underground coal car. Any vehicle for carrying coal or debris. D

**wagon drill**
A drilling machine mounted on a light, wheeled carriage. D

**wake**
A region of reduced total head behind a body situated in the flow. The turbulent wake is another fluid mechanism in which energy is lost. D

**waler**
In coal mining, one who is employed to pick slate and stone from coal at the breaker or pit mouth. D

**walk**
To deviate from the intended course, such as a borehole that is following a course deviating from its intended direction. Also called deviating, wander, wandering. D

**walking beam**
The beam used to impart a reciprocating movement to the drilling column in percussive drilling. Also called rocking beam. D

**walkout**
Act of walking out or leaving; specifically, a labor strike. D

**wall**
(1) The face of a longwall working or stall, commonly called a coal wall. (2) A rib of solid coal between two rooms. (3) The sides of an entry. D

**wall coal**
The middle division of three in a seam, the other two being termed top coal and ground coal. D

**wall friction**
The drag created in the flow of a liquid because of contact with the wall surfaces of its conductor, such as the inside surfaces of a pipe or drill rod or the annular space between a drill string and the walls of a borehole. D

**wallplate**
A horizontal timber supported by posts resting on sills and extending lengthwise on each side of the tunnel. On these wallplates the roof supports rest. D

**walls**
Coal roadways in pillar-and-stall mining. The side of an ore body defining where the ore ceases and the country rock begins. Walls may be definite or indefinite. D

**wander**
An unintentional change in the course of a borehole. D

**wandering coal**
A coal seam that exists only over a small area; an irregular seam of coal. D

**wane**
A defect in a timber or plank. D

**want**
(1) An area in which a bed, usually a coal seam, is missing due to the presence of a normal or lag fault. (2) A portion of a coal seam in which the coal has been washed away and its place filled with clay or sand; a nip. D

**warm spring**
A thermal spring having a temperature lower than 98°F. E

**warning lamp**
A safety lamp fitted with a certain delicate apparatus for indicating very small proportions of firedamp in the atmosphere of a mine. D

**warping**
(1) The bending or twisting of a ceramic article in drying or firing. (2) The gentle bending of the earth's crust without forming pronounced folds or dislocations. D

**warrant**
A general term for the clay floors of coal seams, particularly when hard and tough. D

**warrenite**
A general term for gaseous and liquid bitumens consisting of a mixture of paraffins, isoparaffins, etc. D

**wash**
(1) Applied to the defined bed of waterworn gravels, boulders, and sand in alluvial deposits and containing concentrations of the metal or mineral sought. (2) The wet cleaning of coal or ores. (3) A western miner's term for any loose surface deposits of sand, gravel, boulders, etc. D

**washability**
Coal properties determining the amenability of a coal to improvement in quality by cleaning. D

**washbox**
In coal preparation, the jig box in which coal is stratified and separated into fractions, heavier below and lighter above. D

**washed coal**
(1) Coal from which impurities have been removed by treatment in a liquid medium. (2) Coal produced by a wet-cleaning process. D

**washed gases**
Purified coal gas from which the chemicals benzene and naphthalene have been extracted by scrubbing with oil. D

**washed out**
Said of a coal seam when the bed thins out. D

**washery refuse**
The refuse removed at preparation plants from newly mined coal. D

**washout**
(1) A channel cut into or through a coal seam at some time during or after the formation of the seam, generally filled with sandstone—or more rarely with shale—similar to that of the roof. (2) Barren, thin, or jumbled areas in coal seams in which there is no actual disruption and no vertical displacement of the coal and strata. D

**waste**
Any material which is of no further utility to the particular process involved. D

**waste coal**
Coal obtained as a byproduct from mine waste. D

**waste drainage**
The controlled leakage of air through a waste to insure that large concentrations of mine gases do not accumulate in that waste. D

**waste-heat boiler**
One which uses heat of exit gases from furnaces to produce steam or to heat water. D

**waste, radioactive**
Equipment and materials (from nuclear operations) which are radioactive and for which there is no further use. Generally classified as: high-level, low-level, or intermediate. O

**water barrier**
An area of solid mineral left unworked to protect a mine, or part of a mine, against entry of secondary water. D

**water bed**
A soil or rock layer that is laden with water or through which water percolates; sometimes, a swampy surface area. D

**water boiler**
A research reactor whose core consists of a small metal tank filled with uranium fuel in an aqueous solution. Heat is removed by a cooling coil in the core. Not to be confused with boiling water reactor. O

**water drive**
One of three types of "drives" which generates flush production in an oil well. Where water is present beneath the oil, as in the case of many oil reservoirs, it is under tremendous pressure. When a drill opens the reservoir, the resulting release of pressure enables the underlaying water to drive the oil to the well bore, and in some cases, upward to the surface. In the process, water displaces the oil. M

**water engine**
An engine used exclusively for pumping water. D

**water flooding**
The most widely used secondary recovery technique in oil wells. In 5-

spot water flooding, four injection wells are drilled forming a square with a producing well at the center. The injection operation is carefully controlled to maintain an even advance of the water front. M

**water gas**
A poisonous, flammable, gaseous mixture made principally of carbon monoxide and hydrogen with small amounts of methane, carbon dioxide, and nitrogen; made usually by blowing air and then steam over red-hot coke or coal; used especially formerly as a fuel (as in welding) and, after carbureting, as an illuminant, but chiefly as a source of hydrogen and as a synthesis gas. D

**water infusion**
A technique being used abroad to suppress or prevent the formation of dust, in advance of mining a coal seam. D

**watering method**
Dust control in coal mines by watering or by wetting agents. D

**water of condensation**
Water formed by condensation of water vapor appearing from the atmosphere or air in rock interstices and also of water vapor arising from the magmosphere or from the volcanic focuses of the lithosphere. E

**water of imbibition**
The proportionate amount of water that a rock can contain above the line of water level or saturation. D

**waterpower**
The power of water derived from its gravity or its momentum as applied or applicable to the driving of machinery. D

**water privilege**
The right to the use of the water of a certain stream. D

**waterproofing**
The process of rendering surfaces or materials impervious to water. D

**water quality criteria**
The levels of pollutants that affect the suitability of water for a given use. Generally, water use classification includes: public water supply; recreation; propagation of fish and other aquatic life; agricultural use and industrial use. E

**water quality standard**
A plan for water quality management containing four major elements: the use (recreation, drinking water, fish and wildlife propagation, industrial or agricultural) to be made of the water; criteria to protect those uses; implementation plans (for needed industrial-municipal waste treatment improvements) and enforcement plans, and an antidegradation statement to protect existing high quality waters. E

**water rate**
The weight of dry steam consumed by a steam engine for each horsepower per hour. The result is stated in either horsepower or brake horsepower. E

**water right**
The right to use water for mining, agricultural, or other purposes. D

**watershed**
(1) The area contained within a drainage divide above a specified point on a stream. (2) In water-supply engineering, it is termed a watershed. (3) In river control engineering, it is termed a drainage area, drainage basin, or catchment area. D

**water softening**
Removal of excess calcium and magnesium. D

**water soluble oils**
Oils having the property of forming permanent emulsions or almost clear solutions with water. D

**water table**
(1) The upper limit or surface of the groundwater as it follows approximately the profile of the land surface. (2) The upper limit of the portion of the ground wholly saturated with water whether very near the surface or many feet below it. Also called ground water level. D

**water-table levels**
(1) Levels showing the depth of the water table below the surface. (2) The depth at which water is encountered in trial pits or boreholes. D

**water treatment**
The purification of water to ensure that it is potable. Treatment would also include neutralizing acid water or softening water of more than moderate hardness to render it suitable for use in washing or in steam boilers. D

**water velocity**
The rate, measured in feet per minute, at which water progresses through a conductor. D

**waterwall incinerator**
A solid waste combustion chamber containing tubes in its walls; e.g., when trash is burned, steam is generated in the wall piping system. Z

**water wheel**
Any wheel designed to be rotated by the direct impact or reaction force of water. K

**watt**
The rate of energy transfer equivalent to one ampere under an electrical pressure of one volt. One watt equals 1/746 horsepower, or one joule per second. G

**watt-hour**
The total amount of energy used in one hour by a device that uses one watt of power for continuous operation. Electrical energy is commonly sold by the kilowatt hour (1,000 watt-hours). G

**wattmeter**
An instrument for measuring electric power in watts, the unit of electrical energy, volt times amperes, therefore combining the functions of a voltmeter and an ammeter. D

**wave**
A disturbance which moves through or over the surface of the medium, with speed dependent upon the properties of the medium. D

**wave front**
In seismology, the surface of equal time elapse from the point of detonation to the position of the resulting outgoing signal at any given time after the charge has been detonated. D

**wavelength**
The distance between similar points on successive waves. That of visible light varies between 3,900 angstrom units (violet) and 7,600 angstrom units (red). D

**wave period**
The time interval between the appearance of two consecutive wave segments at a given point, usually expressed in seconds. The wave segments considered must be the same, that is, the crests, troughs, etc. D

**wave propagation**
The radiation, as from an antenna of r-f energy into space, or of sound energy into a conducting medium. D

**wave refraction**
The bending of the wave crests due to variations in the water depth, or to currents. D

**wave spectrum**
A concept used to describe the distribution of energy among waves of different period. Wave speed increases with wavelength. D

**wavy vein**
A vein that alternately enlarges or pinches at short intervals. D

**waxes**
Derivatives of petroleum which are extracted from lubricating oil fractions by chilling, filtering, and solvent washing. There are two types: paraffin and microcrystalline. M

**wax shale**
Another name for oil shale. D

**WBT**
Wet-bulb temperature.

**wear**
A process by which material is removed from one or both of two surfaces moving in contact with one another, for example, abrasion. D

**weather**
(1) To undergo or endure the action of the elements. (2) To wear away, disintegrate, discolor, or deteriorate under atmospheric influences. D

**weathered crude petroleum**
The product resulting from crude petroleum through loss, due to natural causes during storage and handling, of an appreciable quantity of the more volatile components. D

**weathering**
The group of processes, such as the chemical action of air and rainwater and of plants and bacteria, and the mechanical action of changes of temperature, whereby rocks on exposure to the weather change in character, decay, and finally crumble into soil. D

**weathering index**
A measure of the weathering or slacking characteristics of coal. In determining the index, the United States Bureau of Mines applies the following test: a 500 to 1,000 gram sample of coal in lumps approximately 1 to 1-1/2 inches in diameter is air-dried at 30° to 35°C, with humidity at 30 to 35 percent for 24 hours. It is then immersed in water for one hour, the water is drained off, and the coal is again air-dried for 24 hours. The amount of disintegration is determined by sieving on an 8-inch wire mesh sieve with 0.236-inch square openings, and weighing the undersize and oversize. The percentage of the undersize, after passing a blank sieving test, is the weathering or slacking index of the coal. D

**web**
(1) The slice or thickness of coal taken by a cutter loader when cutting along the face. The thickness of web varies from a few inches with plough-type machines up to about 6 feet. The term web tends to be restricted to thin or medium slices of coal. (2) In forging, the thin section of metal remaining at the bottom of a cavity or depression or at the location of the top and bottom punches. D

**WECS**
Wind Energy Conversion System. U

**wedge off**
To deviate or change the course of a borehole by using a deflecting wedge. D

**wedge pyrometer**
An instrument for the approximate measurement of high temperatures. It depends on a wedge of colored glass, the position of which is adjusted until the source of heat is no longer visible when viewed through the glass. D

**wedging crib**
In a circular mine shaft, a steel ring made of segments wedged securely to rock walls for use as a foundation for masonry lining. D

**weeper**
A hole in the ceiling of an underground aqueduct to let water from

above drain through, or a hole in a retaining wall to permit the escape of water from behind. Also called weep hole. D

**weight**
(1) Relative heaviness; ponderability regarded as a property of matter. (2) The force with which a body is attracted toward the earth or a celestial body by gravitation and which is a quantity dependent on the place where it is determined. D

**weight dropping**
A seismic technique by which energy can be sent downward into the earth without the necessity of drilling shotholes. This technique involves lifting a weight, then permitting it to fall and strike the ground. The waves from the impact are then recorded. In areas where drilling is difficult or unduly expensive, this technique may be highly advantageous. D

**weightometer**
An appliance for the continuous weighing of coal or other material in transit on a belt conveyor. D

**weir**
(1) An obstruction placed across a stream for the purpose of diverting the water so as to make it flow through a desired channel, which may be a notch or opening in the weir itself. (2) The term usually applies to rectangular notches in which the water touches only the bottom and ends, the opening being a notch without any upper edge. (3) That part of a dam, embankment, canal bank, etc., which contains gates and over which surplus water flows; specifically called waterweir. D

**weir head**
The depth of water is a measuring weir as measured from the bottom of the notch to the surface of the water upstream of the weir. D

**weir table**
A record or memorandum used to estimate the quantity of water that will flow in a given time over a weir of a given width at different heights of the water. D

**weld**
A union made by welding. D

**welding**
The joining of two metal surfaces which have been heated sufficiently to melt and fuse them together. D

**well**
(1) A shaft or hole sunk into the earth to obtain oil, gas, etc. (2) Commonly used as a synonym for borehole or drill hole, especially by individuals associated with the petroleum drilling industry. D

**well abandoning**
When the oil reservoir is depleted, the site is cleaned up and the well abandoned. The hole is plugged with cement to protect all underground strata, prevent any flow or leakage at the surface and protect water zones. Salvageable equipment is removed and pits used in the operations are filled in and the site regraded. On those sites where practical, grass or other types of vegetation is replanted. M

**well core**
A sample of rock penetrated in a well or other borehole obtained by use of a hollow bit that cuts a circular channel around a central column or core. D

**well log**
A record of the formations penetrated by a borehole and their approximate thickness, as determined by an examination of the cuttings or core recovered. D

**well logging**
A widely used geophysical technique which involves probing of the earth

with instruments lowered into boreholes, their readings being recorded at the surface. Among rock properties currently being logged are electrical resistivity, self-potential, gamma-ray generation (both natural and in response to neutron bombardment), density, magnetic susceptibility, and acoustic velocity. D

**well pressure**
The natural pressure of the oil or gas in a well. It is often several hundred pounds per square inch and sufficient to cause the oil to rise to the surface. The well pressure is not related to the depth of the oil deposit below the surface. D

**well rig**
An assemblage of all mechanisms, including power motors, necessary to drilling, casing, and finishing a tube well. D

**well shooting**
(1) The firing of a charge of nitroglycerin, or other high explosive, in the bottom of a well for the purpose of increasing the flow of water, oil or gas. (2) In seismic work, a method or methods of logging wells so that average velocities, continuous velocities, or interval velocities are obtained by lowering geophones into the hole. Shots are usually fired from surface shot holes, but may be fired in the well itself, or perforating-gun detonations may be used. D

**welt**
Large-scale topographic elevation, elongate in shape, with relatively steep sides. Generally parallel to continental coasts, a welt may rise above water level to form islands, island chains, or even mountain ranges. D

**westerly**
Wind blowing from the west, as in "the prevailing westerlies."

**wet-air oxidation (WAO)**
Process in which the organic materials dissolved or suspended in water are converted to water and carbon dioxide, carried out to 400-600°F and 600-1850 psi. Y

**wet analysis**
A method of estimating the effective diameters of particles smaller than 0.06 millimeter by mixing the sample in a measured volume of water and checking its density at intervals with a sensitive hydrometer. D

**wet-bulb temperature**
The temperature taken on the wet-bulb thermometer, the one whose bulb is kept moist while making determinations of humidity. Because of the cooling that results from evaporation, the wet-bulb thermometer registers a lower temperature than the dry-bulb thermometer. M

**wet cleaning**
A coal-cleaning method that involves the use of washers plus the equipment necessary to dewater and heat-dry the coal. This method is generally used when cleaning the coarser sizes of coal. It is a more expensive method than air cleaning and creates the additional problem of water pollution. Coal can, however, be cleaned more thoroughly by this method than by air cleaning. D

**wet gas**
A gas containing a relatively high proportion of hydrocarbons which are recoverable as liquids. M

**wet gas wells**
Gas wells that produce a gas from which gasoline can be obtained but which do not produce crude oil. D

**wet mining**
In wet mining, water is sprayed into the air at all points where dust is liable to be formed, and no attempt is made to prevent the air from picking

up moisture. It therefore soon becomes saturated and remains so throughout the ventilation circuits. D

**wet natural gas**
Wet natural gas in unprocessed natural gas or partially processed natural gas, produced from strata containing condensable hydrocarbons. The term is subject to varying legal definitions as specified by certain state statutes. DD

**wet scrubber**
In a steel plant, a giant cylindrical shower that removes the stubborn particles of raw material (mostly oxide) which remain behind when the heated air that reduces ore, coke, and limestone to molten iron in the blast furnace swirls up the stack. The dust-laden liquid is pumped to a giant settling basin in which the particles drop to the bottom in a thick sludge. E

**wet separation**
A term used in connection with coal washing or other processes using fluid. D

**wet steam**
An energy source obtained when water is heated by surrounding rock to well above its normal boiling point, but remains liquid because of high pressures underground. When reservoirs are tapped, this superheated water in many wells flows to the surface, the pressure drops and a mixture of steam (about 10 to 20 percent) and water is produced. More abundant than dry steam, but power production is more complicated. Steam must be separated from the water before it can drive the turbine.

**wetting agent**
A substance which, when added to a liquid, increases the ability of the liquid to spread over solid surfaces. Such agents include soaps, synthetic detergents, sulfonated oils, and other chemicals. M

**wheelbase**
The distance between the leading and trailing axles of a vehicle. D

**whim**
A large capstan or vertical drum turned by horsepower or steampower for raising coal, or water, etc., from a mine. Also called whimsey, whim gin, horse gin. D

**whin**
Igneous rock. When parallel to the bedding planes, it is called a whinsill; when cutting across the strata a whin dike.

**whirling mygrometer**
In mining, a hygrometer used to obtain wet-bulb temperatures.

**white coal**
Waterpower; a French designation (houille blanche). D

**white damp**
(1) Carbon monoxide, CO. A gas that may be present in the afterdamp of a gas- or coal-dust explosion, or in the gases given off by a mine fire. (2) One of the constituents of the gases produced by blasting. Rarely found in mines under other circumstances, it is an important constituent of illuminating gas, supports combustion, and is very poisonous. D

**white heat**
A common division of the color scale, generally given as about 1,540°C (2,804°F). D

**white-hot**
Heated to full incandescence so as to emit all the rays of the visible spectrum, in such proportion as to appear dazzling white. D

**whole body counter**
A device used to identify and measure the radiation in the body (body burden) of human beings and animals; it uses

heavy shielding to keep out background radiation and ultrasensitive scintillation detectors and electronic equipment. O

**whr**
Watt-hour. D

**wildcat**
(1) A borehole and/or the act of drilling a borehole in an unproved territory where the prospect of finding anything of value is questionable. It is analogous to a prospect in mining. (2) Specifically applied to a mining or oil company organized to develop unproven ground far from the actual point of discovery. Any risky venture in mining. D

**wildcatter**
An individual or corporation devoted to exploration in areas far removed from points where actual minerals or other substances of value are known to occur. Also called cold noser. D

**wild coal**
Brittle slate interstratified with thin coal seams. Also called rashings. D

**wild gas**
Blast-furnace gas that does not burn steadily or properly. D

**wild well**
A well flowing out of control. D

**williwaw**
Sudden, often violent, movement of cold air from a coastal mountain range, down to sea level. Common in Alaska and British Columbia.

**win**
(1) To extract ore or coal. (2) To mine, to develop, to prepare for mining. D

**wind-driven current**
The ocean current which develops as a result of the changes in density distribution caused by the wind drift. D

**wind gage**
An instrument for measuring the force or velocity of wind; an anemometer. E

**wind generators**
Devices to extract energy from the wind to be used to generate electricity directly. The electrical energy could be stored as hydrogen derived by electrolyzing water or as water in an elevated lake (pumped-hydro), or the energy extracted from the wind could be converted to and stored as compressed air and then used to drive air turbines. The turbines in turn would drive electrical generators whenever there was a demand for electrical energy. EE

**winding**
The operation of hoisting coal, ore, men, or materials in a shaft. The conventional system is to employ two cages actuated by a drum type of winding engine with steel ropes attached at either end of the drum, one over and the other under it, so that as a cage ascends the other descends and they arrive at pit top and bottom simultaneously. D

**windmill anemometer**
An anemometer in which a windmill is driven by the air stream, and its rotation transmitted through gearing to dials or other recording mechanism. In the windmill type, the operation of air measurement involves readings of the dials at the beginning and end of a measured period, and a watch or clock is required. Windmill instruments may be fitted with an extension handle providing a form of remote control, and used to measure the air speed in an otherwise inaccessible spot. D

**wind power**
A source of energy power derived from the possible use of sophisticated windmills which could convert the wind's energy into electricity. Such

systems could be used regionally in areas such as some Rocky Mountain states, parts of New England, Alaska, Hawaii and the Great Plains. Disadvantages to wind power used in large systems on land might be the effect on air current patterns, changing humidity and temperatures. Used on a large scale basis, it could create aesthetic blights worse than present transmission lines. B

**wind pressure**
The pressure on a structure due to wind, which increases with wind velocity approximately in accordance with the formula $p=0.003v^2$, where p is the pressure in pounds per square feet of area affected, v is the wind velocity in miles per hour. D

**wind pump**
A pump operated by the force of the wind rotating a multiblade propeller. E

**wind rose**
A diagram which shows the proportion of winds blowing from each of the main points of the compass at a given locality, recorded over a long period. The prevailing wind with its average strength is thereby revealed at a glance. O

**Winkler system**
A fluidized gasification of coal process widely practiced in Germany on lignites of particle size, 0 to 8 millimeters. D

**winning**
The excavation, loading and removal of coal or ore from the ground; winning follows development. D

**winter oil**
A heavy railway-car and engine oil which has a solidifying point of below -20°F. D

**winter peak**
The greatest load on an electric system during any prescribed demand interval in the winter or heating season, usually between December 1 of a calendar year and March 31 of the next calendar year. K

**winze**
A vertical or inclined opening, or excavation, connecting two levels in a mine, differing from a raise only in construction. A winze is sunk underhand and a raise is put up underhand. When the connection is completed, and one is standing at the top, the opening is referred to as a winze, and when at the bottom, as a raise, or rise. D

**WMO**
World Meteorological Organization.

**wood coal**
Charcoal. Also lignite. D

**wood gas**
Gas produced during production of charcoal by heating wood in absence of air, usually used as a fuel at the production site. D

**wor**
Water-oil ratio. Also abbreviated WOR. D

**work**
(1) Work is normally measured in terms of force times distance, which for a closed system is equivalent to pressure (p) times the change in volume (dV). Work is measured in foot-pounds or in ergs. Since heat and work are both forms of energy, they can be equated. Thus 1 calorie of heat is equivalent to $4.187 \times 10^7$ ergs or 4.187 joules. In other units, 1 Btu of heat is equivalent to 778 ft-lb. In mining terminology, the process of mining coal. (2) A place where industrial labor of any kind is carried on. E

**workable**
A coal seam or ore body of such thickness, grade, and depth, as to make it a good prospect for development. In remote and isolated

locations, other factors would influence its workability, such as access, water supply, transport facilities, etc. D

**work measurement**
The determination of the proper time to allow for the effective performance of a specific task. The proper time is determined by taking into account all the factors affecting the execution of the task. Where machines are involved, the task is also dependent on the effectiveness of the machine and this is determined. D

**work study**
Embodies the techniques of analyzing methods used in performing an operation and of measuring the work involved. Work study insures better use of materials, plant, and manpower, that is, higher productivity. D

**WPI**
Wholesale price index.

**WPPSS**
Washington Public Power Supply System.

**wt**
Weight.

**x**
Symbol for unknown quantity. Also, symbol for multiplication. D

**xeno-**
A combining form from the Greek zenos meaning guest, foreigner, strange, foreign and intrusive. D

**x-ray**
A penetrating form of electromagnetic radiation emitted either when the inner orbital electrons of an excited atom return to their normal state (these are characteristic x-rays), or when a metal target is bombarded with high speed electrons (these are bremsstrahlung). X-rays are always non-nuclear in origin. O

**xylain**
A subvariety of provitrain in which the woody origin of the cellular structure is microscopically visible. D

**xylan**
A carbohydrate component in wood, associated with the cellulose, 15-20% in hardwood, and 3-8% in softwood (Kirk-Othmer, v. 22, p. 364 1970). The general formual is $(C_5H_8O_4)_n$. Y

**xylanthrax**
Wood coal; charcoal: distinct from mineral coal. D

**xylinite**
A variety of provitrinite. The micropetrological constituent, or maceral, of xylain. It consists of wood (xylem or lignified tissues) almost jellified in bulk but still showing faint traces of cell walls and resin contents under the microscope. D

**xyloid coal**
Brown coal or lignite mostly derived from wood. D

**xylovitrain**
Those coal constituents derived from lignified plant tissues and from which all structure has disappeared. Also structureless vitrain. D

**y**
Symbol for ordinate, y-coordinate, or vertical coordinate in a plane Cartesian coordinate system. Also abbreviation for yard(s). D

**yard**
The British standard of length, equal to 36 inches, 3 feet, or 0.9144 meter. D

**yd**
Yard. D

**year**
(1) Measure of time. (2) A solar year (successive intervals between transits of first point of Aries) is 365.2422

mean solar days. (3) A civil year is 365.2425. (4) A sidereal year is 365.2564 days. D

**yellow cake**
The material which results from the first processing (milling) of uranium ore. It is sometimes called "artificial carnotite" and is about 53% uranium, a mixture of $UO_2$ and $UO_3$. V

**yellow heat**
A division of the color scale, generally given as about 1,090°C (1,994°F). D

**yield**
(1) The total energy released in a nuclear explosion. It is usually expressed in equivalent tons of TNT (the quantity of TNT required to produce a corresponding amount of energy). Low yield is generally considered to be less than 20 kilotons; low intermediate yield from 20 to 200 kilotons; intermediate yield from 200 kilotons to 1 megaton. There is no standardized term to cover yields from 1 megaton upward. (2) In mining, the tonnage of coal per yard advance of a face, or per acre of coal seam worked. (3) The amount of product obtained from any operation expressed as a percentage of the feed material. O

**Z**
Symbol for unknown quantity. Symbol for atomic number, and gram-equivalent weight. D

**Zeiss konimeter**
A portable dust-sampling instrument. D

**zenith**
Point in celestial sphere directly above observer. D

**zephyr**
Gentle breeze, often used to describe a breeze from the west.

**zero group**
The group of inert gases, having a valence of 0, in the periodic system. D

**zero-point energy**
Energy remaining in a substance at the absolute 0 of temperature. D

**zero-power reactor**
An experimental reactor operated at such low power levels that a coolant is not needed and little radioactivity is produced. O

**zinc**
A lustrous, bluish-white metallic element in group II of the periodic system. D

**Z number**
The atomic number of an element. D

**zoic**
In geology, containing fossils, or yielding evidence of plant or animal life; said of rocks. D

**zone rate schedule**
A rate schedule restricted in its availability to a particular service area. L

**zoogene**
In geology, of, pertaining to, consisting of, resulting from, or indicative of animal life or structure.

# APPENDIX

## LISTING OF FIGURES AND TABLES

**Section 1** **WORLD ENERGY OVERVIEW**



**Section 2** **FOSSIL FUELS**

## Oil Shale



## Coal

## SOURCE MATERIAL FOR FIGURES AND TABLES

| | |
|---|---|
| Source 1 | American Gas Association. |
| Source 2 | Argonne National Laboratory, Biomedical and Environmental Research Program. |
| Source 3 | Atomic Industrial Forum, Inc. |
| Source 4 | Battelle Pacific Northwest Laboratories. |
| Source 5 | Cameron Engineering Inc. |
| Source 6 | Congressional Budget Office, Congress of the United States. |
| Source 7 | ESCOE Report, Coal Technologies Market Analysis, 1980. |
| Source 8 | ESCOE Report, Synthetic Fuels Summary, 1980. |
| Source 9 | Exxon Company, U.S.A. Energy Outlook, 1980-2000, 1981. |
| Source 10 | Carl Hall, Biomass as an Alternative Fuel, Government Institutes, Inc., 1981. |
| Source 11 | Henry R. Linden, Perspectives on U.S. and World Energy Problems, Gas Research Institutes, 1980. |
| Source 12 | National Coal Association. |
| Source 13 | Proceedings of the 5th Energy Technology Conference, Government Institutes, Inc., 1978. |
| Source 14 | Proceedings of the 6th Energy Technology Conference, Government Institutes, Inc., 1979. |
| Source 15 | Proceedings of the 7th Energy Technology Conference, Government Institutes, Inc., 1980. |
| Source 16 | Proceedings of the Synfuels Industry Development Seminar, Government Institutes, Inc., 1980. |

Source 17 — Ocean Energy Programs, Applied Physics Laboratory, The Johns Hopkins University.

Source 18 — U.S. Department of Energy, The Nuclear Industry, WASH-1174 UC-2.

Source 19 — U.S. Department of Energy.

Source 20 — U.S. Department of Energy, Secretary's Annual Report to Congress, DOE/S-0010(80), January 1980.

Source 21 — Passive Solar Heating, U.S. Department of Energy, 1980.

Source 22 — U.S. Department of Housing and Urban Development, Solar Hot Water and Your Home, HUD-PDR 466(2), February 1980.

Source 23 — Bureau of Mines, U.S. Department of Interior.

Source 24 — U.S. Department of Interior.

Source 25 — U.S. Bureau of Mines.

Source 26 — U.S. Department of the Navy, Energy Fact Book, AD-A069 138, May 1979.

Source 27 — World Energy Resources 1985-2020: World Energy Conference.

Source 28 — Kirk-Othmer, Encyclopedia of Chemical Technology, 2nd Edition, Vol. 15 (John Wiley and Sons).

Source 29 — Government Institutes, Inc., 1981.

# WORLD ENERGY OVERVIEW

SECTION 1

**Figure 1: Annual World Energy Production of Individual Energy Sources. (Source 11)**

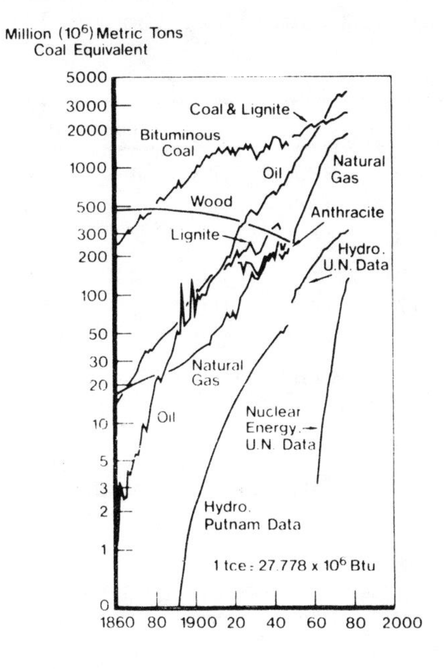


**Figure 2: Total World Primary Energy Demand. (Source 14)**

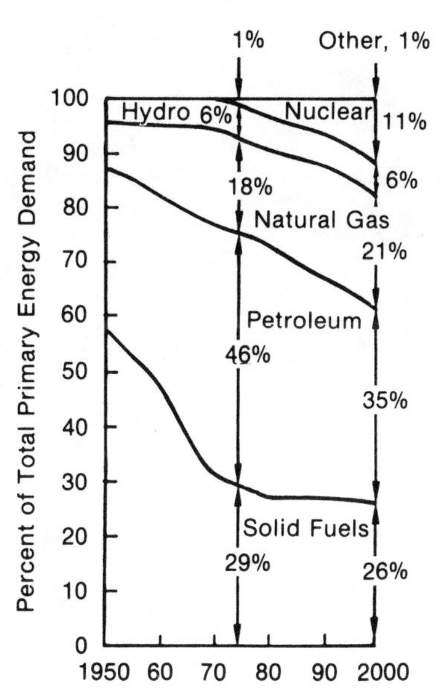

**Table 1: Change in Mix of Commercial Fuel Used by Selected Countries. (Source 11)**

| | OIL, % | | | GAS, % | | | COAL, % | | |
|---|---|---|---|---|---|---|---|---|---|
| | 1950 | 1965 | 1978 | 1950 | 1965 | 1978 | 1950 | 1965 | 1978 |
| FRANCE | 17.2 | 46.3 | 62.8 | 0.3 | 4.4 | 12.0 | 79.9 | 45.3 | 19.8 |
| W. GERMANY | 3.1 | 39.6 | 49.2 | 0.0 | 1.8 | 17.1 | 95.9 | 57.7 | 31.7 |
| JAPAN | 4.3 | 56.6 | 74.3 | 0.2 | 1.5 | 5.2 | 85.3 | 37.0 | 17.4 |
| SWEDEN | 35.9 | 75.2 | 78.0 | 0.0 | 0.0 | 0.0 | 49.5 | 8.9 | 4.4 |
| SWITZERLAND | 27.6 | 72.0 | 76.0 | 0.0 | 0.0 | 4.4 | 49.4 | 10.6 | 1.3 |
| U.K. | 9.1 | 31.5 | 40.3 | 0.0 | 0.4 | 20.7 | 90.8 | 67.2 | 37.2 |
| U.S. | 38.9 | 41.7 | 47.1 | 19.9 | 33.0 | 29.8 | 40.1 | 23.9 | 20.8 |

**Table 2: U.S. Energy Supply. (Source 8)**

U.S. Energy Supply
(Billion barrels oil equivalent)

| | Estimated Resources | Proven Reserves |
|---|---|---|
| Crude Oil | 50 - 370 | 28 |
| Natural Gas | 56 - 210 | 36 |
| Unconventional Gas | | |
| - Coal seams | 70 | — |
| - Tight sands | — | 38 |
| - Devonian shale | — | 2.9 |
| - Geopressured brine | 570 - 17,000 | — |
| Tar Sands | 26 - 37 | 2.5 |
| Heavy Oil | 55 | 1.8 |
| Coal | 4,500 - 17,000 | 1,200 |
| Peat | 29 - 250 | — |
| Shale Oil | 1,026 - 6,000 | 74 |

## Table 3: U.S. Consumption of Energy by Source. (Source 19)

| | Percent of Total | | | | | | Total | Percent |
|---|---|---|---|---|---|---|---|---|
| Year | Coal | Gas | Oil | Hydro | Nuclear | Other* | Quads | Change |
| 1972 | 17.4 | 31.7 | 46.0 | 4.1 | 0.8 | — | 71.6 | |
| 1973 | 17.8 | 30.2 | 46.7 | 4.0 | 1.2 | 0.1 | 74.6 | +4.2 |
| 1974 | 17.7 | 29.9 | 46.0 | 4.5 | 1.7 | 0.2 | 72.8 | -2.5 |
| 1975 | 18.1 | 28.2 | 46.3 | 4.5 | 2.7 | 0.2 | 70.7 | -2.8 |
| 1976 | 18.4 | 27.3 | 47.2 | 4.1 | 2.8 | 0.2 | 74.5 | +5.4 |
| 1977 | 18.4 | 26.0 | 48.6 | 3.3 | 3.5 | 0.2 | 76.5 | +2.7 |
| 1978 | 17.9 | 25.5 | 48.5 | 4.0 | 3.8 | 0.3 | 78.4 | +2.5 |
| 1979 | 19.5 | 25.4 | 47.4 | 4.0 | 3.5 | 0.2 | 78.2 | -0.3 |

**************************************************************************************

## Table 4: U.S. Energy Demand by Primary Energy Source. (Source 6)

| Energy Source | 1979 | 1985 (Projected) | 1990 (Projected) |
|---|---|---|---|
| | | Amount | |
| Oil | 18.7 | 19.5 | 19.9 |
| Natural Gas | 9.8 | 9.5 | 9.7 |
| Coal | 7.6 | 10.0 | 12.2 |
| Nuclear Energy | 1.5 | 2.6 | 3.9 |
| Hydropower | 1.5 | 1.5 | 1.5 |
| Other | 0.2 | 0.3 | 0.6 |
| Total | 39.3 | 43.4 | 47.8 |
| | | Percent Distribution | |
| Oil | 47.6 | 44.9 | 41.6 |
| Natural Gas | 24.9 | 21.9 | 20.3 |
| Coal | 19.3 | 23.0 | 25.5 |
| Nuclear Energy | 3.8 | 6.0 | 8.2 |
| Hydropower | 3.8 | 3.5 | 3.1 |
| Other | 0.5 | 0.7 | 1.3 |
| Total | 100.0 | 100.0 | 100.0 |

**Figure 3:** **U.S. Electric Utility Demand by Fuel Consumed. (Source 9)**

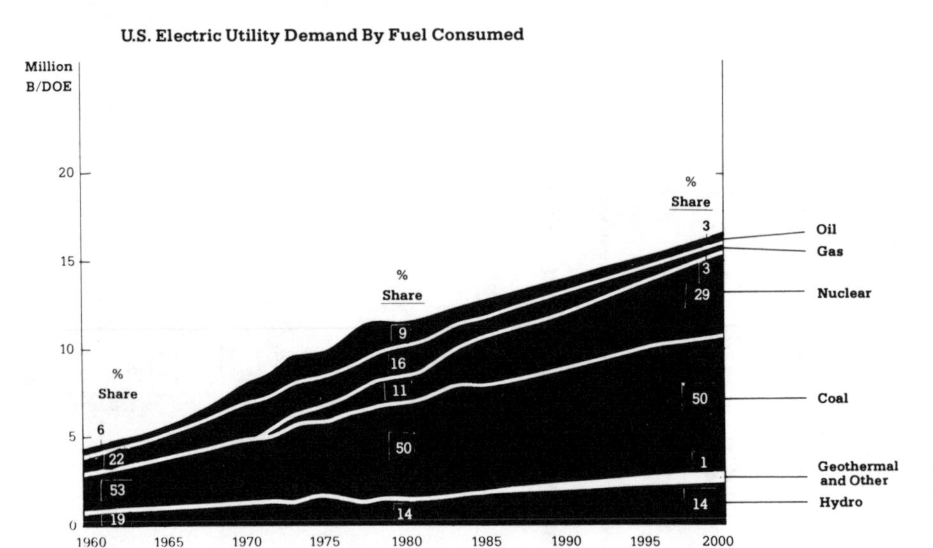


**Figure 4:** **U.S. Electric Utility Demand by Consuming Sector. (Source 9)**

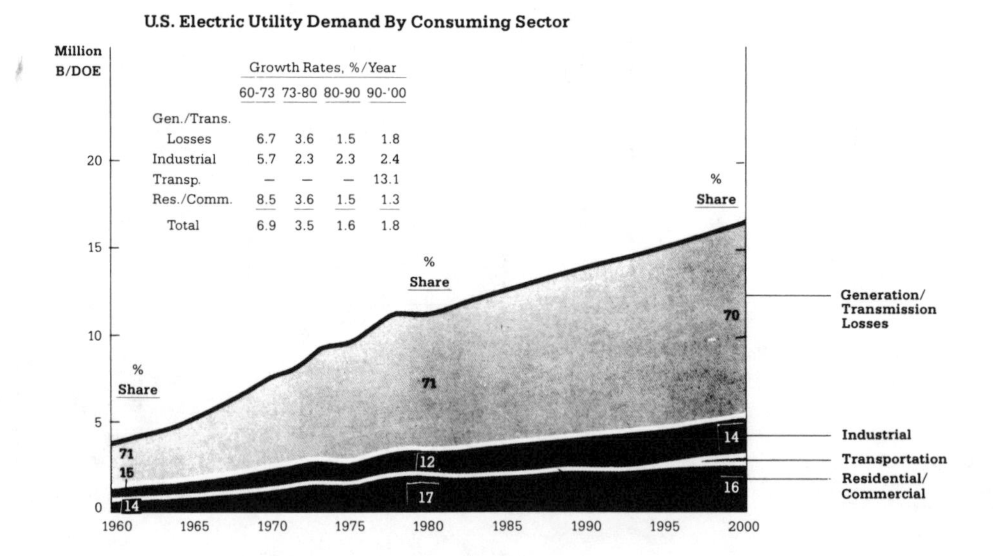

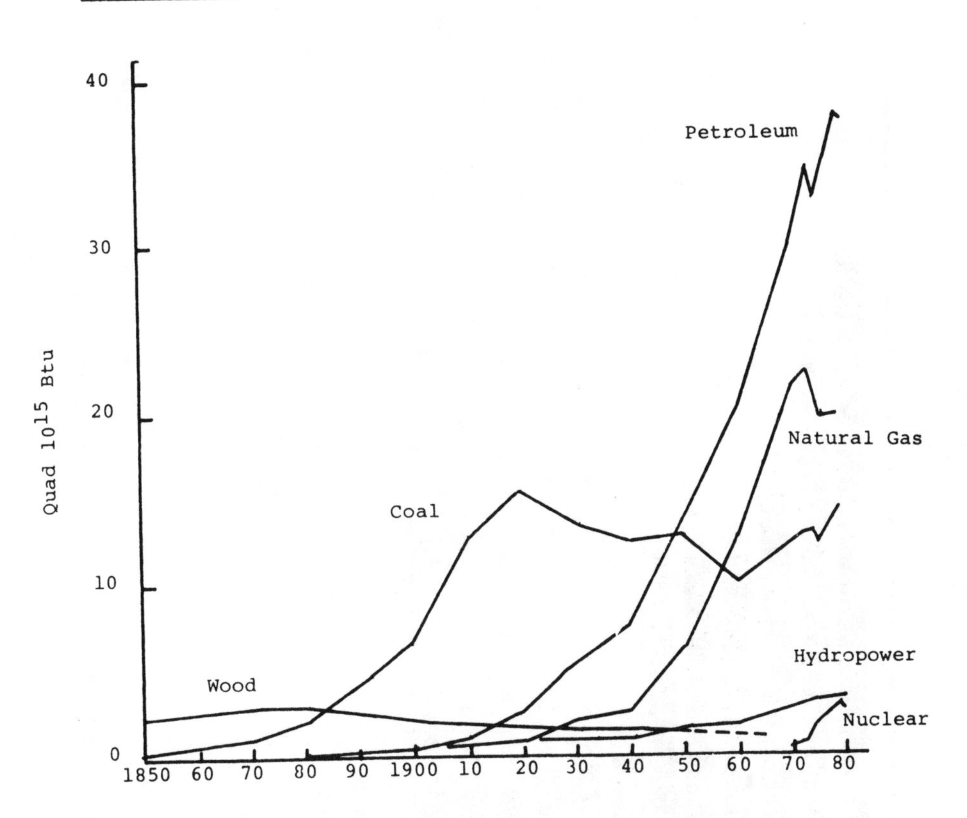


**Figure 5:** U.S. Energy Consumption Patterns (Source 7).

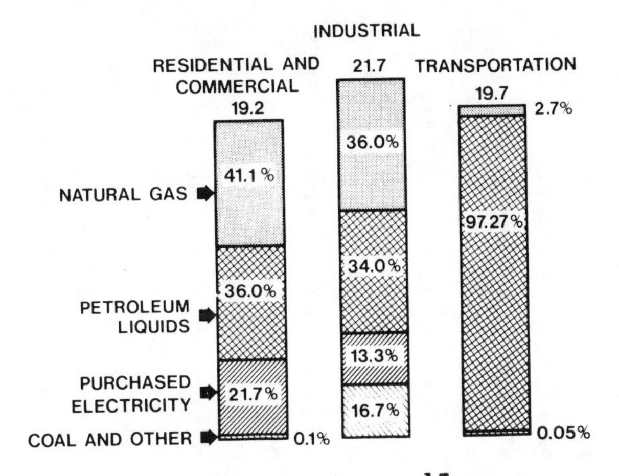


**Figure 6:** U.S. Energy End Use, $10^{15}$ Btu (Source 11).

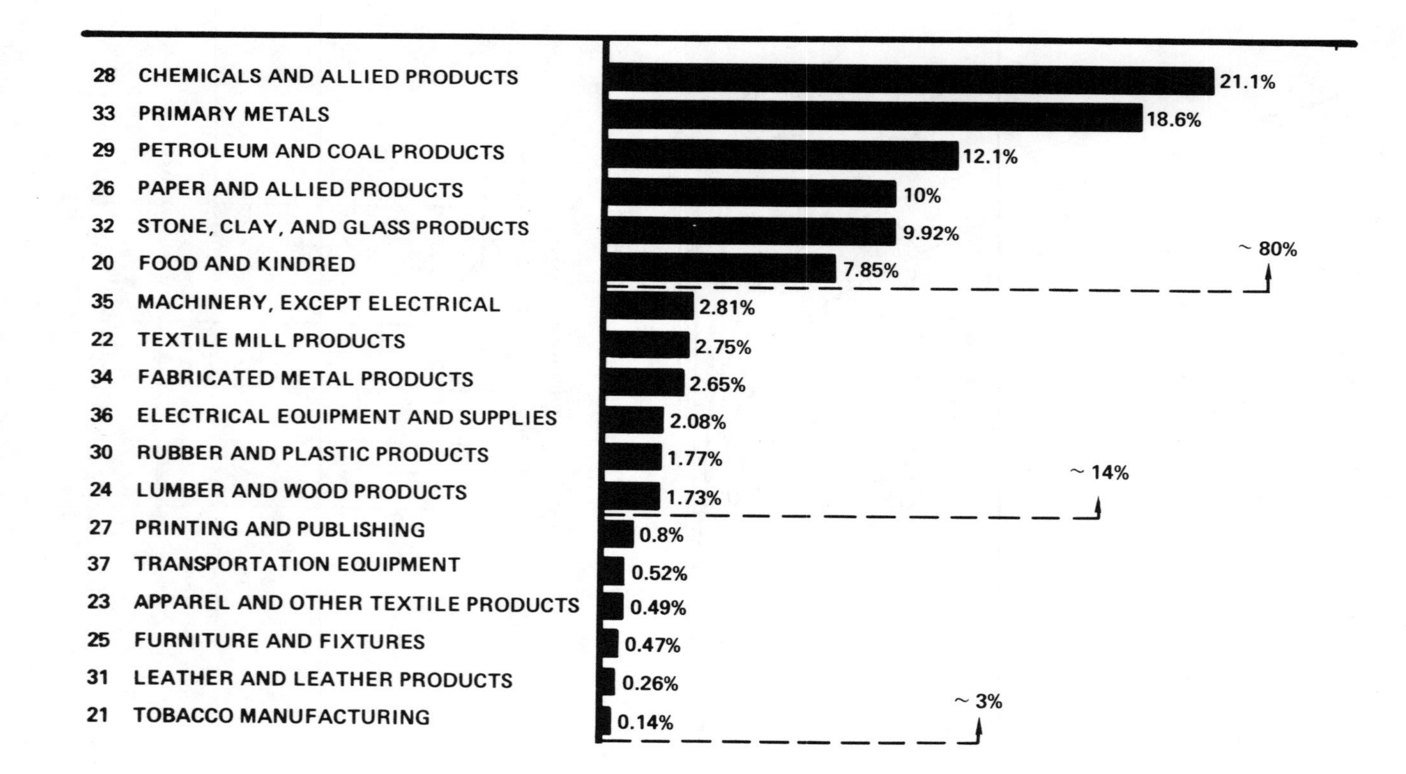


Figure 7: Nationwide Industrial Energy Consumption. (Source 14)

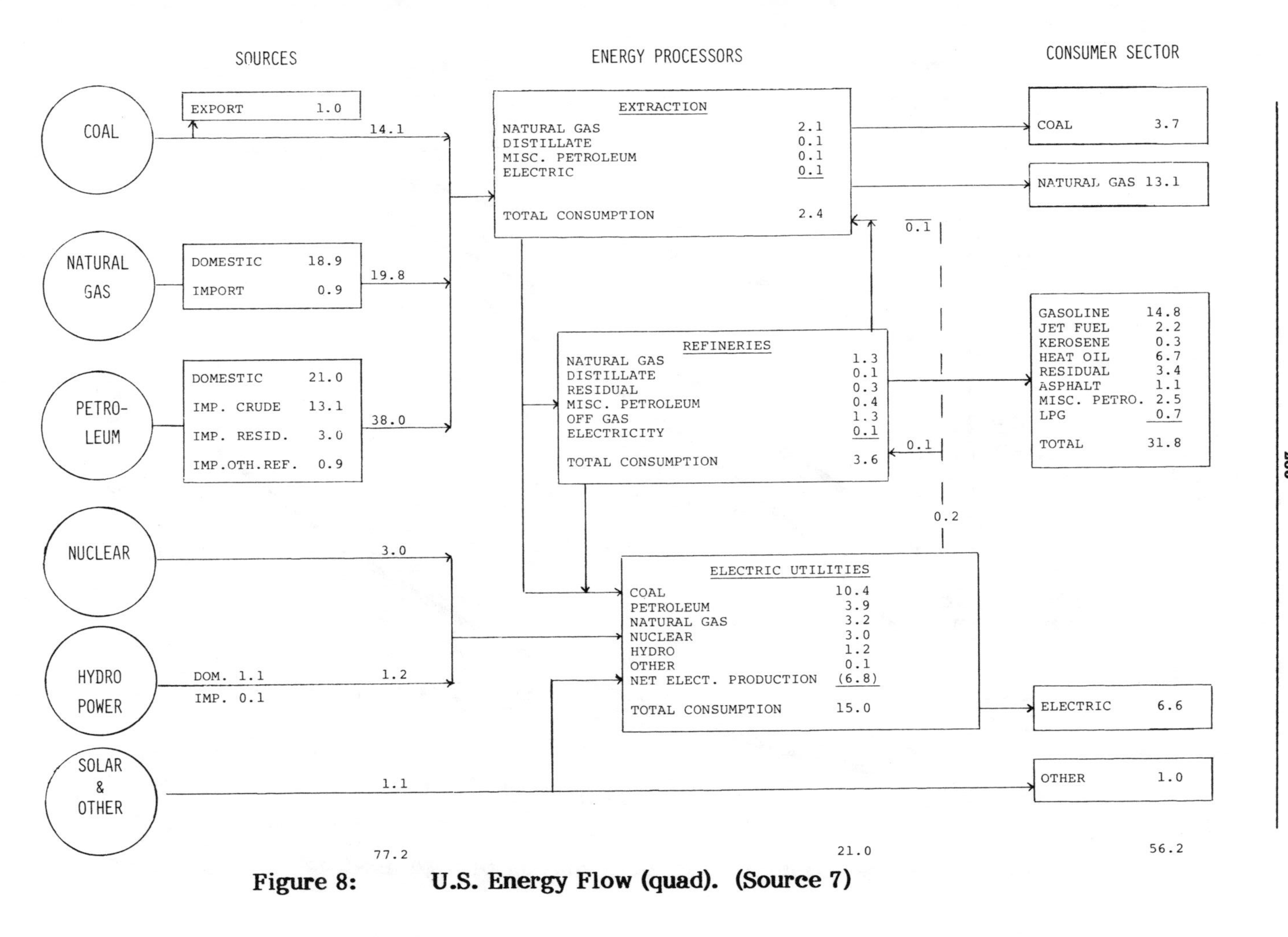


**Figure 8: U.S. Energy Flow (quad). (Source 7)**

**Figure 9: Proposed World Energy Production Timetable. (Source 10)**

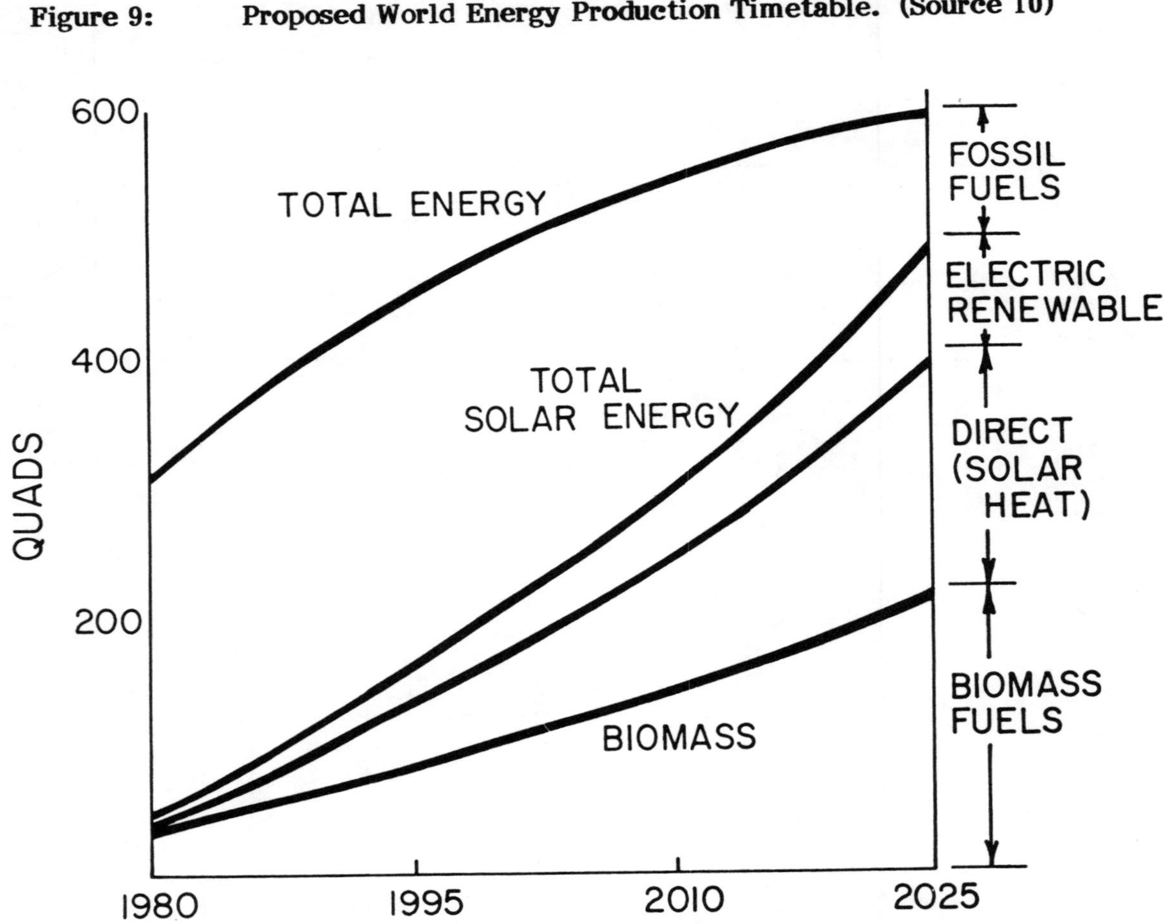

**Figure 10: Proposed U.S. Energy Production Timetable. (Source 10)**

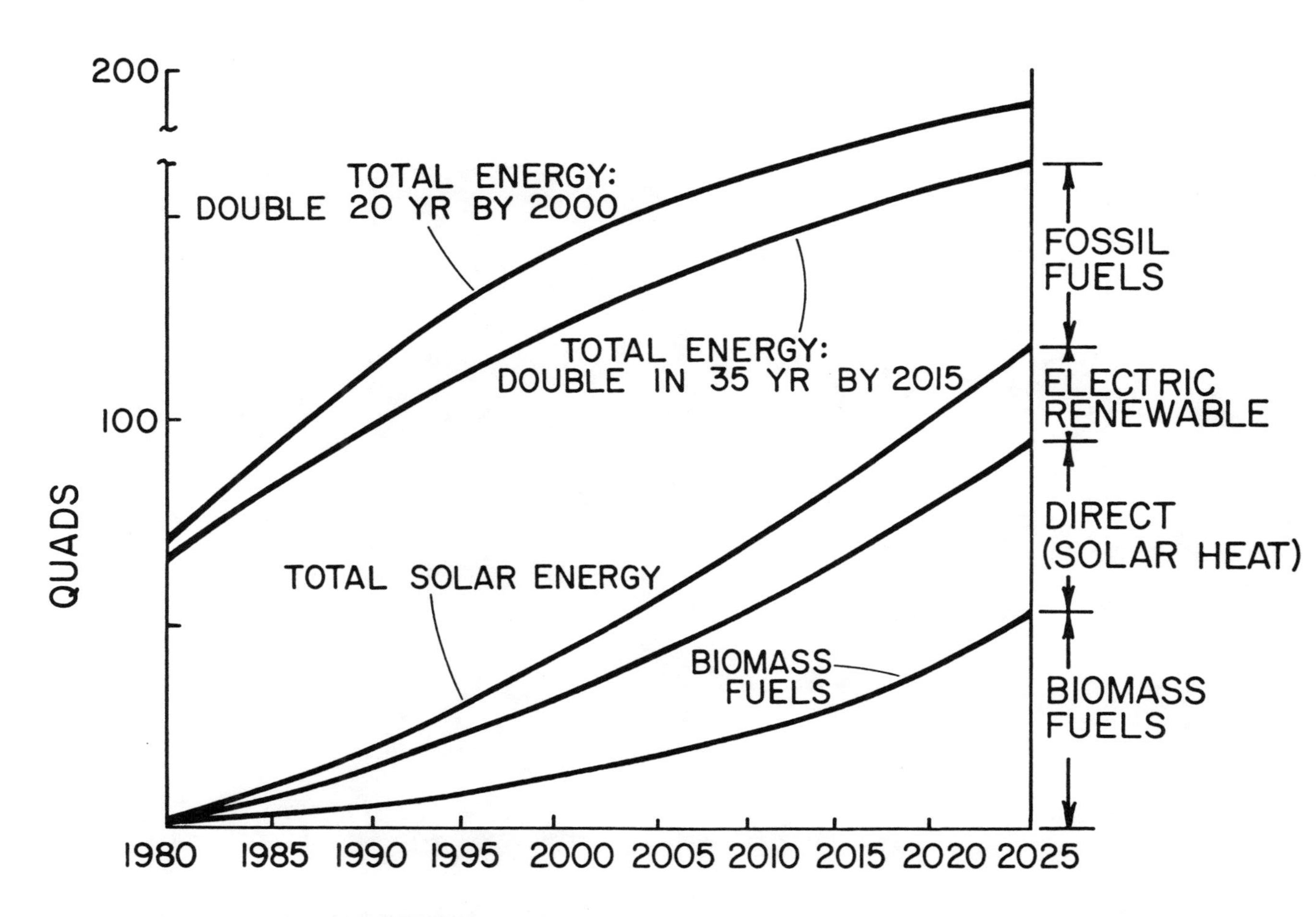

# FOSSIL FUELS

SECTION 2

| | Proved & Currently Recoverable. $10^{18}$ Btu | Estim. Total Remaining Recoverable. $10^{18}$ Btu |
|---|---|---|
| Natural Gas | 2.3 - 2.7 | 9.3 - 9.7 |
| Natural Gas Liquids | 0.3 (Estim.) | 1.0 |
| Crude Oil | 3.1 - 3.8 | 8.3 - 12.0 |
| Syncrude | 1.5 | 13.5 |
| Coal | 17.4 - 18.7 | 126.4 - 141.5 |
| Total | 24.6 - 27.0 | 158.4 - 177.7 |

Figure 11: Total World Fossil Fuel Resources. (Source 11)

| | Proved & Currently Recoverable $10^{18}$ Btu | Estim. Total Remaining Recoverable. $10^{18}$ Btu |
|---|---|---|
| Dry Natural Gas & Natural Gas Liquids | 0.23 | 0.81 - 1.38 |
| Crude Oil | 0.16 | 0.79 - 2.13 |
| Coal | 4.78 | 20.70 - 35.74 |
| Shale Oil & Bitumens | 0.44 | 6.04 |
| Total | 5.61 | 28.3 - 45.3 |

Figure 12: U.S. Fossil Fuel Resources. (Source 11)

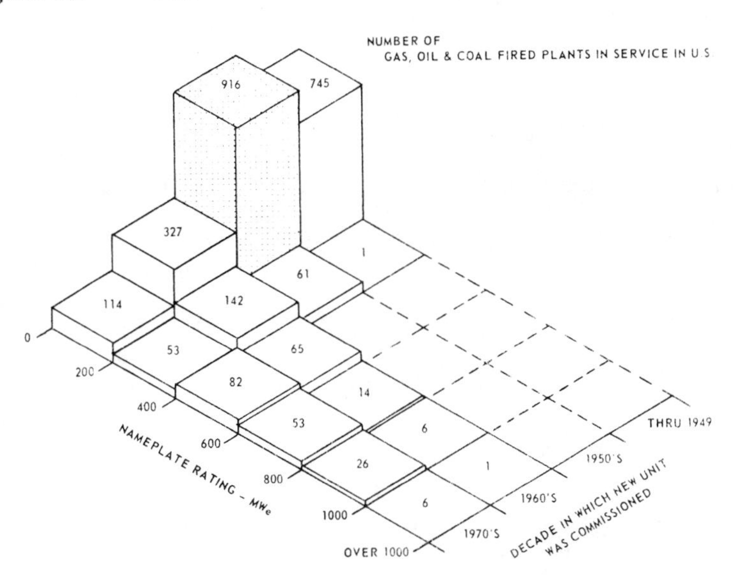


Figure 13: Number of Gas, Oil and Coal-Fired Plants in U.S. (Source 14)

| | Range of carbon values, percent |
|---|---|
| Anthracite | 90-93 |
| Bituminous | 79-90 |
| Sub-bituminous | 50-70 |
| Petroleum | 82-87 |
| Natural gas ($CH_4$) | 75 |
| Lignites | 45-55 |
| Peat | 45-60 |

**Table 5: Carbon Content of Fossil Fuels and Peat Sources (Source 10)**

| Resource | Years |
|---|---|
| Coals | 150 to 210 x $10^6$ ($400 \times 10^6$)[1] |
| Sub-bituminous and bituminous | 100 x $10^6$ |
| Lignite | 20 to 60 x $10^6$ |
| Peat | 1 x $10^6$ |
| Oil, gas | 1 x $10^6$ |

[1] Up to 400 million years may have been involved in the development of some anthracite coals.

**Table 6: Approximate Number of Years to Produce Fossil Fuels In Earth (Source 10)**

# OIL / TAR SANDS

## Table 7: World Crude Oil Supply. (Source 6)

(In millions of barrels per day)

| | 1978 | 1985 (Projected) | 1990 (Projected) |
|---|---|---|---|
| Industrialized Countries | | | |
| United States | 10.3 | 9.4 | 8.5 |
| North Sea (Britain & Norway) | 1.5 | 3:7 | 3.7 |
| Canada | 1.6 | 1.5 | 1.9 |
| Other | 0.4 | 0.5 | 0.5 |
| OPEC Countries | | | |
| Saudi Arabia | 8.3 | 8.5 | 8.5 |
| Other OPEC | 22.0 | 21.6 | 22.8 |
| Non-OPEC LDCs | | | |
| Mexico | 1.3 | 4.5 | 5.5 |
| Other Latin America | 1.2 | 1.4 | 1.2 |
| Africa | 1.0 | 1.8 | 2.3 |
| Asia | 0.9 | 1.8 | 1.9 |
| Non-OPEC Middle East | 0.6 | 0.8 | 0.8 |
| Soviet Bloc and China, Net Exports a/ | 2.0 | -1.0 | -1.8 |
| Total | 51.1 | 54.5 | 55.8 |

a/ Includes natural gas.

## Table 8: Proven Reserves of Crude Oil, January 1, 1980. (Source 6)

| Country | Reserves, Billion Barrels |
|---|---|
| Saudi Arabia | 163.35 |
| U.S.S.R. | 67.00 |
| Mexico | 31.25 |
| United States | 26.50 |

## Figure 14: World Petroleum Production. (Source 11)

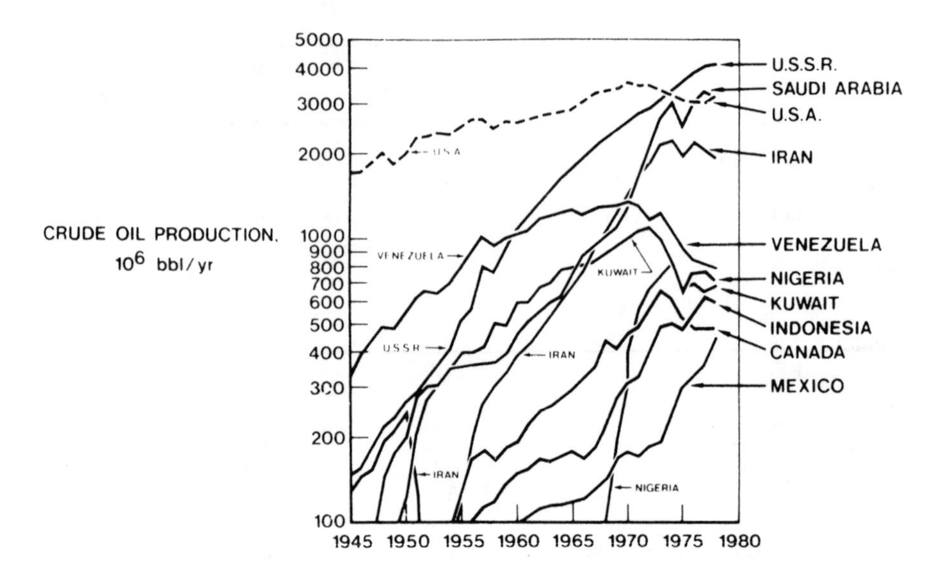


## Table 9: Projected World Oil Balance. (Source 29)

| | 1985 | | | 1990 | | |
|---|---|---|---|---|---|---|
| | Production | Consumption | Imports (-) Exports (+) | Production | Consumption | Imports (-) Exports (+) |
| Industrialized Countries | | | | | | |
| United States | 9.4 | 19.5 | -10.1 | 8.5 | 19.9 | -11.4 |
| Canada | 1.5 | 2.0 | -0.5 | 1.9 | 2.2 | -0.3 |
| Japan | --- | 6.5 | -6.5 | --- | 7.4 | -7.4 |
| Western Europe | 4.2 | 15.9 | -11.7 | 4.2 | 17.3 | -13.1 |
| Subtotal, Industrialized Less Developed Countries | 15.1 | 43.9 | -28.8 | 14.6 | 46.8 | -32.2 |
| Less Developed Countries | | | | | | |
| OPEC | 30.1 | 4.3 | +25.8 | 31.3 | 6.3 | +25.0 |
| Non-OPEC | 10.3 | 10.7 | -0.4 | 11.7 | 13.3 | -1.6 |
| Subtotal, LDCs | 40.4 | 15.0 | +25.4 | 43.0 | 19.6 | +23.4 |
| Subtotal, Non-Communist World | 55.5 | 58.9 | -3.4 | 57.6 | 66.4 | -8.8 |
| Soviet Bloc and China, Net Imports | | | -1.0 | | | -1.8 |
| Total Imports (-) | | | -4.4 | | | -10.6 |

SOURCE: Projected by the Congressional Budget Office, using price assumption of $30 per barrel in the fourth quarter of 1979, rising at an annual rate 2 percent greater than the general price level.

**Table 10: OPEC Crude Oil Supply. (Source 6)**

(In millions of barrels per day)

| Country | 1978 | 1985 (Projected) | 1990 (Projected) |
|---|---|---|---|
| Algeria | 1.2 | 1.0 | 1.0 |
| Ecuador | 0.2 | 0.3 | 0.3 |
| Gabon | 0.2 | 0.3 | 0.2 |
| Indonesia | 1.6 | 1.7 | 1.7 |
| Iran | 5.1 | 3.5 | 4.0 |
| Iraq | 2.6 | 4.0 | 5.0 |
| Kuwait | 2.1 | 1.6 | 1.6 |
| Libya | 2.0 | 2.0 | 2.0 |
| Nigeria | 1.9 | 2.5 | 2.0 |
| Qatar | 0.5 | 0.5 | 0.5 |
| Saudi Arabia | 8.4 | 8.5 | 8.5 |
| United Arab Emirates | 1.8 | 2.0 | 2.0 |
| Venezuela | 2.2 | 2.2 | 2.5 |
| Total | 29.8 | 30.1 | 31.3 |

**Table 11: Crude Oil Supply In Non-OPEC Less Developed Countries. (Source 6)**

1978, 1985, and 1990 (In millions of barrels per day)

| Area | 1978 | 1985 (Projected) | 1990 (Projected) |
|---|---|---|---|
| Africa | | | |
| Angola | 0.1 | 0.3 | 0.3 |
| Egypt | 0.5 | 1.0 | 1.0 |
| Tunisia | 0.1 | 0.1 | 0.2 |
| Other | 0.1 | 0.4 | 0.8 |
| Subtotal | 0.8 | 1.8 | 2.3 |
| Asia | | | |
| Australia | 0.4 | 0.5 | 0.5 |
| Brunei | 0.2 | 0.3 | 0.3 |
| India | 0.2 | 0.4 | 0.4 |
| Malaysia | 0.2 | 0.4 | 0.3 |
| Other | 0.1 | 0.2 | 0.4 |
| Subtotal | 1.1 | 1.8 | 1.9 |
| Latin America | | | |
| Argentina | 0.4 | 0.5 | 0.6 |
| Brazil | 0.2 | 0.3 | 0.2 |
| Mexico | 1.2 | 4.5 | 5.5 |
| Other | 0.6 | 0.6 | 0.4 |
| Subtotal | 2.4 | 5.9 | 6.7 |
| Middle East | | | |
| Oman | 0.3 | 0.4 | 0.4 |
| Syria | 0.2 | 0.3 | 0.3 |
| Other | 0.1 | 0.1 | 0.1 |

## Table 12: World Crude Oil Demand. (Source 6)

(In millions of barrels per day)

| Area | 1978 | 1985 (Projected) | 1990 (Projected) | Average Annual Rate of Change (Percent) 1978-1985 | 1985-1990 |
|---|---|---|---|---|---|
| Industrialized Countries | | | | | |
| United States | 18.3 | 19.5 | 19.9 | 0.9 | 0.4 |
| Western Europe | 14.6 | 15.9 | 17.3 | 1.2 | 2.0 |
| Japan | 5.4 | 6.5 | 7.4 | 2.7 | 2.6 |
| Canada | 1.8 | 2.0 | 2.2 | 1.5 | 1.5 |
| OPEC | 2.5 | 4.3 | 6.3 | 8.0 | 8.0 |
| Non-OPEC LDCs | 8.5 | 10.7 | 13.3 | 3.3 | 4.4 |
| Total | 51.1 | 58.9 | 66.4 | 2.1 | 2.4 |

## Figure 15: U.S. Oil Demand. (Source 9)

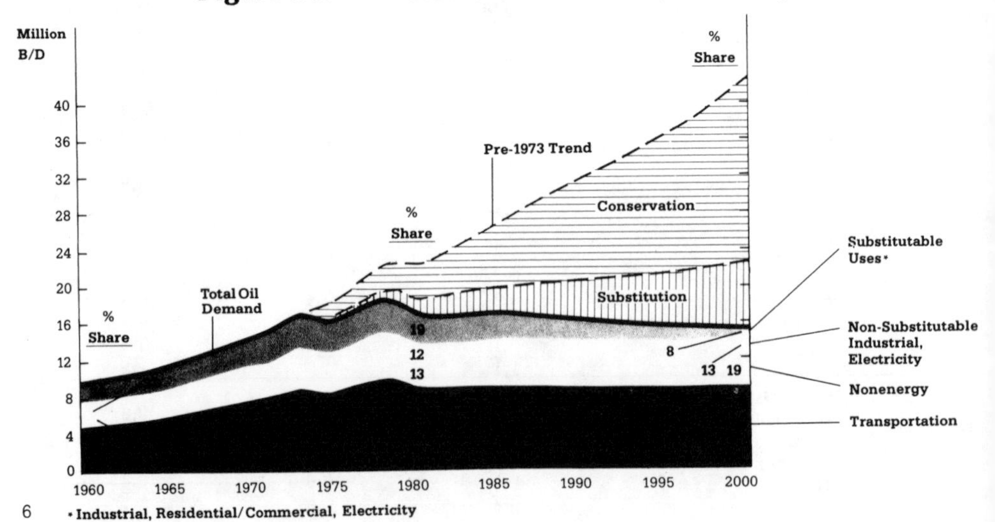


* Industrial, Residential/Commercial, Electricity

Figure 16: U.S. Oil Supply. (Source 9)

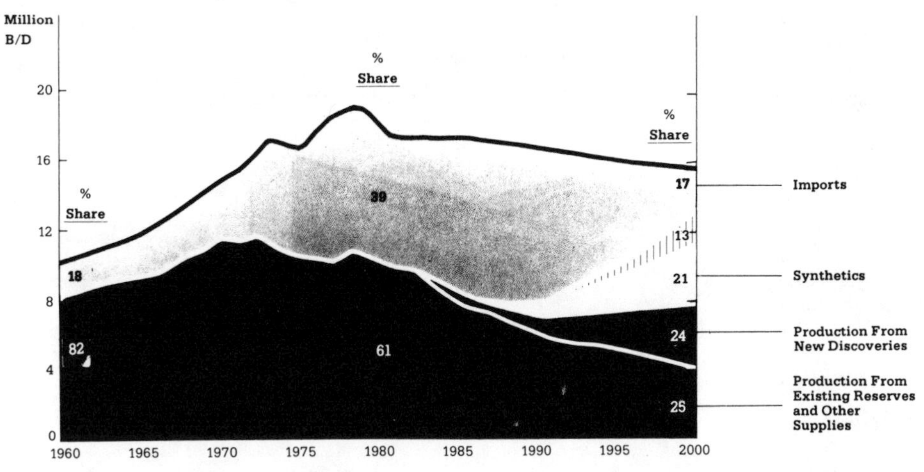


## Table 13: Crude Oil Production from Larger Fields in the U.S. (Source 8)

| State | Production thousands of barrels | Estimated Number of Wells | Average Production per well Barrels/day |
|---|---|---|---|
| Alaska | 505,462 | 377 | 4,060 |
| Alabama | 3,144 | 477 | 20 |
| Arkansas | 3,000 | 2,160 | 4 |
| California | 304,745 | 35,227 | 26 |
| Colorado | 19,306 | 445 | 131 |
| Florida | 35,904 | 100 | 1,088 |
| Illinois | 11,088 | 11,395 | 3 |
| Kansas | 6,042 | 4,880 | 4 |
| Louisiana | 199,032 | 12,823 | 47 |
| Mississippi | 9,979 | 515 | 59 |
| Montana | 9,621 | 1,156 | 25 |
| New Mexico | 36,607 | 3,860 | 29 |
| North Dakota | 3,651 | 369 | 30 |
| Oklahoma | 47,504 | 18,578 | 8 |
| Texas | 686,440 | 60,407 | 34 |
| Utah | 23,299 | 1,018 | 69 |
| Wyoming | 53,235 | 2,949 | 55 |

## Table 14: Gross U.S. Imports of Crude Oil and Petroleum Products. (Source 20)

(millions of barrels per day)

| OPEC | 1979[1] |
|---|---|
| Saudi Arabia | 1.3 |
| Iran | 0.3 |
| Nigeria | 1.0 |
| Libya | 0.7 |
| Venezuela | 0.7 |
| Algeria | 0.6 |
| Indonesia | 0.4 |
| United Arab Emirates | 0.3 |
| Other | 0.2 |
| Total OPEC | 5.5 |
| Non-OPEC | |
| Bahamas | 0.1 |
| Canada | 0.5 |
| Mexico | 0.4 |
| Netherlands Antilles | 0.2 |
| Puerto Rico | 0.1 |
| Trinidad/Tobago | 0.2 |
| Virgin Islands | 0.4 |
| Other | 0.7 |
| Total non-OPEC | 2.7 |
| Total imports | 8.2 |

[1] These figures are gross imports based on data from the first eleven months of the year. Imports from Iran were suspended by the President on November 12, 1979. No imports for the Strategic Petroleum Reserve are included. Imported petroleum products are counted on a par with crude oil. The sources of the crude refined into products overseas and then imported into the United States are not known. To get net oil imports, subtract U.S. petroleum exports of about 450,000 barrels per day.

Note: Totals may not add due to rounding.

## Table 15: Petroleum: Energy Equivalents. (Source 10)

Equivalent: 100% recovered if burned)

1 TJ = 156 bbl of oil (42 gal) = $10^{12}$ J = $10^9$ Btu
1 lb crude oil = 18,000 - 19,500 Btu/lb
1 bbl crude oil = 5,800,000 Btu
1 metric tonne oil = 7.33 bbl = 44.86 GJ = $5.8 \times 10^6$ Btu
1 metric tonne of refined gasoline = 7.7 bbl oil
1 bbl crude oil = 5500 $ft^3$ natural gas
1 bbl crude oil = 158 $m^3$ natural gas
1 quad = $10^{15}$ Btu = 500,000 bbl/day of oil for 1 year
= $172.4 \times 10^6$ bbl oil
1 bbl = 0.16 $m^3$
1 bbl of natural gas liquids = 4,200,000 Btu/lb
1 tonne of average crude oil = 44.86 GJ
1 bbl oil has the energy content = 2 bbl methanol
1 million barrels per day of oil = 2.1 quads per year

**Figure 17: Basic Refinery Operations. (Source 28)**

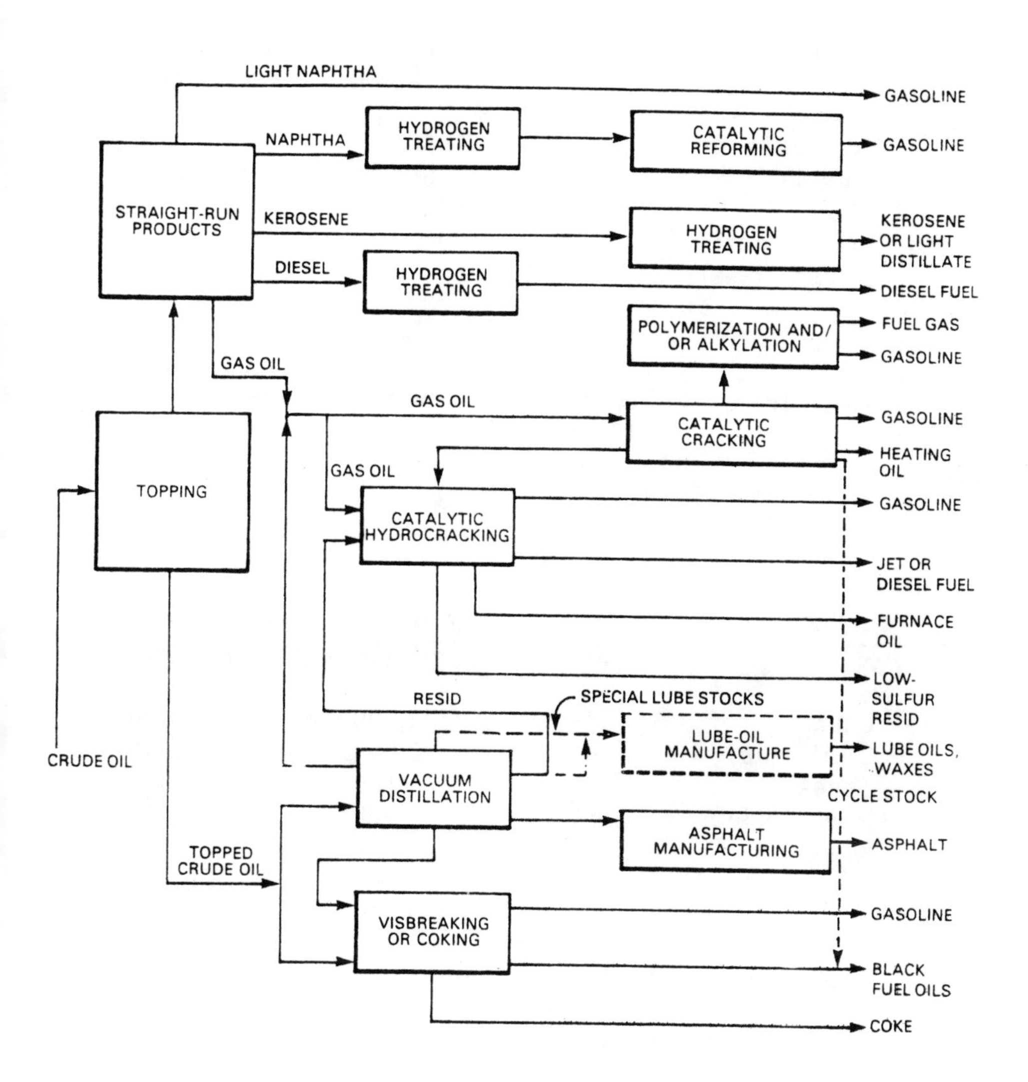

| Country | Estimated Reserves (billion barrels bitumen) |
|---|---|
| Venezuela | 1050 |
| Canada | 970 |
| USSR | 144 |
| United States | 30 |
| Others | 6 |
| | 2200 |

**Table 16: International Tar Sand Resources (Source 29)**

| State | Estimated Reserves (million barrels bitumen) | | |
|---|---|---|---|
| Utah | 23,000 | - | 30,000 |
| California | 1,000 | - | 3,000 |
| Texas | 100 | - | 3,000 |
| New Mexico | 50 | - | 60 |
| Kentucky | 30 | - | 40 |
| | 24,180 | - | 36,100 |

**Table 17: Domestic Tar Sand Resources (Source 29)**

# OIL SHALE

Figure 18: Reported World Recoverable Oil Shale. (Source 27)

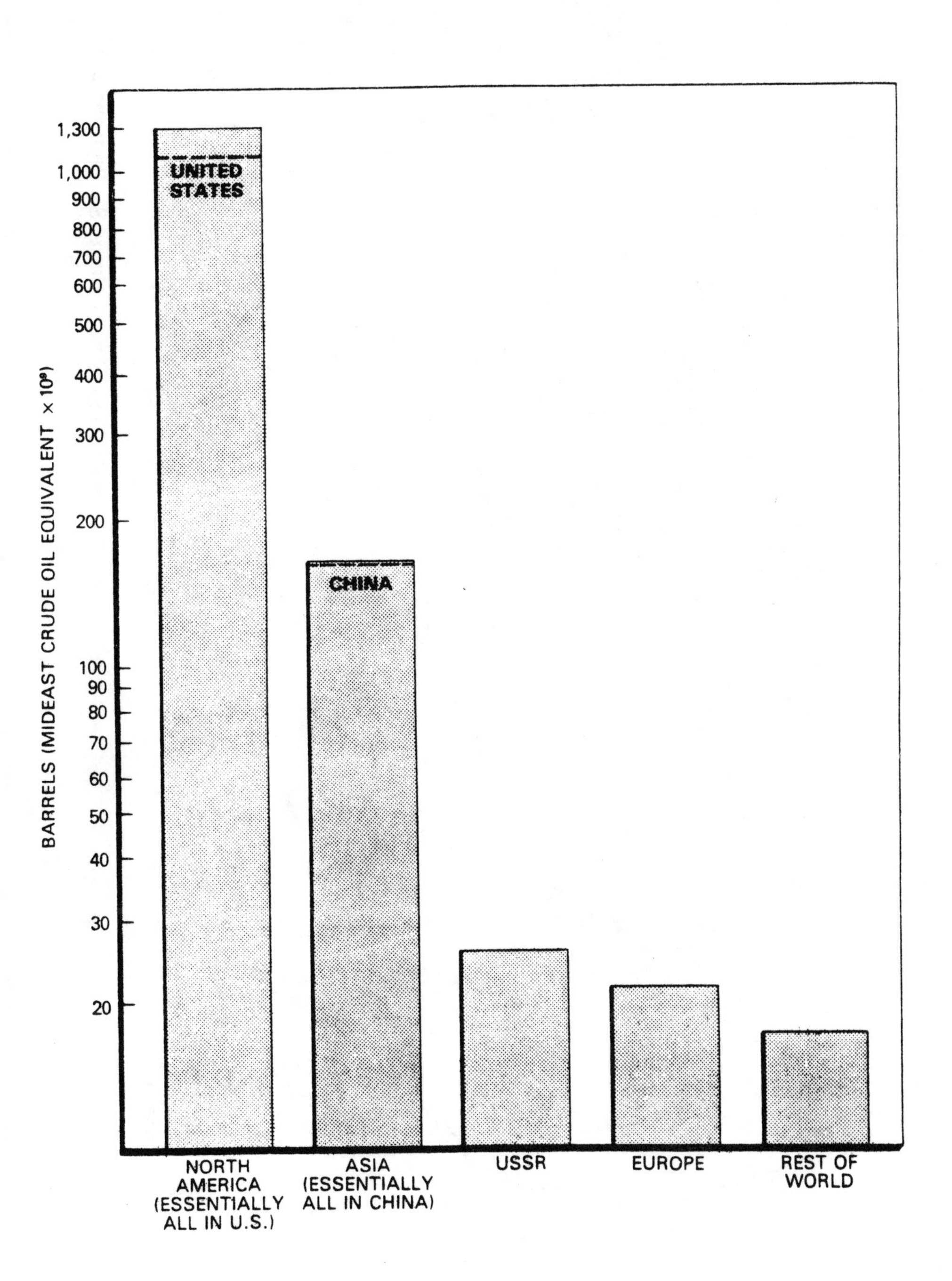

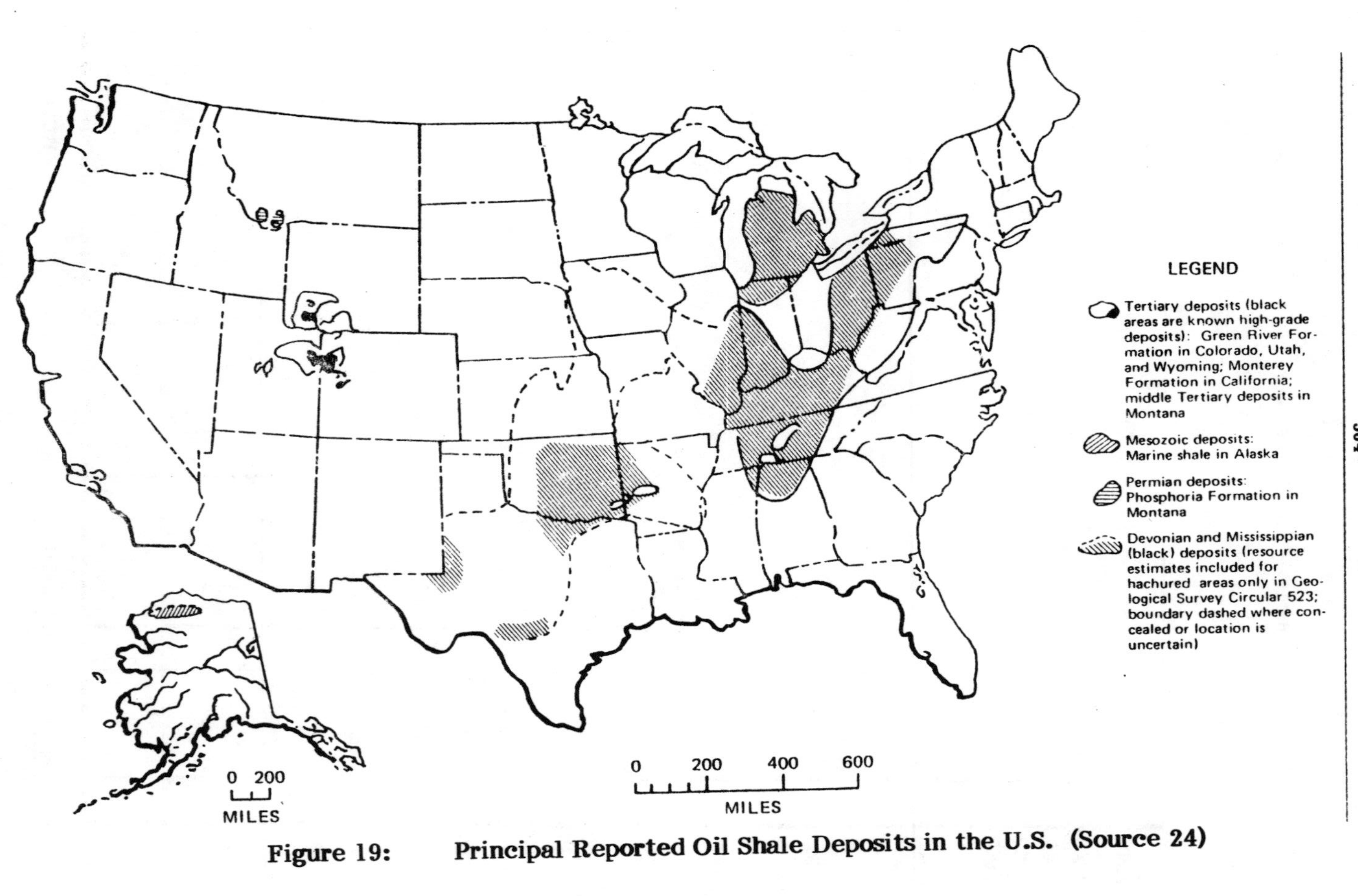


**Figure 19: Principal Reported Oil Shale Deposits in the U.S. (Source 24)**

| Location | Range of Shale Oil Yields, Gallons Per Ton | | |
|---|---|---|---|
| | 5 - 10 | 10 - 25 | Over 25 |
| Colorado, Utah, and Wyoming (the Green River formation) | 4,000 | 2,800 | 1,200 |
| Central and Eastern States (includes Antrim, Chattanooga, Devonian, and other shales | 2,000 | 1,000 | (?) |
| Alaska | Large | 200 | 250 |
| Other deposits | 134,000 | 22,500 | (?) |
| Total | 140,000 | 26,000 | 2,000(?) |

**Figure 20: Potential Shale Oil in Oil Shale Deposits of the U.S. (Source 6)**

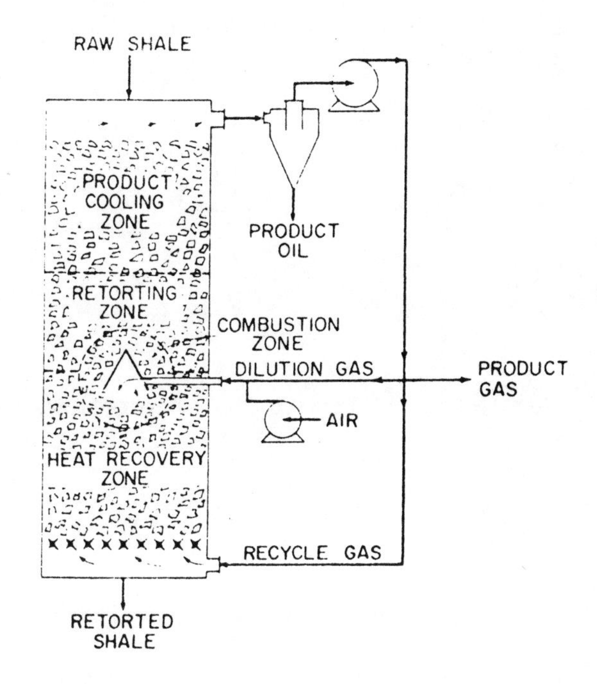


**Figure 21: Gas Combustion Shale Retorting Process. (Source 5)**

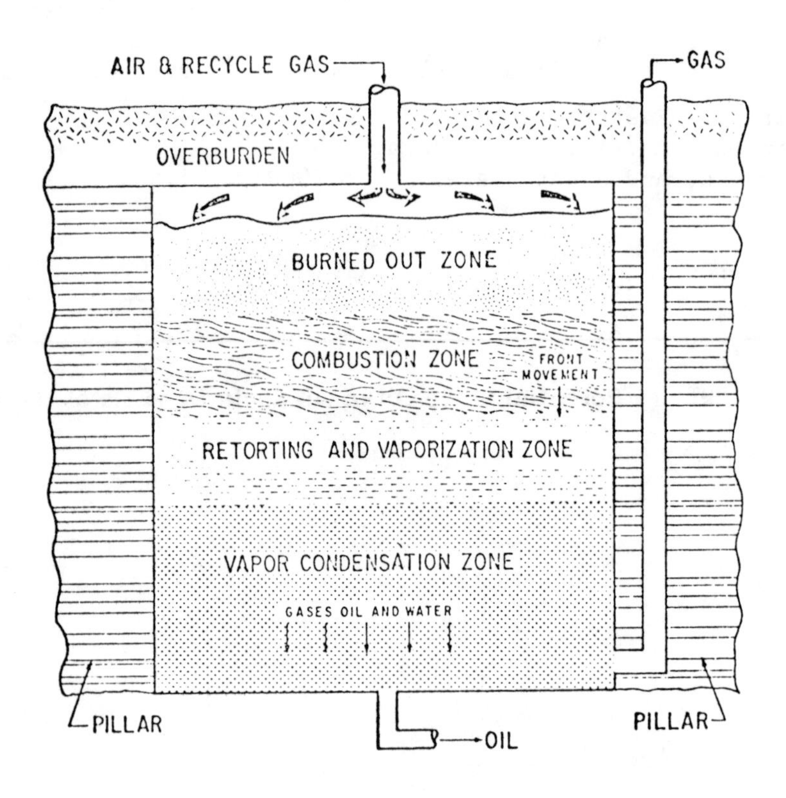


**Figure 22: In Situ Shale Retorting. (Source 8)**

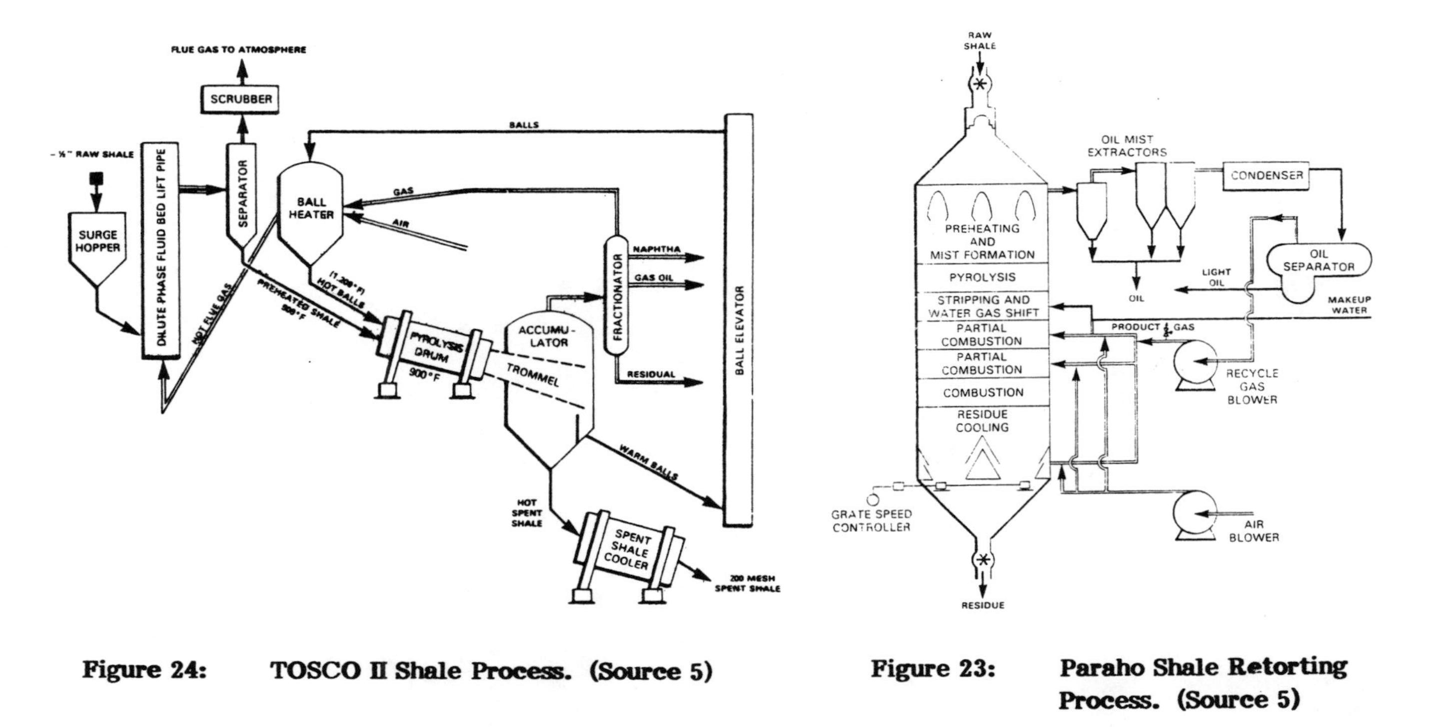


**Figure 24: TOSCO II Shale Process. (Source 5)**

**Figure 23: Paraho Shale Retorting Process. (Source 5)**

# COAL

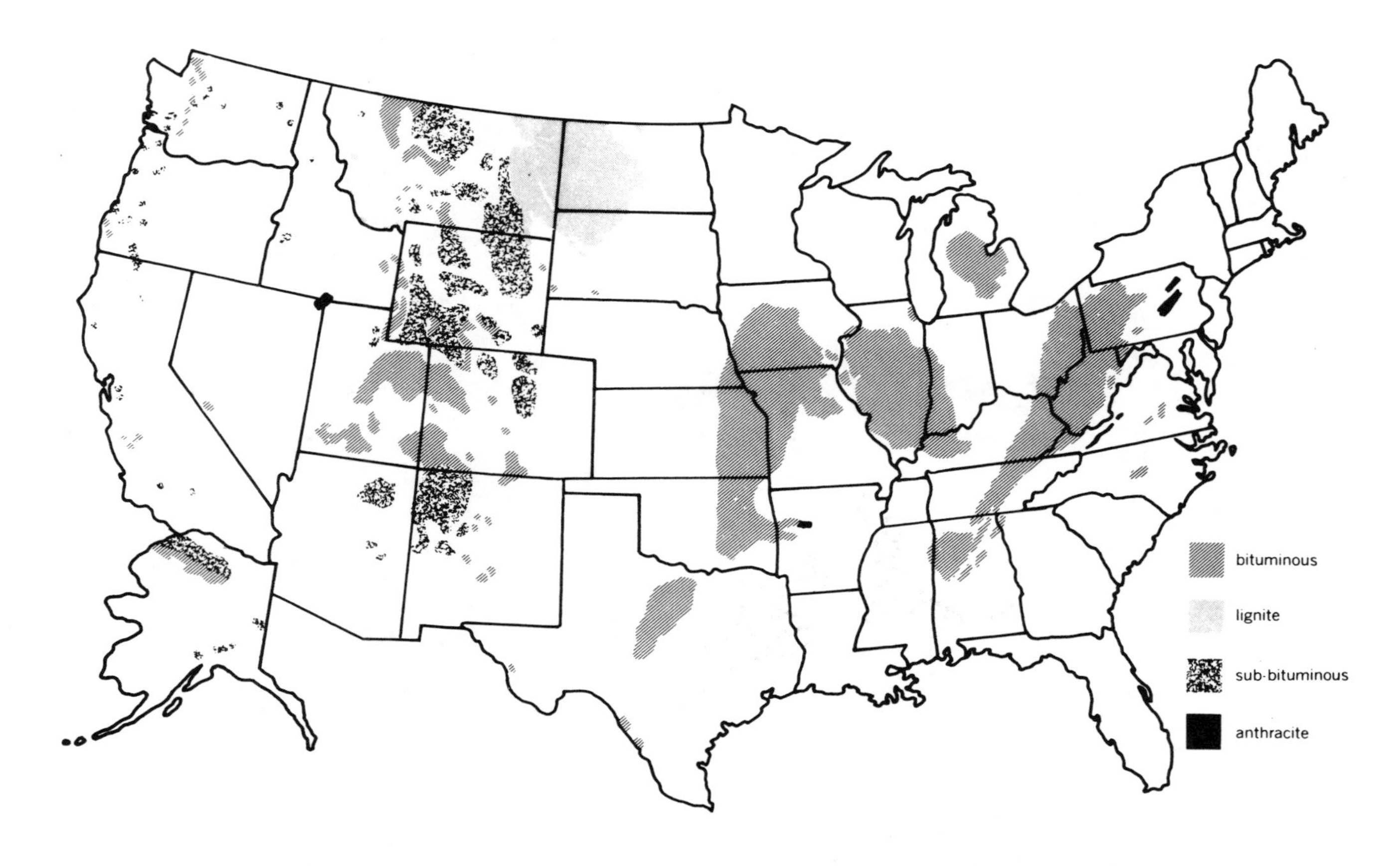


**Figure 25: Coal Areas in the United States. (Source 12)**

1978 WORLD COAL PRODUCTION (million tons)

| Country | Anthracite | Bituminous & Subbituminous | Lignite & Peat | Total |
|---|---|---|---|---|
| U.S.S.R. | 87 | 530 | 186 | 803 |
| China | ---- | (No details available) | ---- | 680 |
| U.S. | 5 | 629 | 36 | 670 |
| Poland | ---- | 212 | 45 | 257 |
| W. Germany | 6 | 87 | 136 | 229 |
| Remainder | 78 | 604 | 604 | 1,286 |
| Total | | | | 3,925 |

**Table 18: World Coal Production. (Source 19)**

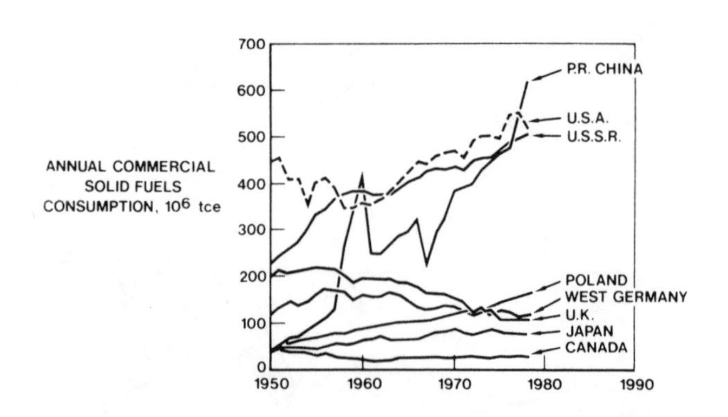


**Figure 26: World Coal Consumption. (Source 11)**

## Figure 27: U.S. Coal Supply. (Source 9)

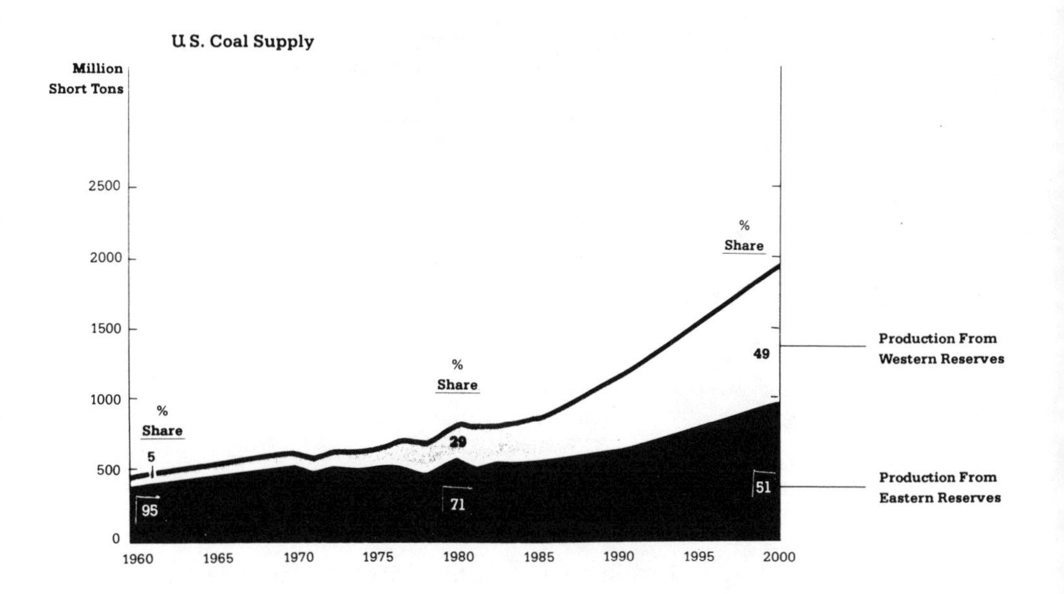


## Figure 28: U.S. Coal Demand. (Source 9)

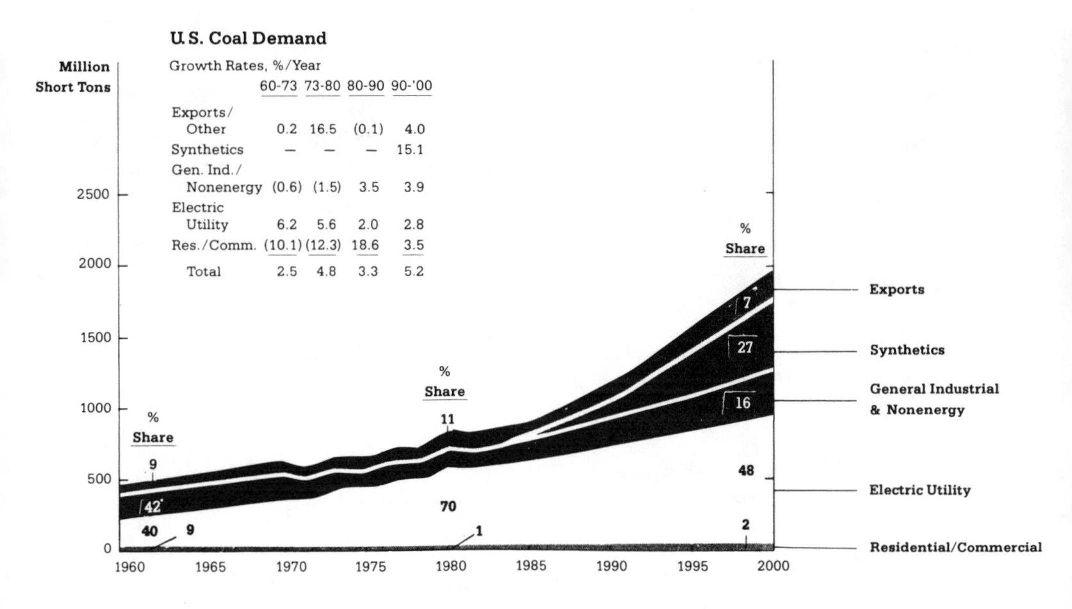


| | 60-73 | 73-80 | 80-90 | 90-'00 |
|---|---|---|---|---|
| Exports/ Other | 0.2 | 16.5 | (0.1) | 4.0 |
| Synthetics | — | — | — | 15.1 |
| Gen. Ind./ Nonenergy | (0.6) | (1.5) | 3.5 | 3.9 |
| Electric Utility | 6.2 | 5.6 | 2.0 | 2.8 |
| Res./Comm. | (10.1) | (12.3) | 18.6 | 3.5 |
| Total | 2.5 | 4.8 | 3.3 | 5.2 |

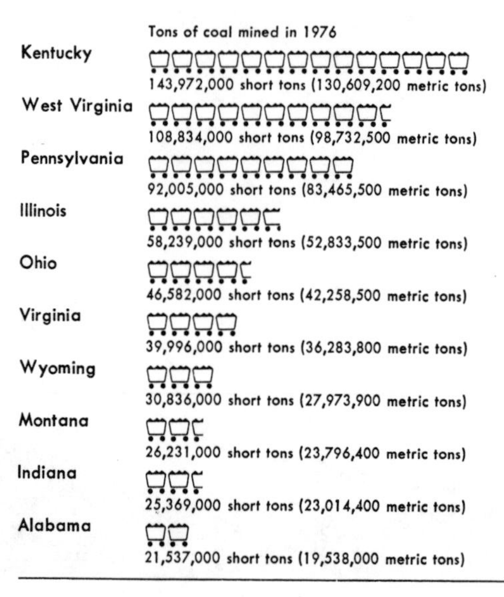


Figure 29: Leading Coal-Mining States. (Source 19)

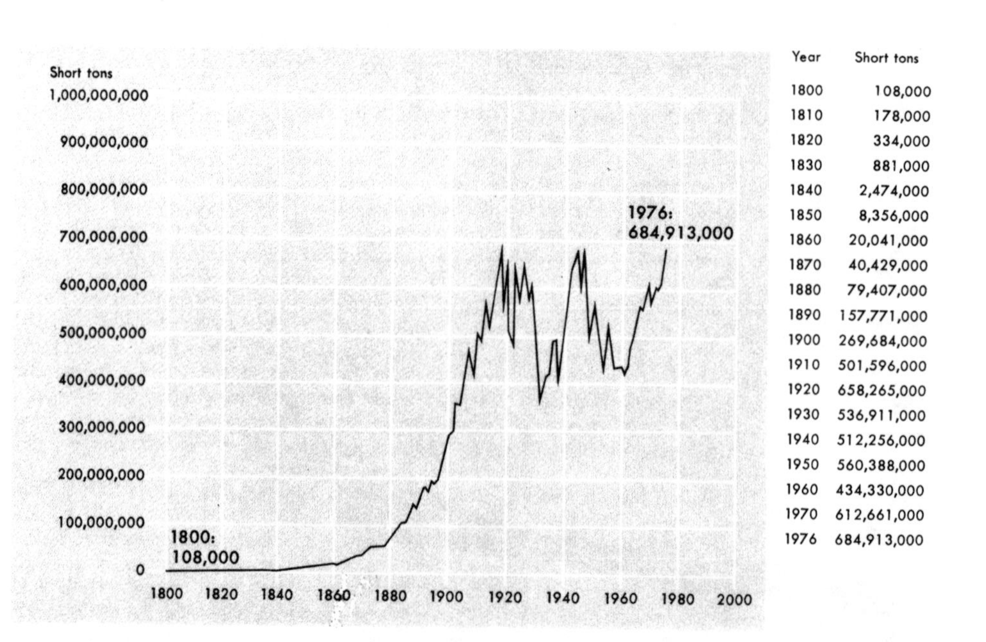


| Year | Short tons |
|---|---|
| 1800 | 108,000 |
| 1810 | 178,000 |
| 1820 | 334,000 |
| 1830 | 881,000 |
| 1840 | 2,474,000 |
| 1850 | 8,356,000 |
| 1860 | 20,041,000 |
| 1870 | 40,429,000 |
| 1880 | 79,407,000 |
| 1890 | 157,771,000 |
| 1900 | 269,684,000 |
| 1910 | 501,596,000 |
| 1920 | 658,265,000 |
| 1930 | 536,911,000 |
| 1940 | 512,256,000 |
| 1950 | 560,388,000 |
| 1960 | 434,330,000 |
| 1970 | 612,661,000 |
| 1976 | 684,913,000 |

Figure 30: Coal Production in the U.S. (Source 19)

## Table 19: U.S. Coal Reserve Base. (Source 25)

(millions of tons)

| State | Anthracite | Bituminous | Subbituminous | Lignite | Total |
|---|---|---|---|---|---|
| Alabama | — | 2,010 | — | 1,085 | 3,095 |
| Alaska | — | 700 | 5,445 | 15 | 6,160 |
| Arizona | — | 325 | — | — | 325 |
| Arkansas | 95 | 270 | — | 25 | 390 |
| Colorado | 25 | 9,145 | 4,120 | 2,965 | 16,255 |
| Georgia | — | 1 | — | — | 1 |
| Idaho | — | 5 | — | — | 5 |
| Illinois | — | 67,970 | — | — | 67,970 |
| Indiana | — | 10,715 | — | — | 10,715 |
| Iowa | — | 2,200 | — | — | 2,200 |
| Kansas | — | 1,000 | — | — | 1000 |
| Kentucky | — | 26,000 | — | — | 26,000 |
| Louisiana | — | — | — | — | — |
| Maryland | — | 1,050 | — | — | 1,050 |
| Michigan | — | 130 | — | — | 130 |
| Missouri | — | 5,015 | — | — | 5,015 |
| Montana | — | 1,385 | 103,420 | 15,770 | 120,575 |
| New Mexico | 2 | 1,860 | 2,735 | — | 4,597 |
| North Carolina | — | 35 | — | — | 35 |
| North Dakota | — | — | — | 10,145 | 10,145 |
| Ohio | — | 19,230 | — | — | 19,230 |
| Oklahoma | — | 1,620 | — | — | 1,620 |
| Oregon | — | — | 20 | — | 20 |
| Pennsylvania | 7,110 | 23,730 | — | — | 30,840 |
| South Dakota | — | — | — | 425 | 425 |
| Tennessee | — | 965 | — | — | 965 |
| Texas | — | — | — | 3,180 | 3,180 |
| Utah | — | 6,550 | 1 | — | 6,551 |
| Virginia | 140 | 4,165 | — | — | 4,305 |
| Washington | — | 255 | 1,320 | 10 | 1,585 |
| West Virginia | — | 38,610 | — | — | 38,610 |
| Wyoming | — | 4,005 | 51,370 | — | 55,375 |
| Total | 7,372 | 228,946 | 168,431 | 33,620 | 438,369 |

Equivalents: (100% recovered if burned)

1 TJ = 36 tons of coal = $10^{12}$ J = $10^9$ Btu
1 lb coal = 12,000 Btu = 4 kWh
1 metric ton coal = 4.9 bbl of oil
1 ton dry coal = 25,000,000 - 26,000,000 Btu.
1 ton anthracite = 33.5 GJ
1 ton bituminous = 29.3 GJ
1 ton sub-bituminous = 25.1 GJ
1 ton brown coal & lignite = 14.7 GJ
1 ton peat = 8.3 GJ
1 quad = $10^{15}$ Btu = 40 x $10^6$ tons of bituminous coal
= 50 x $10^6$ tons of sub-bituminous coal
= 62 x $10^6$ tons of lignite

Conversions (Approximate):

Of coal to synthetic crude oil at 480°C, gives 40-45% oil by weight.
Of coal to low Btu gas, 85-90% conversion.
2000 lb of coal produces 4.5 bbl of synthetic oil.
2000 lb of coal produces 126 gal of liquid fuel (gasoline).
2000 lb of coal produces 2-3 bbl (42 gal) of high octane gasoline.
2000 lb of coal produces 50,000 $ft^3$ (Lurgi process) gas (1000Btu/$ft^3$).
2000 lb of coal produces 2200 kWh .
0.87 lb of coal produces 1 kWh (1968).
2000 lb of coal produces 1500 lb coke, 9500-11,500 $ft^3$ of (550 Btu/ft),(50% $H_2$ at 1000°C ).
Manufactured (coal) gas, 8-10 gal coal tars, 3-3.5 gal of light oil, and 28 lb chemicals.
2000 lb of coal produces 3000 ft gas (950 Btu/ft), which is 65% $CH_4$.
2000 lb of shale yields an average of 25-35 gal of liquid fuel (10-100 gal/ton).
2000 lb of oil shale rock yields 1-2 bbl of oil.

**Table 20: Coal: Energy Equivalents and Conversion (Source 10)**

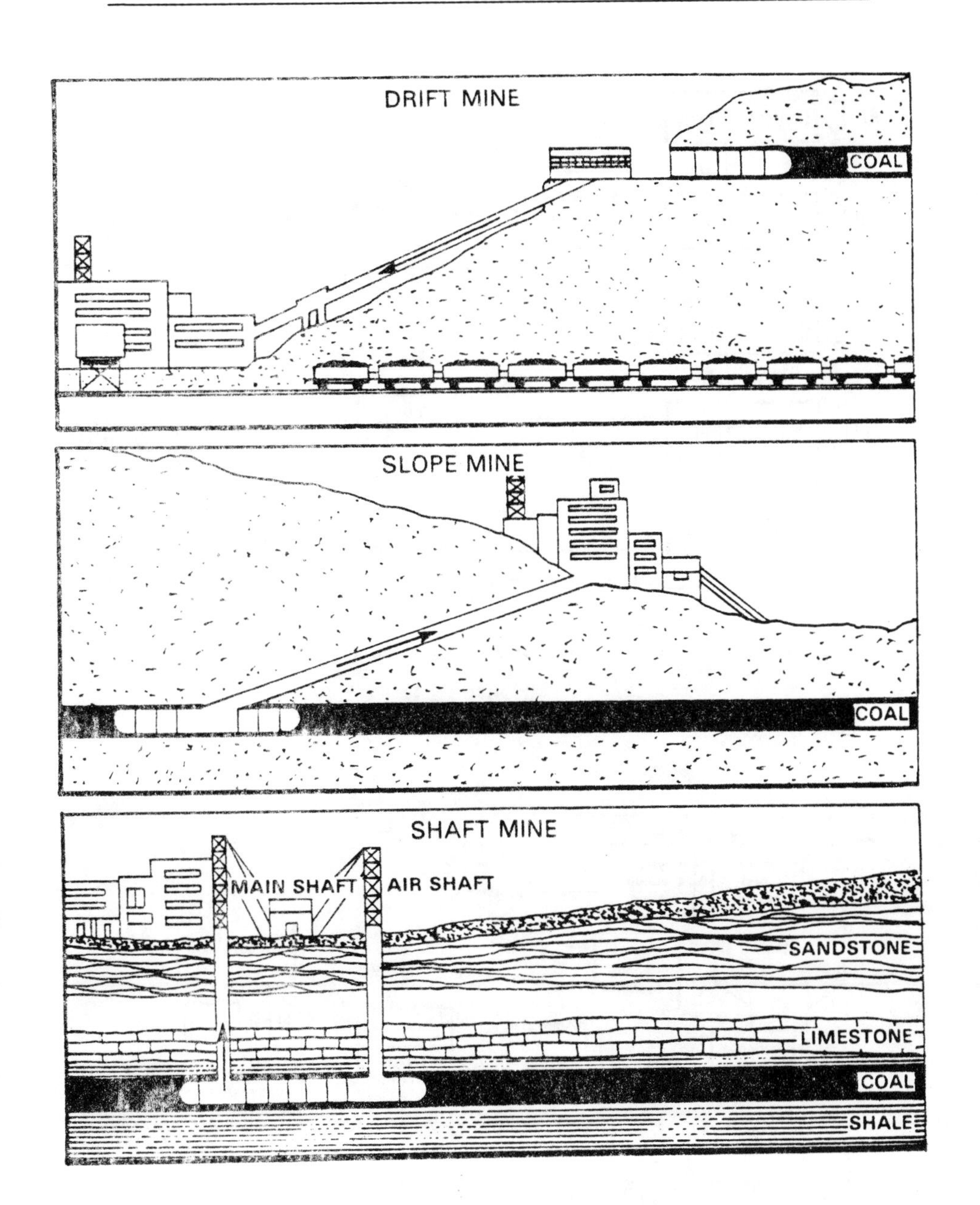


Figure 31: Underground Mines. (Source 12)

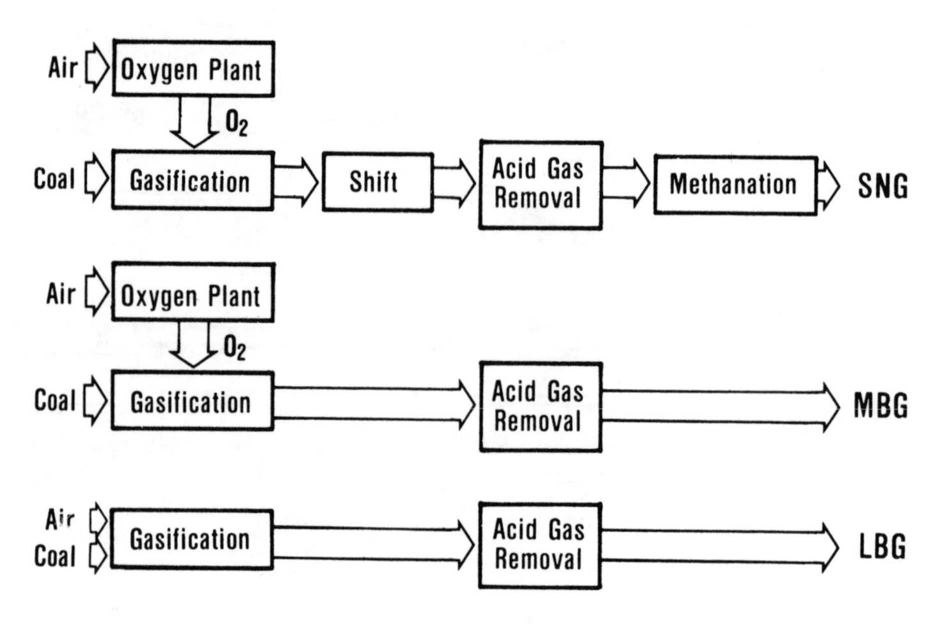


Figure 32: First Generation Coal Gasification Processes. (Source 14)

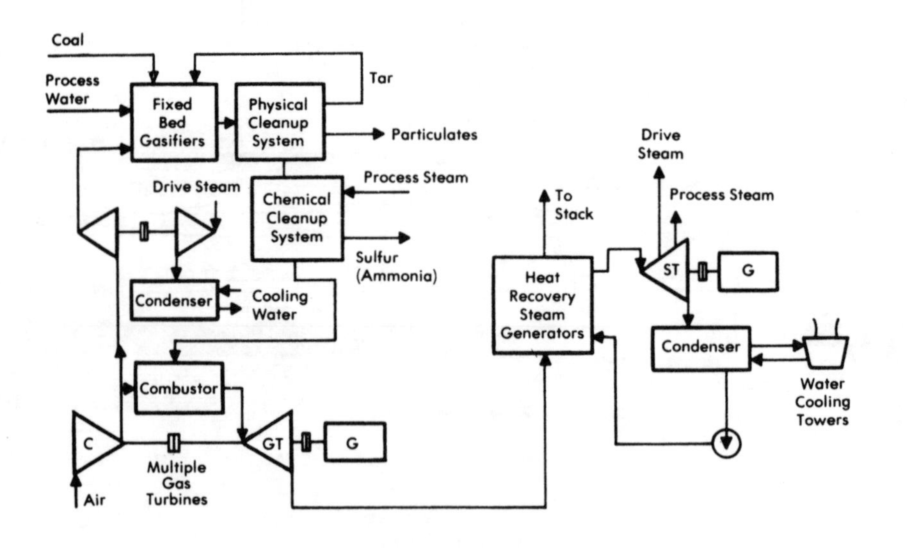


Figure 33: Integrated Gasification Combined Cycle. (Source 14)

Figure 34: Atmospheric Fluidized Bed Combustion. (Source 2)

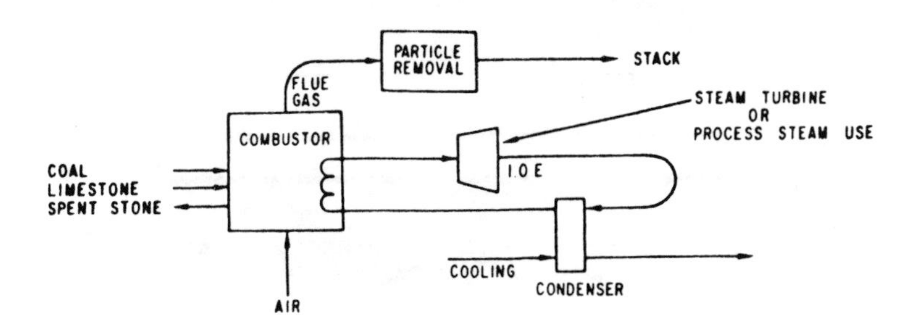


Figure 35: Pressurized Fluidized Bed Combustion. (Source 2)

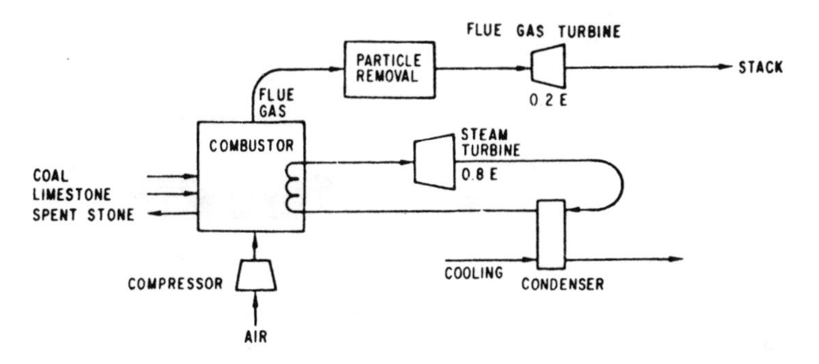


Figure 36: Pressurized Adiabatic Fluidized Bed Combustion. (Source 2)

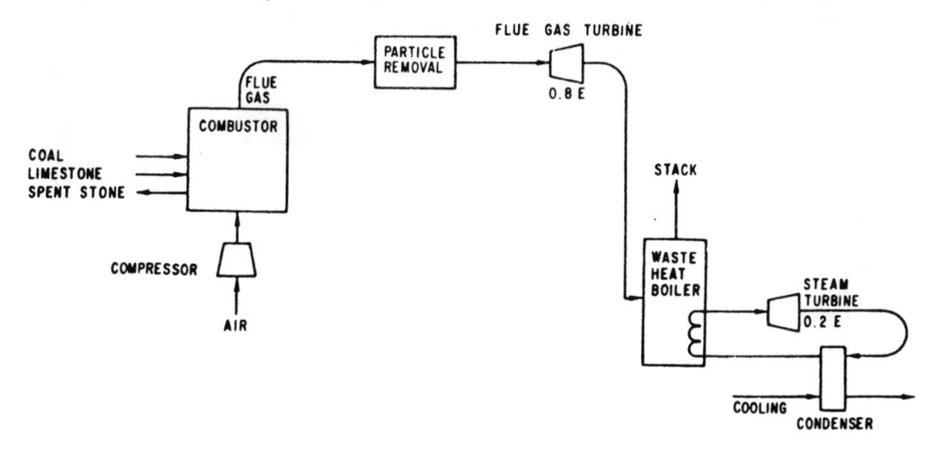

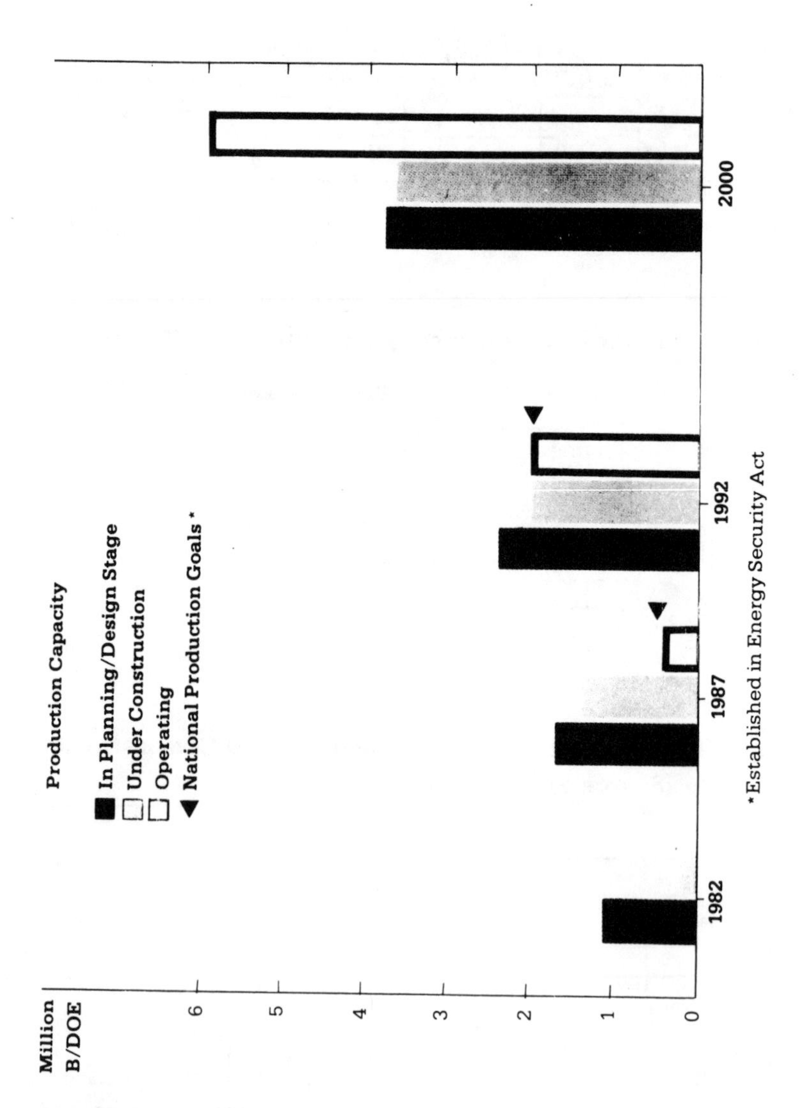


**Figure 37: Synthetic Fuels Production Capacity. (Source 9)**

Figure 38:
Lurgi Pressure Gasifier. (Source 16)

Figure 39:
Koppers-Totzek Two-Headed Gasifier. (Source 16)

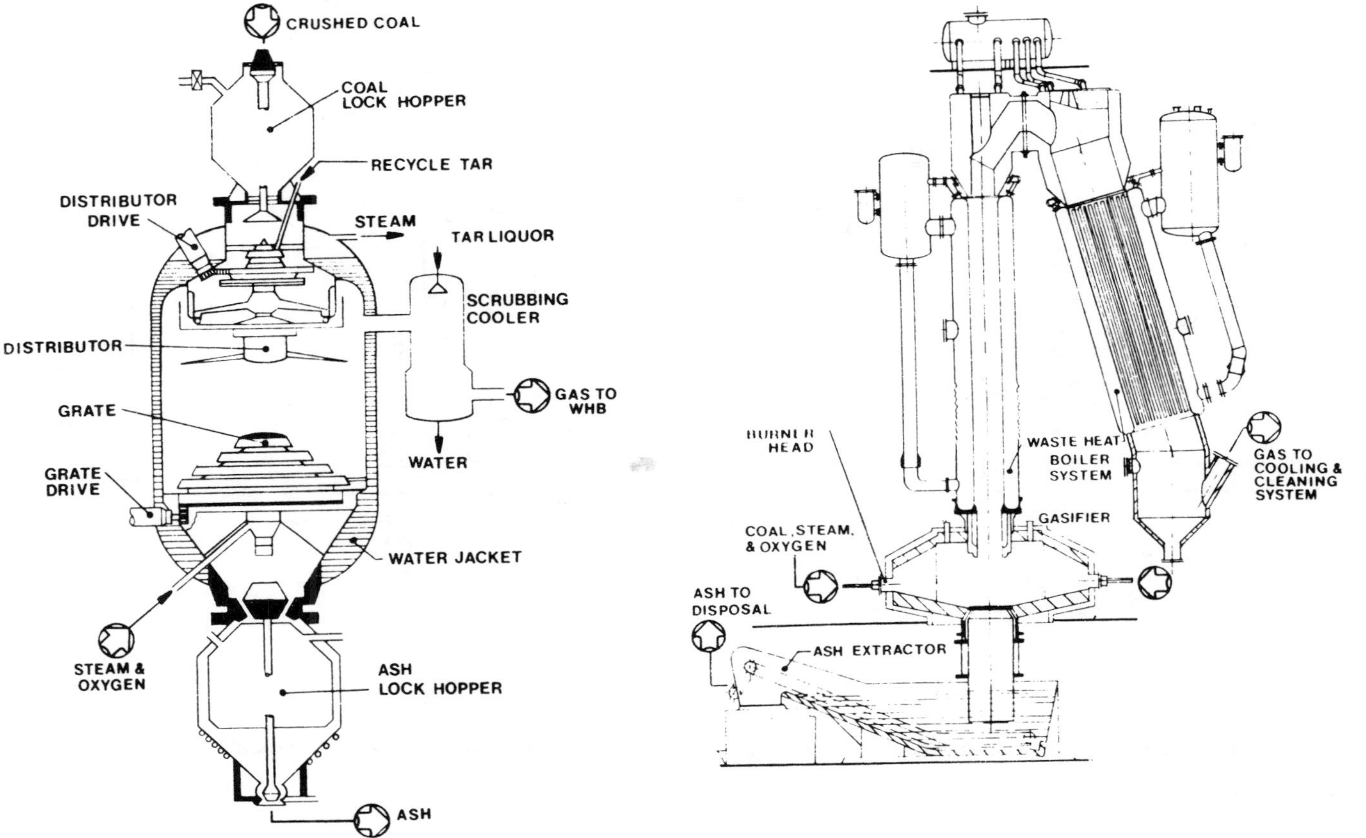

**Figure 40: Texaco Gasifier. (Source 16)**

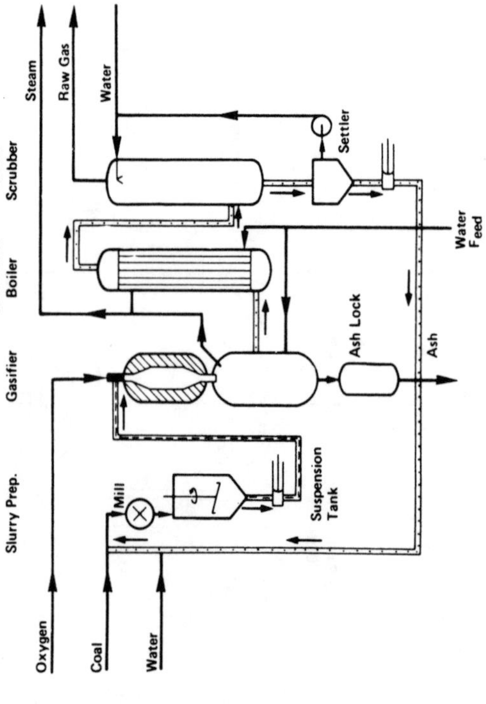

**Figure 41: Winkler Gasifier. (Source 16)**

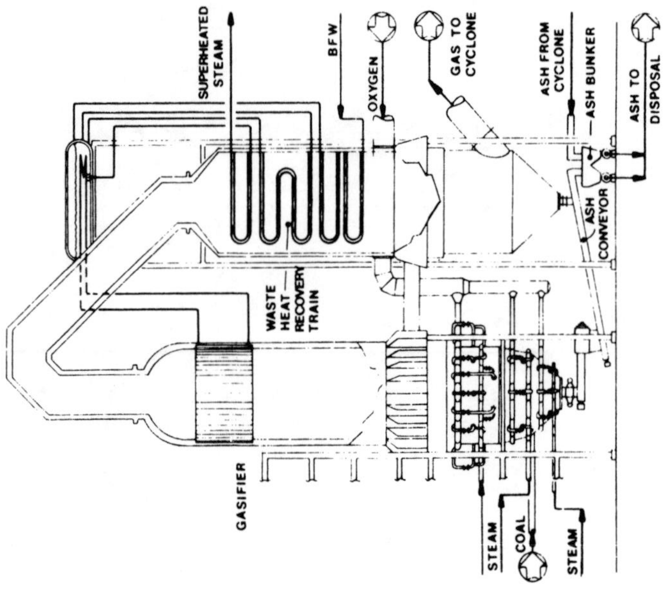

## Figure 42: Classification of Coal Liquefaction Processes. (Source 14)

1. DIRECT LIQUEFACTION
   - SRC-I
   - SRC-II
   - H-COAL
   - EXXON DONOR SOLVENT (EDS)
2. TWO-STAGE LIQUEFACTION
   - CONOCO CSF
3. PYROLYSIS AND HYDROPYROLYSIS
   - CITIES SERVICE
   - ROCKETDYNE
   - COED
4. INDIRECT LIQUEFACTION
   - FISCHER-TROPSCH
   - METHANOL AND MOBIL-M

## Figure 43: SRC–I Process. (Source 14)

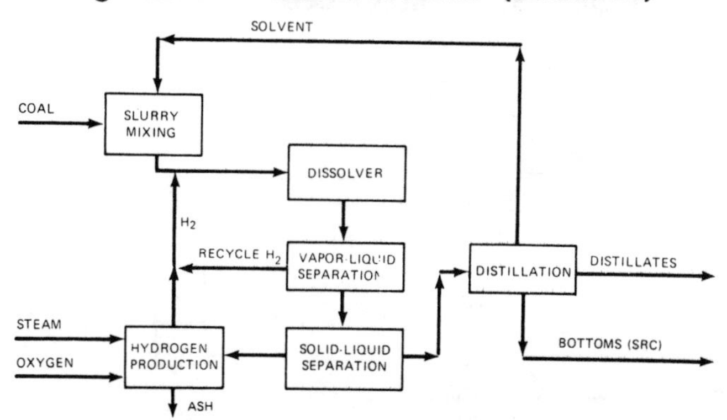


## Figure 44: SRC–II Process. (Source 14)

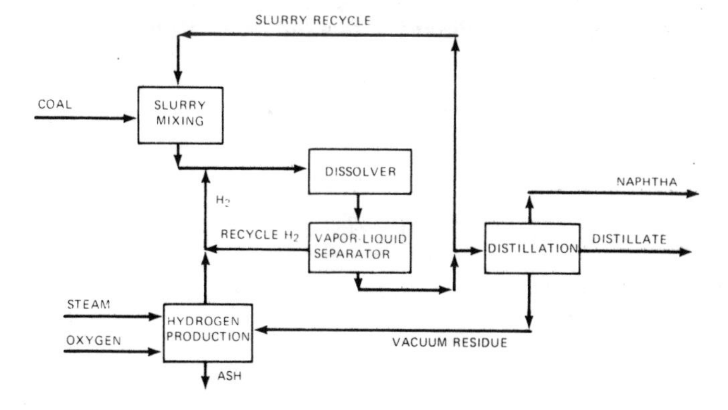

**Figure 45: H–Coal Process. (Source 14)**

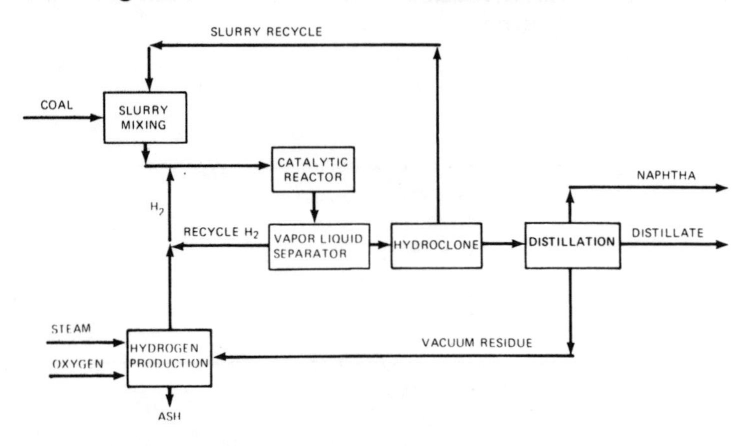


**Figure 46: EDS Process. (Source 14)**

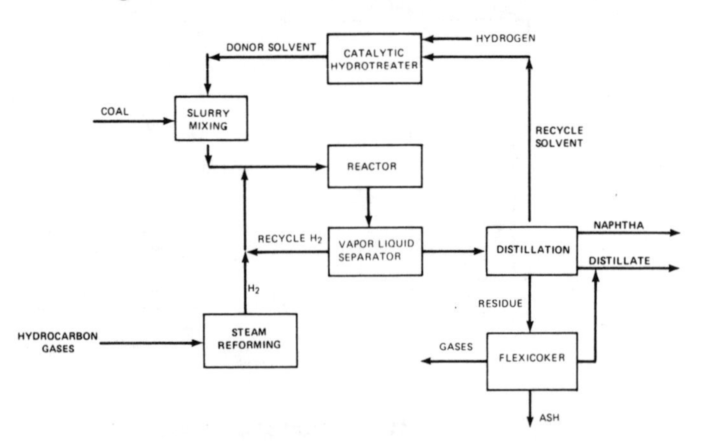


**Figure 47: Indirect Liquefaction. (Source 14)**

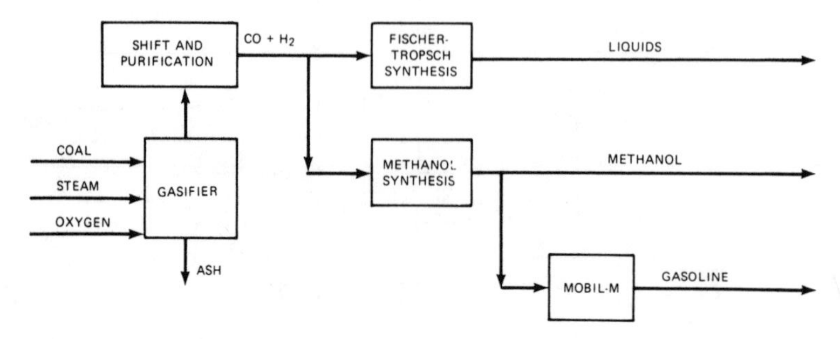

**Figure 48: Two-Stage Liquefaction (CSF). (Source 14)**

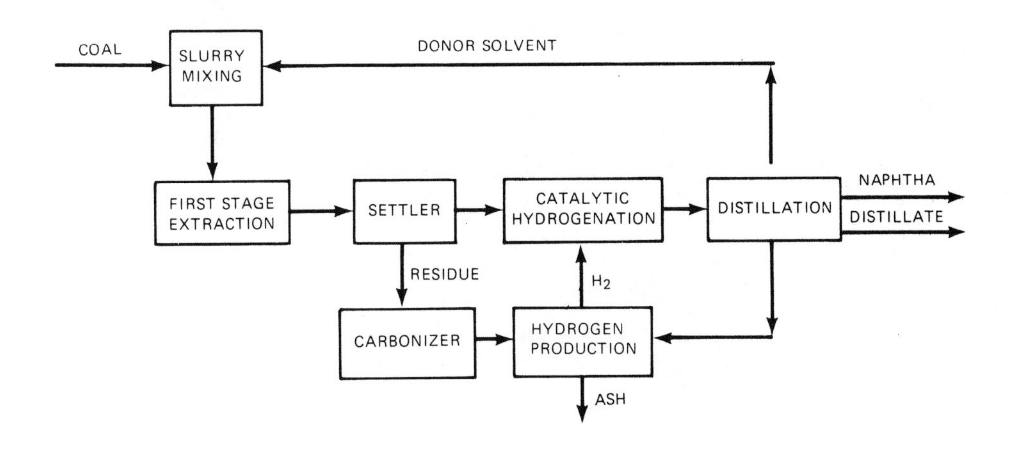


**Figure 49: Rapid Hydropyrolysis. (Source 14)**

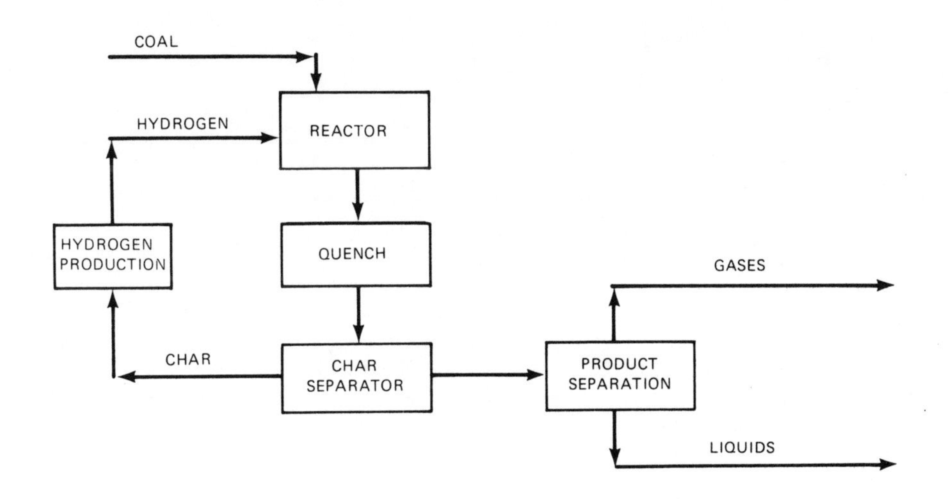

# GAS

**Figure 50: World Natural Gas Production. (Source 11)**

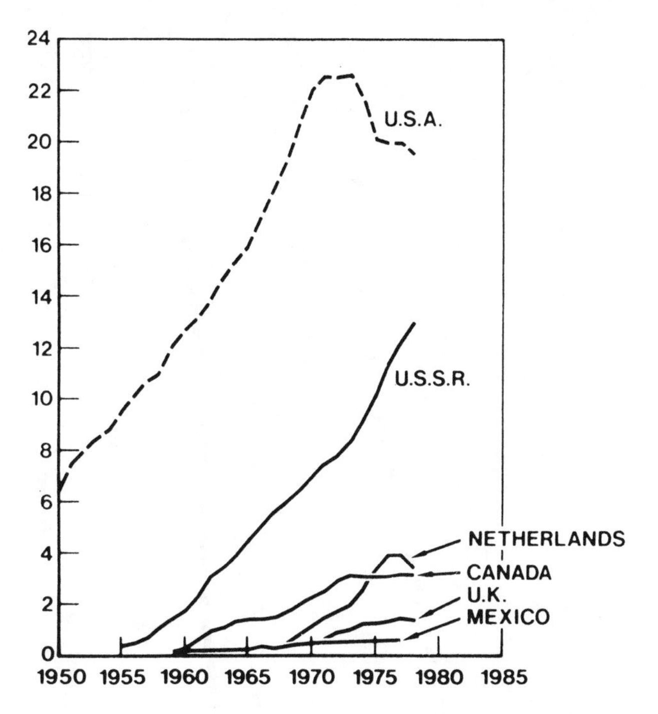


**Figure 51: U.S. Gas Demand. (Source 9)**

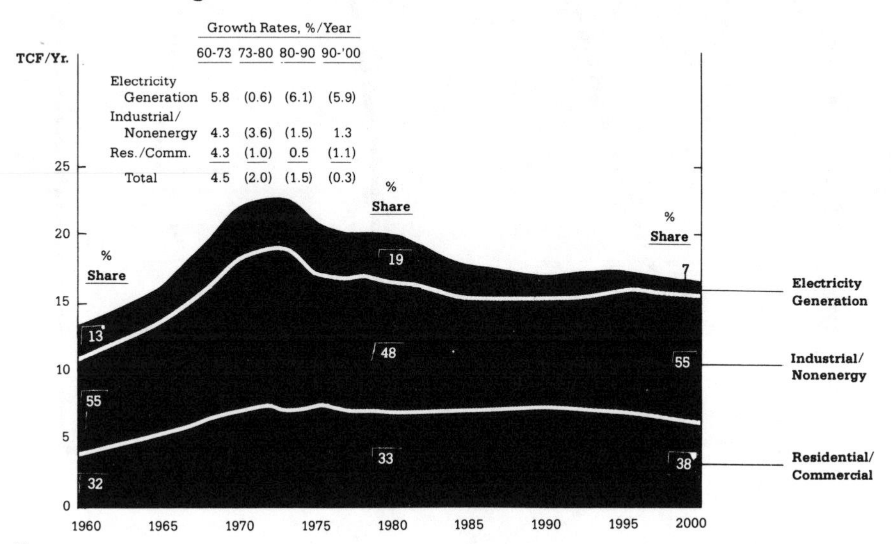

**Figure 52:** U.S. Natural Gas Reserves and Production. (Source 1)
(Lower 48 States)

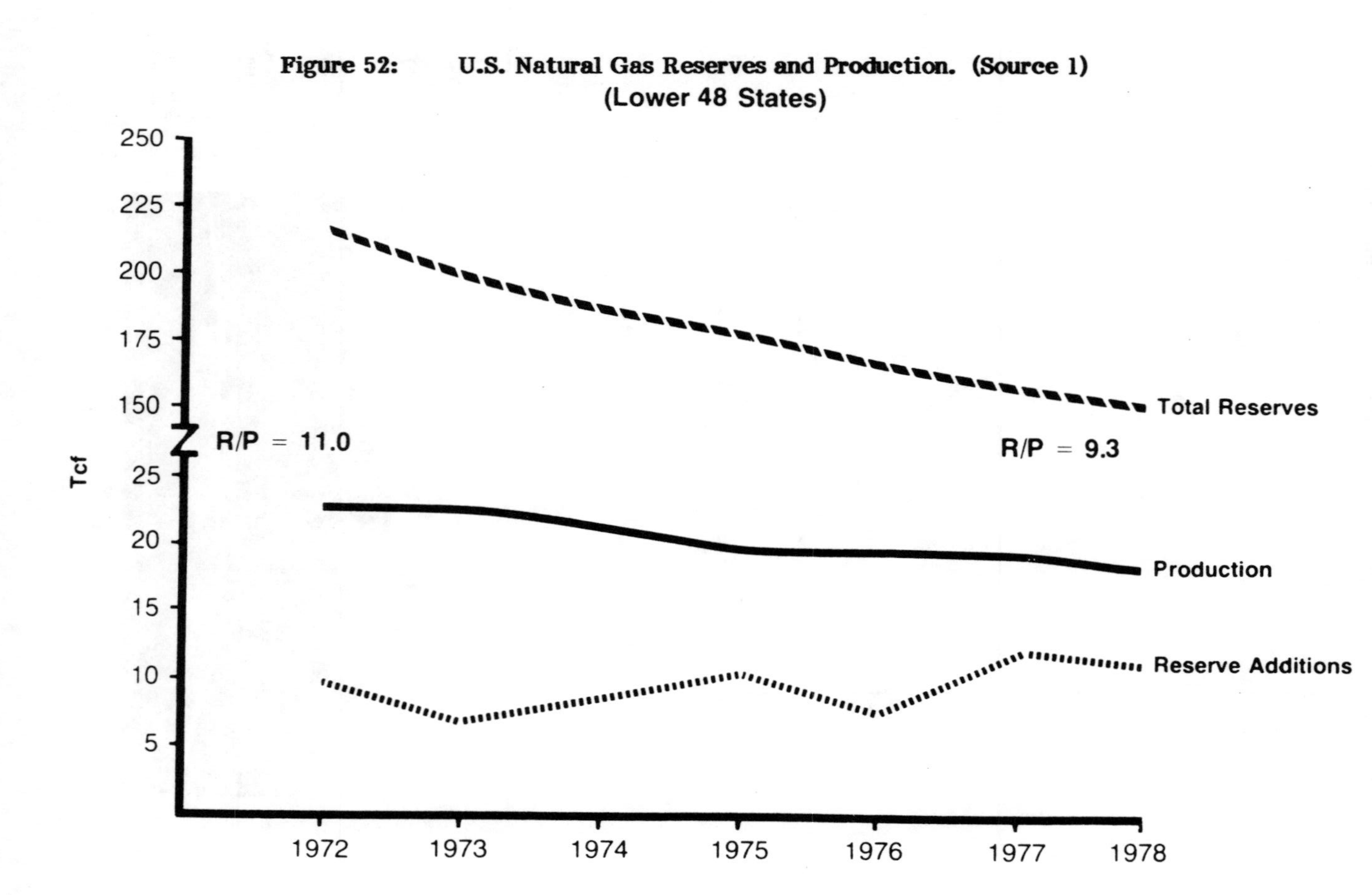

# Table 21: Natural Gas Supply and Production: Lower 48 States. (Source 1)

| Year | Annual Gross Additions to Proved Reserves | Cumulative Discoveries | Preliminary Annual Net Production* | Cumulative Net Production | Proved Reserves | Gas in Underground Storage** | Proved Reserves/ Annual Production Ratio |
|---|---|---|---|---|---|---|---|
| | Tcf at 60°F and 14.73 psia | | | | | | |
| 1945 | — | 233.18 | — | 86.35 | 146.99 | 0.15 | |
| 1946 | 17.63 | 250.81 | 4.92 | 91.26 | 159.70 | 0.15 | 32.46 |
| 1947 | 10.92 | 261.74 | 5.60 | 96.86 | 165.03 | 0.15 | 29.47 |
| 1948 | 13.82 | 275.56 | 5.98 | 102.84 | 172.93 | 0.20 | 28.92 |
| 1949 | 12.61 | 288.16 | 6.21 | 109.05 | 179.40 | 0.29 | 28.89 |
| 1950 | 11.99 | 300.15 | 6.86 | 115.91 | 184.58 | 0.34 | 26.91 |
| 1951 | 15.97 | 316.12 | 7.92 | 123.83 | 192.76 | 0.47 | 24.34 |
| 1952 | 14.27 | 330.38 | 8.59 | 132.42 | 198.63 | 0.67 | 23.12 |
| 1953 | 20.34 | 350.73 | 9.19 | 141.61 | 210.30 | 1.18 | 22.88 |
| 1954 | 9.55 | 360.27 | 9.38 | 150.99 | 210.56 | 1.27 | 22.45 |
| 1955 | 21.90 | 382.17 | 10.06 | 161.05 | 222.48 | 1.36 | 22.12 |
| 1956 | 24.72 | 406.89 | 10.85 | 171.90 | 236.48 | 1.49 | 21.80 |
| 1957 | 20.01 | 426.89 | 11.44 | 183.34 | 245.23 | 1.67 | 21.44 |
| 1958 | 18.90 | 445.79 | 11.42 | 194.76 | 252.76 | 1.73 | 22.13 |
| 1959 | 20.56 | 466.35 | 12.37 | 207.13 | 261.11 | 1.89 | 21.11 |
| 1960 | 13.84 | 480.20 | 13.02 | 220.15 | 262.22 | 2.17 | 20.14 |
| 1961 | 16.35 | 496.55 | 13.38 | 233.53 | 265.35 | 2.33 | 19.83 |
| 1962 | 18.77 | 515.32 | 13.64 | 247.16 | 270.64 | 2.49 | 19.84 |
| 1963 | 18.10 | 533.42 | 14.54 | 261.70 | 274.46 | 2.74 | 18.88 |
| 1964 | 20.11 | 553.52 | 15.34 | 277.04 | 279.42 | 2.94 | 18.22 |
| 1965 | 21.16 | 574.68 | 16.24 | 293.29 | 284.48 | 3.09 | 17.52 |
| 1966 | 19.25 | 593.93 | 17.48 | 310.77 | 286.39 | 3.22 | 16.38 |
| 1967 | 21.09 | 615.02 | 18.36 | 329.13 | 289.27 | 3.38 | 15.76 |
| 1968 | 12.04 | 627.06 | 19.33 | 348.45 | 282.10 | 3.49 | 14.59 |
| 1969 | 8.34 | 635.40 | 20.64 | 369.09 | 269.91 | 3.60 | 13.08 |
| 1970 | 11.12 | 646.52 | 21.82 | 390.91 | 259.62 | 4.00 | 11.90 |
| 1971 | 9.44 | 655.96 | 21.92 | 412.83 | 247.44 | 4.31 | 11.29 |
| 1972 | 9.40 | 665.36 | 22.37 | 435.20 | 234.63 | 4.47 | 10.49 |
| 1973 | 6.51 | 671.86 | 22.47 | 457.67 | 218.31 | 4.12 | 9.72 |
| 1974 | 8.31 | 680.17 | 21.17 | 478.85 | 205.27 | 3.94 | 9.69 |
| 1975 | 10.14 | 690.32 | 19.56 | 498.41 | 196.15 | 4.24 | 10.02 |
| 1976 | 7.45 | 697.77 | 19.32 | 517.73 | 184.10 | 4.05 | 9.53 |
| 1977 | 11.76 | 709.53 | 19.26 | 536.99 | 177.05 | 4.50 | 9.19 |
| 1978 | 10.59 | 720.12 | 19.10 | 556.69 | 168.69 | 4.65 | 8.83 |

Note: The small inconsistencies between annual and cumulative discoveries and between annual and cumulative production are caused by rounding of values.

*Preliminary net production is reported annually by the Committee on Natural Gas Reserves. Adjustments to these preliminary numbers are included in the revisions which are reported in the following year. Actual annual production and cumulative production are tabulated for the years 1965 through 1974 in "Reserves of Crude Oil, Natural Gas Liquids and Natural Gas in the United States and Canada as of December 31, 1975."

**Proved recoverable gas contained in underground storage reservoirs (first reported on a recoverable basis in 1973).

## Table 22: Natural Gas Supply and Production: Total United States. (Source 1)

| Year | Annual Gross Additions to Proved Reserves | Cumulative Discoveries | Preliminary Annual Net Production* | Cumulative Net Production | Proved Reserves | Gas in Underground Storage** | Proved Reserves/ Annual Production Ratio |
|---|---|---|---|---|---|---|---|
| | Tcf at 60°F and 14.73 psia | | | | | | |
| 1945 | — | 233.18 | — | 86.35 | 146.99 | 0.15 | — |
| 1946 | 17.63 | 250.81 | 4.92 | 91.26 | 159.70 | 0.15 | 32.46 |
| 1947 | 10.92 | 261.74 | 5.60 | 96.86 | 165.03 | 0.15 | 29.47 |
| 1948 | 13.82 | 275.56 | 5.98 | 102.84 | 172.93 | 0.20 | 28.92 |
| 1949 | 12.61 | 288.16 | 6.21 | 109.05 | 179.40 | 0.29 | 28.89 |
| 1950 | 11.99 | 300.15 | 6.86 | 115.91 | 184.58 | 0.34 | 26.91 |
| 1951 | 15.97 | 316.12 | 7.92 | 123.83 | 192.76 | 0.47 | 24.34 |
| 1952 | 14.27 | 330.38 | 8.59 | 132.42 | 198.63 | 0.67 | 23.12 |
| 1953 | 20.34 | 350.73 | 9.19 | 141.61 | 210.30 | 1.18 | 22.88 |
| 1954 | 9.55 | 360.27 | 9.38 | 150.99 | 210.56 | 1.27 | 22.45 |
| 1955 | 21.90 | 382.17 | 10.06 | 161.05 | 222.48 | 1.36 | 22.12 |
| 1956 | 24.72 | 406.89 | 10.85 | 171.90 | 236.48 | 1.49 | 21.80 |
| 1957 | 20.01 | 426.89 | 11.44 | 183.34 | 245.23 | 1.67 | 21.44 |
| 1958 | 18.90 | 445.79 | 11.42 | 194.76 | 252.76 | 1.73 | 22.13 |
| 1959 | 20.62 | 466.41 | 12.37 | 207.13 | 261.17 | 1.89 | 21.11 |
| 1960 | 13.89 | 480.31 | 13.02 | 220.15 | 262.33 | 2.17 | 20.15 |
| 1961 | 17.17 | 497.47 | 13.38 | 233.53 | 266.27 | 2.33 | 19.90 |
| 1962 | 19.48 | 516.96 | 13.64 | 247.17 | 272.28 | 2.49 | 19.96 |
| 1963 | 18.16 | 535.12 | 14.55 | 261.71 | 276.15 | 2.74 | 18.98 |
| 1964 | 20.25 | 555.37 | 15.35 | 277.06 | 281.25 | 2.94 | 18.32 |
| 1965 | 21.32 | 576.69 | 16.25 | 293.31 | 286.47 | 3.09 | 17.63 |
| 1966 | 20.22 | 596.91 | 17.49 | 310.81 | 289.33 | 3.22 | 16.54 |
| 1967 | 21.80 | 618.72 | 18.38 | 329.19 | 292.91 | 3.38 | 15.94 |
| 1968 | 13.70 | 632.41 | 19.37 | 348.56 | 287.35 | 3.49 | 14.83 |
| 1969 | 8.38 | 640.79 | 20.72 | 369.28 | 275.11 | 3.60 | 13.28 |
| 1970 | 37.20 | 677.99 | 21.96 | 391.24 | 290.75 | 4.00 | 13.24 |
| 1971 | 9.83 | 687.81 | 22.08 | 413.32 | 278.81 | 4.31 | 12.63 |
| 1972 | 9.63 | 697.45 | 22.51 | 435.83 | 266.08 | 4.47 | 11.82 |
| 1973 | 6.83 | 704.27 | 22.61 | 458.44 | 249.95 | 4.12 | 11.05 |
| 1974 | 8.68 | 712.95 | 21.32 | 479.76 | 237.13 | 3.94 | 11.12 |
| 1975 | 10.48 | 723.43 | 19.72 | 499.48 | 228.20 | 4.24 | 11.57 |
| 1976 | 7.56 | 730.99 | 19.54 | 519.02 | 216.03 | 4.05 | 11.06 |
| 1977 | 11.85 | 742.84 | 19.45 | 538.47 | 208.88 | 4.50 | 10.74 |
| 1978 | 10.59 | 753.43 | 19.31 | 557.78 | 200.30 | 4.65 | 10.37 |

Note: The small inconsistencies between annual and cumulative discoveries and between annual and cumulative production are caused by rounding of values.

*Preliminary net production is reported annually by the Committee on Natural Gas Reserves. Adjustments to these preliminary numbers are included in the revisions which are reported in the following year. Actual annual production and cumulative production are tabulated for the years 1965 through 1974 in "Reserves of Crude Oil, Natural Gas Liquids and Natural Gas in the United States and Canada as of December 31, 1975."

**Proved recoverable gas contained in underground storage reservoirs (first reported on a recoverable basis in 1973).

## Table 23: Lower 48 Production Forecasts. (Source 1)

| Source | 1980 | 1981 | 1982 | 1983 | 1984 | 1985 | 1986 | 1987 | 1988 | 1989 | 1990 |
|---|---|---|---|---|---|---|---|---|---|---|---|
| I. Oil and Gas Co.'s | | | | | | | | | | | |
| AMOCO | 19.1 | 18.5 | 17.6 | 17.1 | 16.6 | 16.3 | | | | | 15.2 |
| ARCO | 18.3 | 17.8 | 16.4 | 17.0 | 16.7 | 16.4 | 16.2 | 16.0 | 15.6 | 15.5 | 15.2 |
| Exxon | 18.2 | | | | | 14.4 | | | | | 13.4 |
| Gulf | 19.0 | 19.0 | 19.2 | 19.4 | 19.5 | 19.2 | 19.2 | 19.2 | 19.2 | 19.2 | 19.2 |
| Mobil | 19.5 | | | | | 18.1 | | | | | 16.3 |
| Shell | 17.9 | 17.4 | 16.9 | 16.5 | 16.1 | 15.7 | 15.2 | 14.7 | 14.4 | 13.8 | 13.2 |
| Medium Size Producer | 19.1 | 18.6 | 17.7 | 16.8 | 15.9 | 15.2 | 15.2 | 15.2 | 15.2 | 15.2 | 15.1 |
| II. Gas Transmission Co. | | | | | | | | | | | |
| Michigan-Wisconsin | 18.1 | 17.4 | 16.9 | 16.4 | 16.0 | 15.5 | 15.1 | 14.7 | 14.4 | 14.0 | 13.6 |
| Tenn. Gas. Trans. | 18.9 | 18.7 | 18.2 | 17.7 | 17.0 | 15.9 | 14.9 | 13.9 | 12.8 | 11.9 | 10.9 |
| Texas Eastern | 19.2 | 18.9 | 18.5 | 18.1 | 17.6 | 17.1 | 16.6 | 16.1 | 15.6E | 15.1E | 14.6E |
| Pipeline A | 18.0 | 17.5 | 17.3 | 16.9 | 16.8 | 16.7 | 16.4 | 16.2 | 16.1 | 16.0 | 15.9 |
| Pipeline B | 18.8 | | | | | 16.6 | | | | | 14.6 |
| III. Government Agencies | | | | | | | | | | | |
| DOE/EIA (Series C Med./Med.) | 18.5 | | | | | 16.8 | | | | | 15.7 |
| GAO | 18.1 | | | | | 17.6 | | | | | 17.1 |
| IV. A.G.A. - TERA | 17.3 | | | | | 16.9 | | | | | 18.0 |
| Total Average | 18.5 | | | | | 16.6 | | | | | 15.1 |

E = Extrapolated based on data through 1987

| Source | Estimated Resource in Place (Tcf) |
|---|---|
| Enhanced Gas Recovery: | |
| • Natural gas from coal seams | 300-800 |
| • Devonian Shale | 500-600 |
| • Rocky Mtn. Tight Formations | 600 |
| • Geopressured Resources | 3,000-50,000 |
| SNG from Biomass: | |
| • Land, renewable per year at | 7-11 |
| • Marine, renewable per year at | 20 |
| High Btu Oil Shale Gasification | 464 |
| Peat Gasification | 1,443 |

Table 24: Potential Unconventional Gas from New Technologies. (Source 1)

(In Trillion Cubic Feet)

| Actual 1979 | | Scenarios: Self Sufficiency | North American Focus | Moderate World Imports | World Conventional Gas Emphasis |
|---|---|---|---|---|---|
| 19.9 | Lower 48 | 12-14 | 12-14 | 12-14 | 12-14 |
| 0.2 | SNG | 0.3 | 0.3 | 0.3 | 0.1 |
| — | Alaska | 3.0 | 3.0 | 1.5 | 3.0 |
| 1.0 | Canada | 1.0 | 2.0 | 2.0 | 2.0 |
| — | Mexico | 0.1 | 2.0 | 2.0 | 2.0 |
| 0.2 | LNG | 0.7 | 0.7 | 2.5 | 4.0 |
| — | Coal Gas | 3.5 | 3.5 | 1.5-2.5 | 1.5-2.5 |
| — | Tight Formations | 1.5-5.0 | 1.5-4.0 | 1.5-3.0 | 1.5-3.0 |
| — | Nonconventional | 1.0-2.5 | 1.0-2.5 | 1.0-2.5 | 1.0-2.5 |
| 21.3 | Totals | 23.1-30.1 | 26.0-32.0 | 24.3-30.3 | 27.1-33.1 |

Table 25: Gas Supply in Year 2000. (Source 1)

Figure 53: Projected United States Gas Supply. (Source 10)

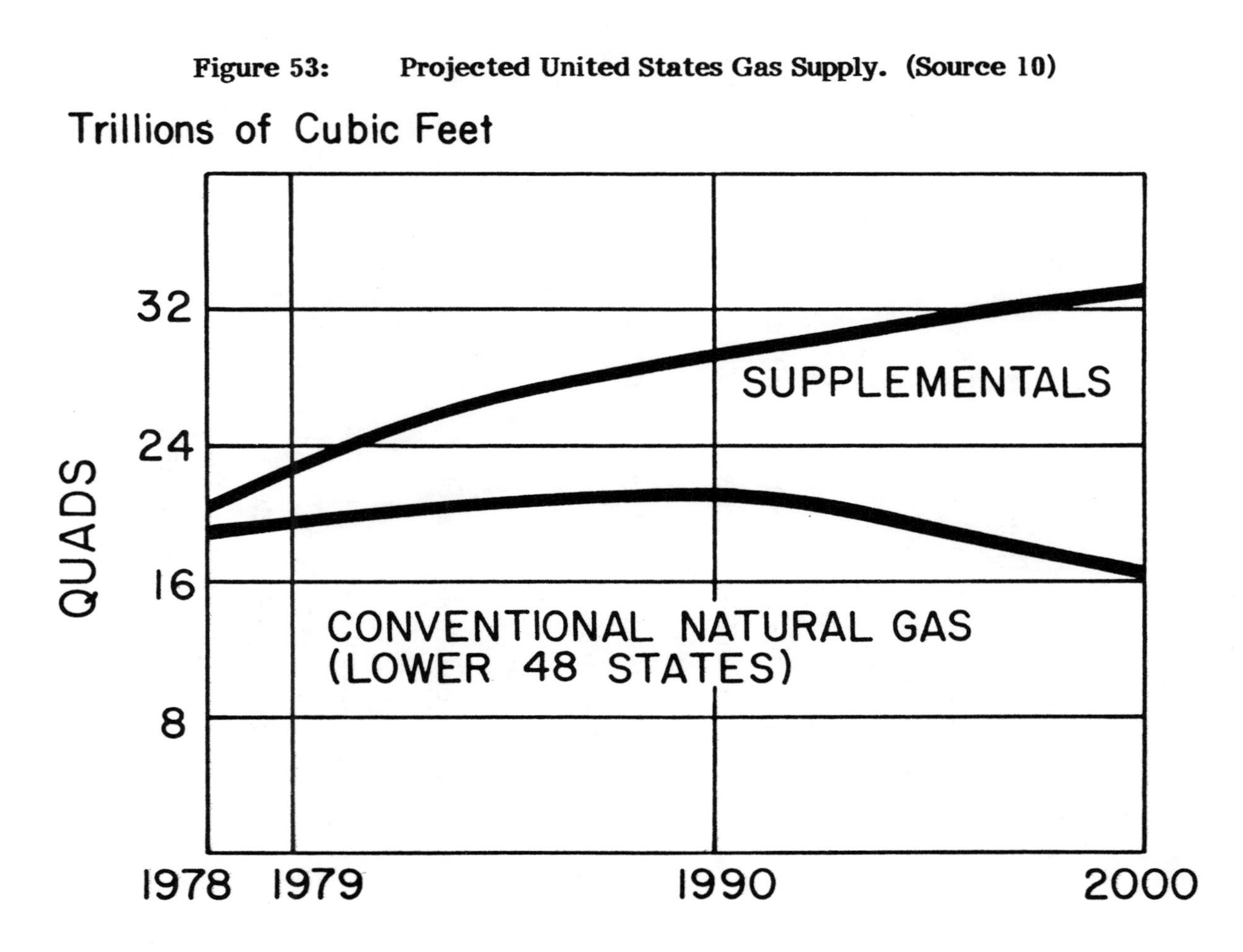

## Table 26: Natural Gas: Energy Equivalents and Conversions (Source 10)

Equivalent:

1 TJ = 1 x $10^6$ $ft^3$ = 1 Mcf = $10^{12}$J = $10^9$ Btu

1 $ft^3$ = 0.028 $m^3$ = 1000 Btu = 1.055 MJ

1 $m^3$ = 35.15 $ft^3$ = 35.315 Btu = 37.6 MJ

1 ton natural gas = 32,000 ft

1 $quad^{15}$ = 10 Btu = $10^{12}$ natural gas

1,167 million $m^3$ = 1 MT crude oil

1 Trillion cubic feet (Tcf) = 160 million bbl oil

1 Tcf = 293 million megawatts (100%) = 42 million tons coal

1 MMcf = 1,000,000 Btu

135 - 160 $ft^3$ of natural gas (methane) = 1 gal gasoline

52 Mcf = 1 tonne LNG

Conversion (Approximate):

20 cu. ft. converts to 1 lb methanol

1 Tcf converts to 88 million megawatts

10 cu. ft. converts to kWh electricity

11 $ft^3$ of natural gas (methane) to produce one brake hp-hr

Natural gas to methanol, 65% energy efficiency (conversion)

# NUCLEAR

SECTION 3

**Figure 54: Nuclear Power Plants in the United States. (Source 3)**

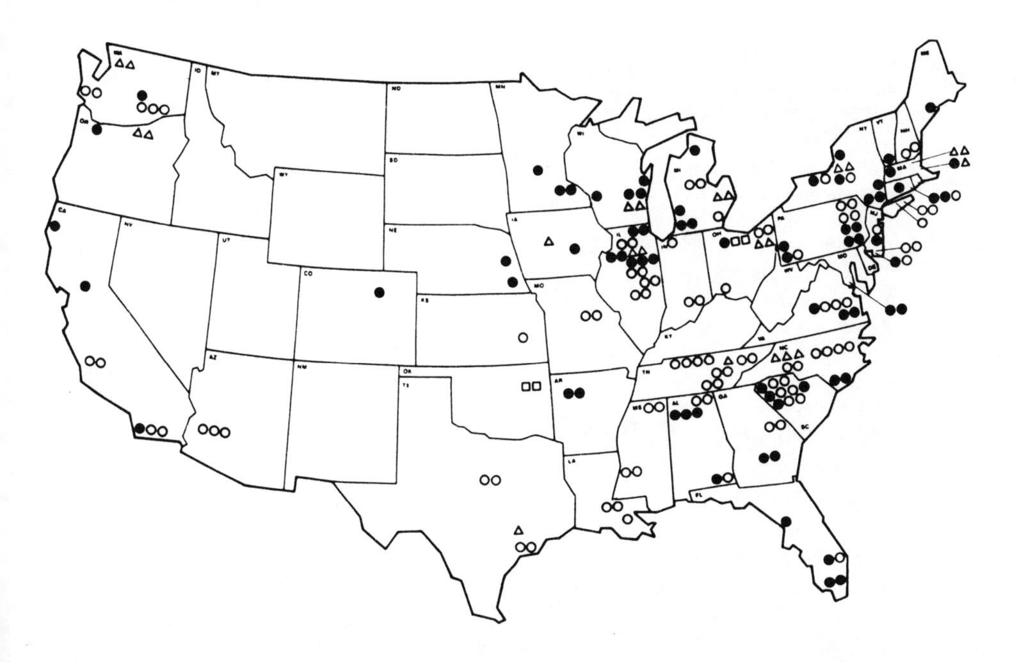

**Key**

- ● With Operating Licenses
- ○ With Construction Permits
- □ Limited Work Authorizations
- △ On Order
- ∧ Letters of Intent/Options

| | | |
|---|---|---|
| 72 | Reactors with operating licenses ....... | 52,366 MWe |
| 91 | Reactors with construction permits ..... | 100,052 MWe |
| 4 | Reactors with limited work authorizations | 4,112 MWe |
| 25 | Reactors on order (including 2 units not sited on map) .................. | 29,094 MWe |
| 192 | Total ........................... | 185,624 MWe |

January 1, 1980

## Table 27: Statistical Summary of Reactors. (Source 20)

| | Operable | Being built | Planned | Shut down or dismantled |
|---|---|---|---|---|
| **I. CIVILIAN REACTORS** | | | | |
| **1. Power Reactors** | | | | |
| A. Central-Station Electric Power | 71 | 94 | 28 | 8 |
| B. Dual-Purpose Plants | 1 | 2 | | |
| C. Propulsion (Maritime) | | | | 1 |
| **2. Experimental Power-Reactor Systems** | | | | |
| A. Electric-Power Systems | 1 | | | 23 |
| B. Auxiliary Power (SNAP) | | | | 9 |
| C. Space Propulsion (Rover) | | | | 21 |
| **3. Test, Research, and University Reactors** | | | | |
| A. General Irradiation Test | 3 | 1 | | 3 |
| B. High-Power Research and Test | 9 | | | 4 |
| C. Safety Research and Test | 3 | 1 | | 8 |
| D. General Research | 24 | 1 | 1 | 40 |
| E. University Research and Teaching | 54 | 1 | | 9 |
| **II. PRODUCTION REACTORS** | | | | |
| **1. Materials Production** | 3 | | | 10 |
| **2. Process Development** | 4 | | | 1 |
| **III. MILITARY REACTORS** | | | | |
| **1. Defense Power-Reactor Applications** | | | | |
| A. Remote Installations | 1 | | | 5 |
| B. Propulsion (Naval) | 121 | 28 | | 5 |
| **2. Developmental Power** | | | | |
| A. Electric-Power Experiments and Prototypes | | | | 3 |
| B. Propulsion Experiments and Prototypes | 7 | | | 7 |
| **3. Test and Research** | | | | |
| A. Test | | | | 4 |
| B. Research | 6 | | | 3 |
| **IV. REACTORS FOR EXPORT** | | | | |
| **1. Power Reactors** | | | | |
| A. Central-Station Electric Power | 21 | 28 | 14 | 2 |
| B. Propulsion | 1 | | | |
| **2. Test, Research, and Teaching** | | | | |
| A. General Irradiation Test | 4 | | | |
| B. General Research | 30 | | | 1 |
| C. University Research and Teaching | 26 | | | |

**Figure 55: Light Water Reactor Fuel Cycle. (Source 26)**

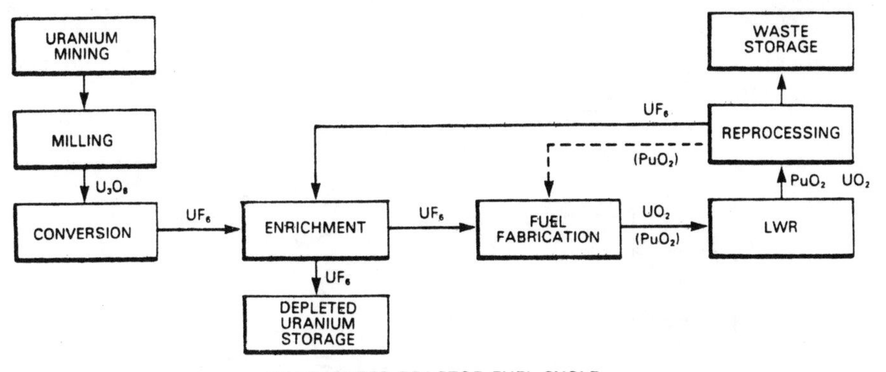


LIGHT WATER REACTOR FUEL CYCLE

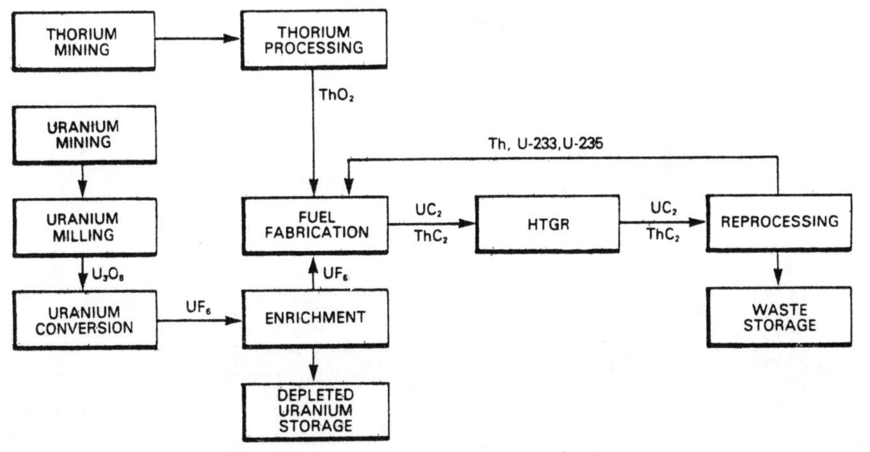


HIGH TEMPERATURE, GAS-COOLED REACTOR FUEL CYCLE

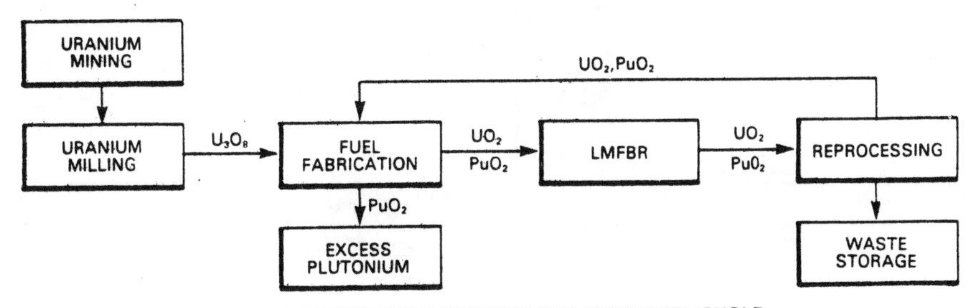


LIQUID METAL FAST BREEDER REACTOR FUEL CYCLE

**Figure 56: Schematic Diagrams for Boiling Water and Pressurized Water Reactors. (Source 18)**

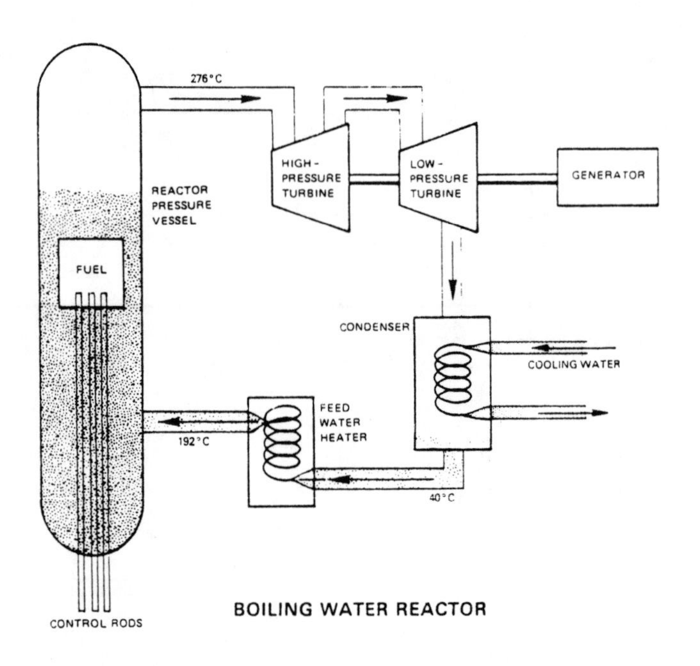


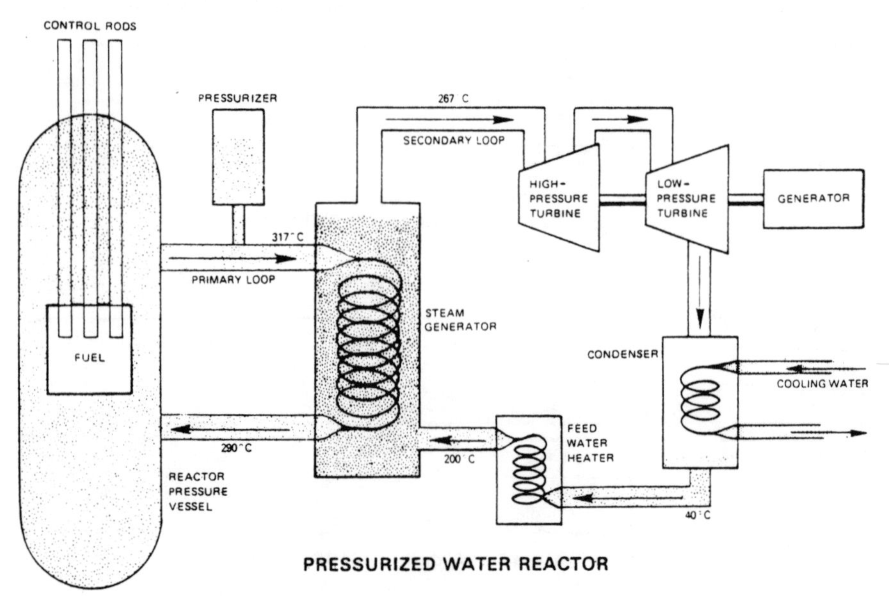

**Figure 57:** **Schematic Diagrams for HTGR and LMFBR Reactors. (Source 18)**

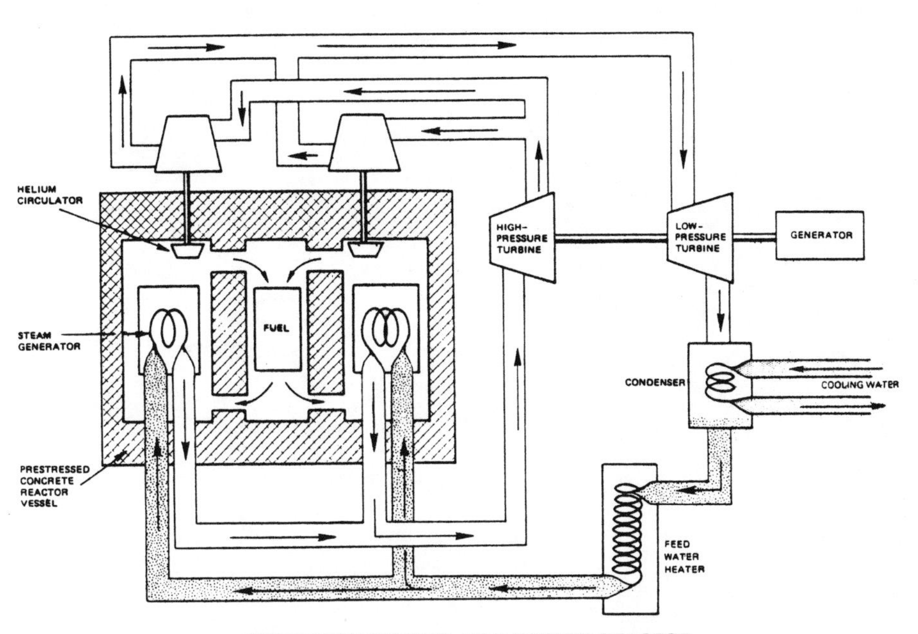


HIGH TEMPERATURE, GAS-COOLED REACTOR

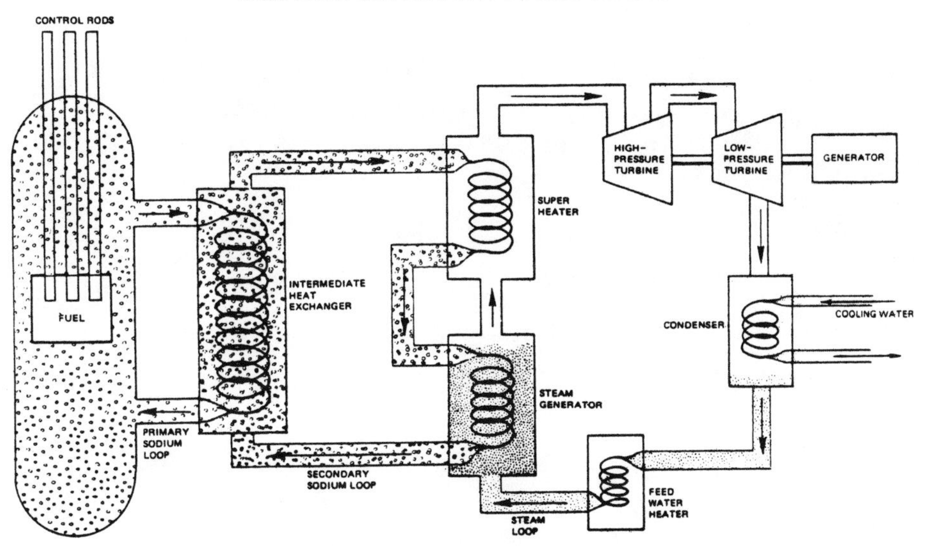


LIQUID METAL FAST BREEDER REACTOR

# RENEWABLES

SECTION 4

# Figure 58: Renewable Energy Sources. (Source 9)

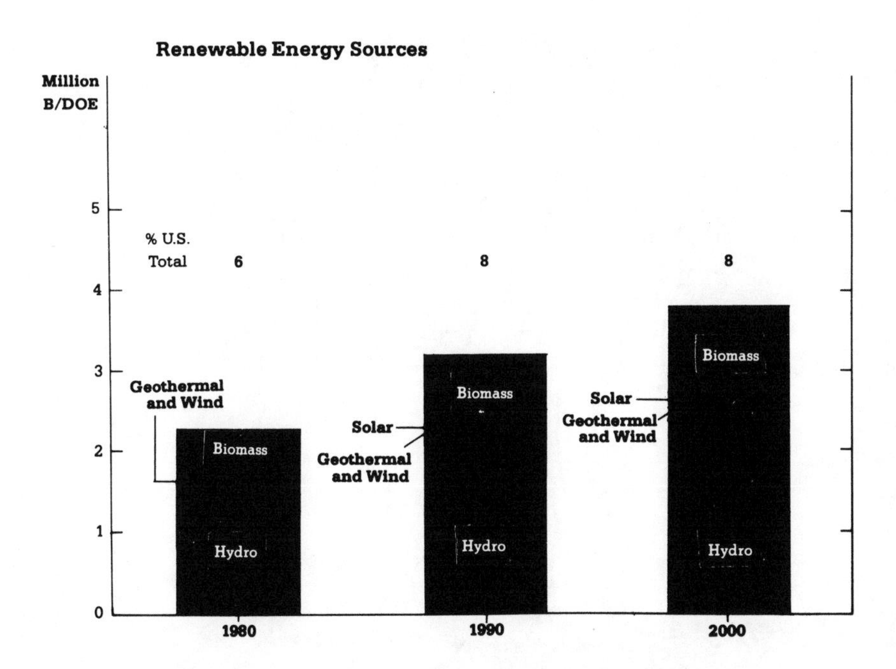

**Figure 59: Annual Heating Degree Days (Base 65° F). (Source 29)**

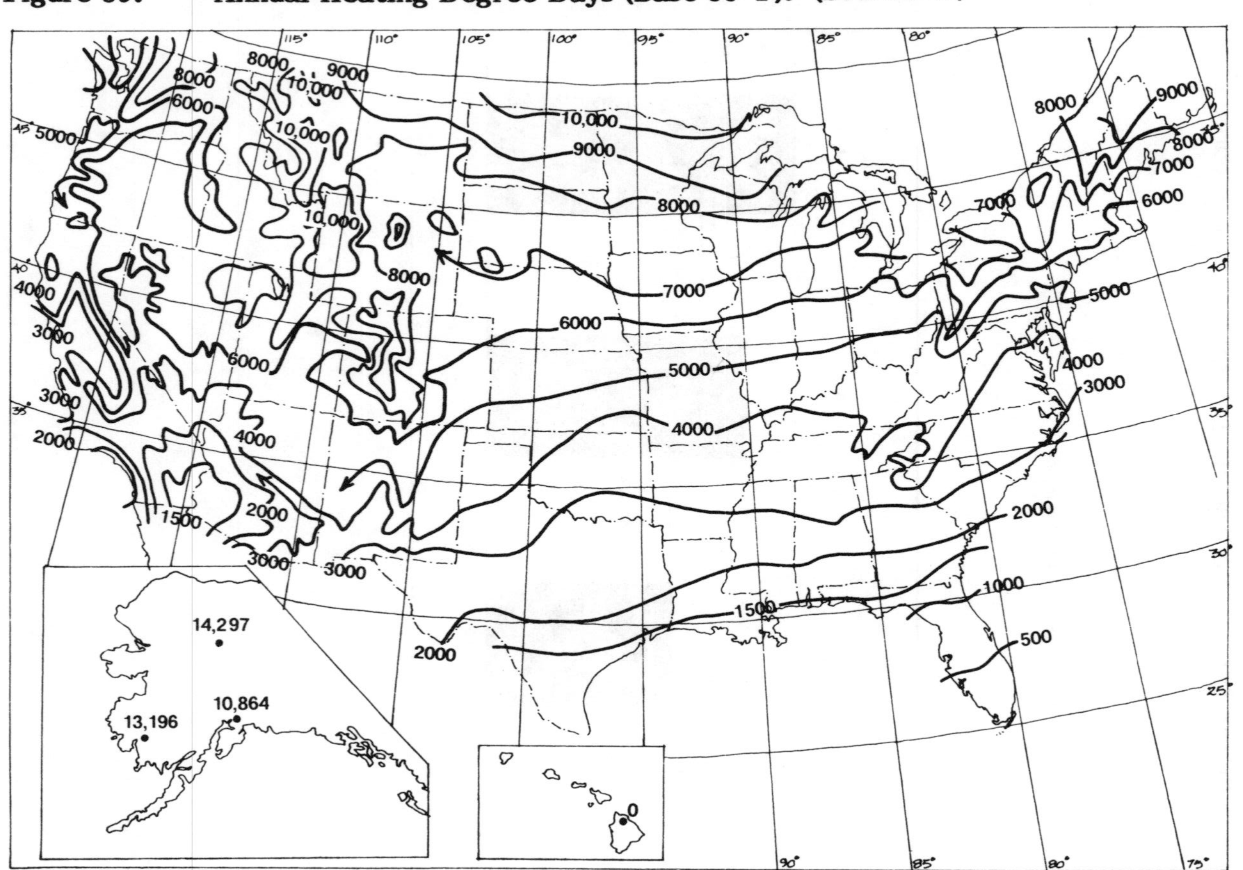

Figure 60: Annual Cooling Degree Hours (Base 85° F). (Source 29)

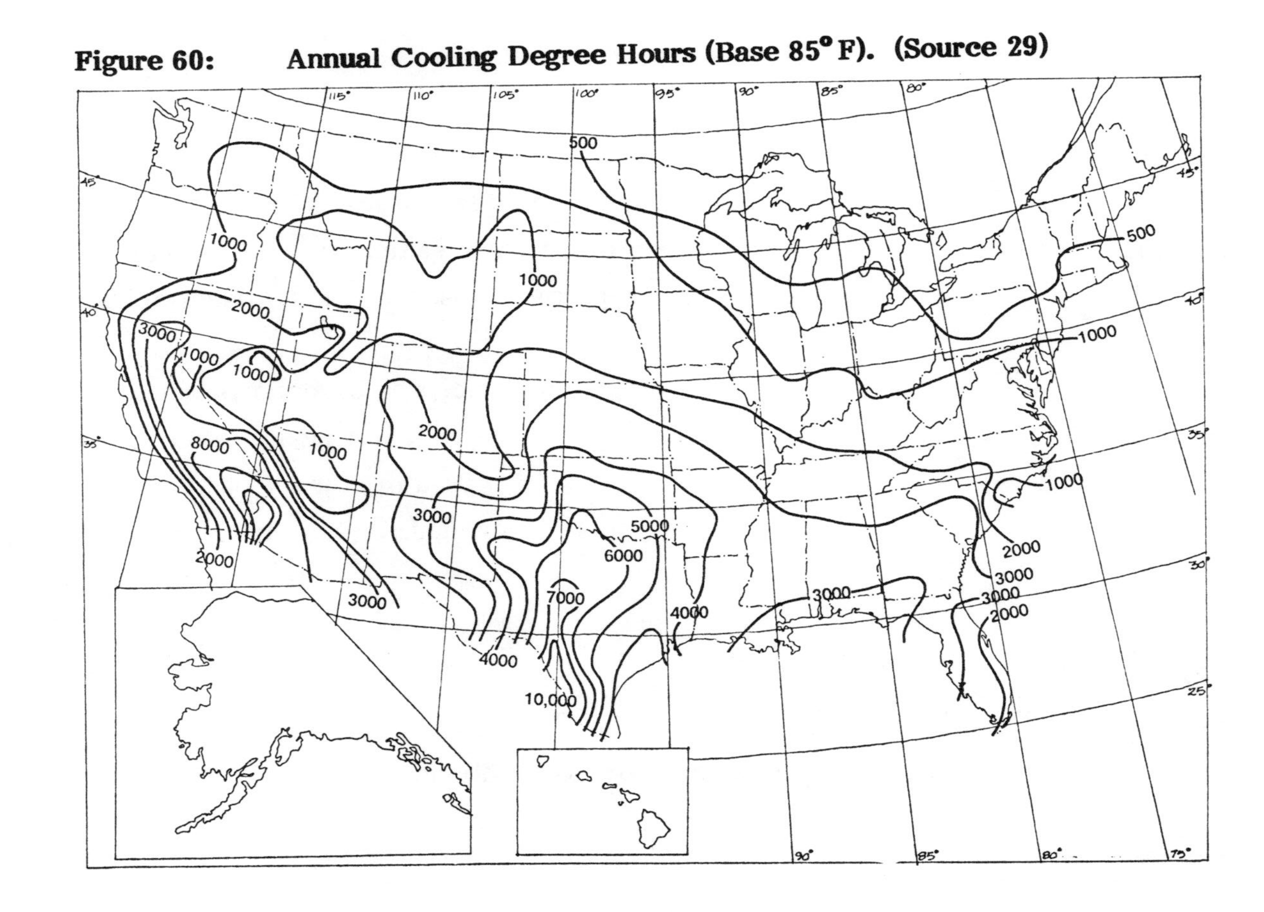

**Table 28: Energy Flow from the Sun. (Source 10)**

| | Watts (x $10^{12}$) |
|---|---|
| Incoming solar radiation | 174,000 |
| Primary disposition of solar radiation | |
| Direct conversion to heat | 82,000 |
| Reflected short wave | 52,000 |
| Evaporation, precipitation, etc. | 40,000 |
| Secondary disposition or use of solar radiation | |
| Weather effects: waves, winds, air flow | 370 |
| Photosynthesis, stored in plants | 40 |

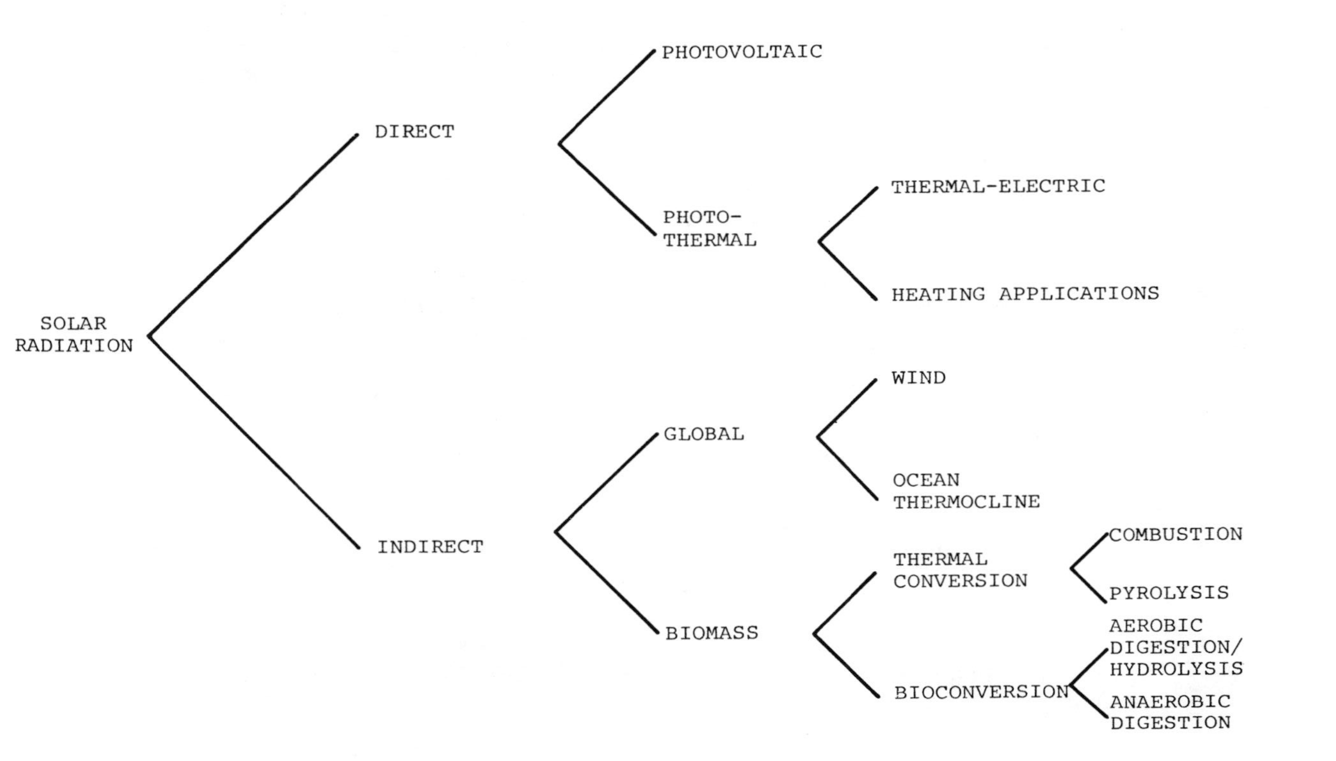


**Figure 61: Categorization of Solar Energy Technologies. (Source 15)**

**Table 29: Summary of Solar Energy Technology Products. (Source 15)**

| Technology | Product | Product Qualities |
|---|---|---|
| Photovoltaic | Electricity | Periodic, non-uniform output; low voltage; moderate, fluctuating current; d.c. |
| Thermal-Electric | Electricity | Periodic; central power station outputs* |
| Solar-Heat-concentrated | Heat | Periodic; small heated volume; temperature range to about 3,000° F |
| Solar Heat-diffuse | Heat | Periodic; water or air medium; temperature range to about 100° F |
| Wind | Mechanical Energy; Electricity | Fluctuating; site sensitive; unit output small to about 4 MWe |
| Ocean Thermocline | Electricity Combustible | Offshore site; central power station outputs*; Hydrogen; low impurities levels, 60, 960 BTU/lb |

*Central power station outputs are 60 hz electricity that can be conditioned for high or low voltage transmission. Power levels are of the order 100-1000MWe. Current outputs for network base-loading or load following capability are implied.

| *Solar technology* | *1977* | *2000 Base case*[2] | *2000 Maximum practical case* |
|---|---|---|---|
| | | *(in quads*[1]*)* | |
| Biomass | 1.8 | 3.1 | 5.4 |
| Hydro | 2.4 | 3.9 | 4.3 |
| high-head | (2.4) | (3.5) | (3.5) |
| low-head | (small) | (0.4) | (0.8) |
| Residential/commercial heating, hot water, cooling | small | 0.9 | 2.0 |
| Passive design | small | 0.2 | 1.0 |
| Industrial and agricultural process heat | — | 1.0 | 2.6 |
| Wind | — | 0.6 | 1.7 |
| Photovoltaics | — | 0.1 | 1.0 |
| Solar thermal electric | — | 0.1 | 0.4 |
| Ocean thermal energy conversion | — | — | 0.1 |
| Total | 4.2 | 9.9 | 18.5 |

[1] 1 quad per year = $10^{15}$ Btu or approximately 500,000 barrels of oil per day.

[2] Based on *Domestic Policy Review of Solar Energy* $25-oil scenario

**Table 30: Potential Fuel Displacement of Solar Technologies (Source 20)**

| Use | Year 1980 | 2000 | 2020 |
|---|---|---|---|
| | (quads, $10^{15}$ Btu) | | |
| Direct heating & cooling | 0.15 | 2.0 | 15 |
| Agricultural applications | 0.03 | 0.6 | 3 |
| Industrial applications | 0.02 | 0.4 | 2 |
| Equivalent fuel energy (wind photovoltaic, solar, thermal and ocean thermal) | 0.07 | 5.0 | 15 |
| Fuels from biomass | 0.5 | 3.0 | 10 |
| Total solar energy | ∿1Q | ∿10Q | ∿45Q |
| U.S. energy demand | ∿ 100Q | ∿150Q | ∿180Q |

**Table 31: Estimates of Heat, Electric Power, and Fuels to be Supplied by Solar Energy in the U.S. (Source 10)**

**Figure 62: Solar Energy Flow to Various Applications. (Source 10)**

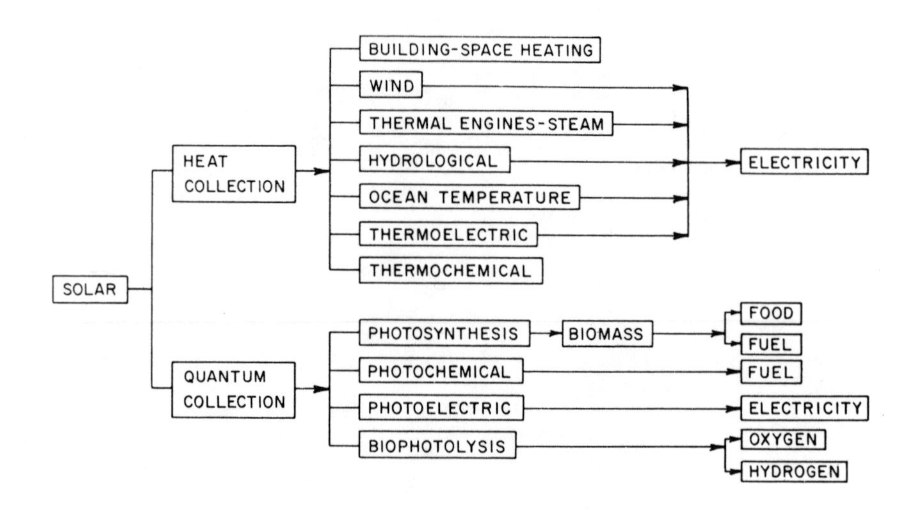


**Figure 63: Solar Energy to Electricity. (Source 10)**

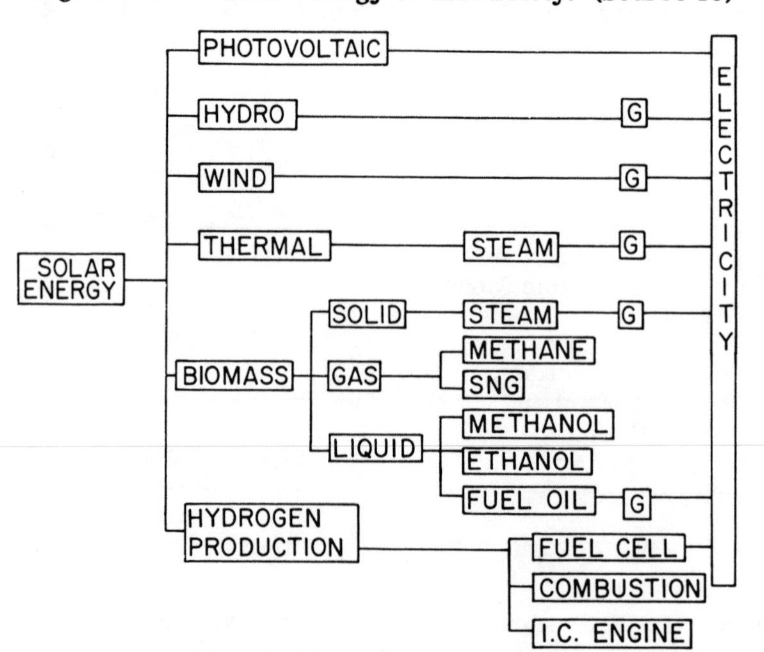

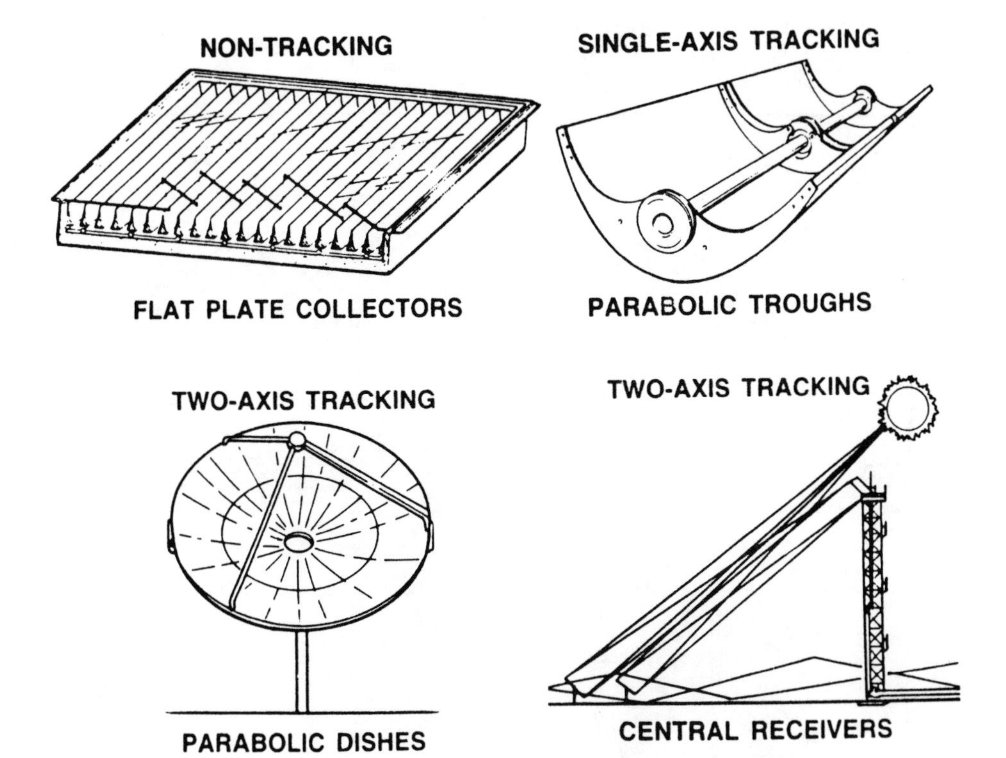


Figure 64: Collector Types. (Source 15)

Figure 65: Solar Hot Water Systems: Operational Description. (Source 22)

# SOLAR HOT WATER SYSTEMS:

# Basics of Solar Energy Utilization

## Collector Orientation and Tilt

Solar collectors must be orientea and tilted within prescribed limits to receive the optimum level of solar radiation for system operation and performance.

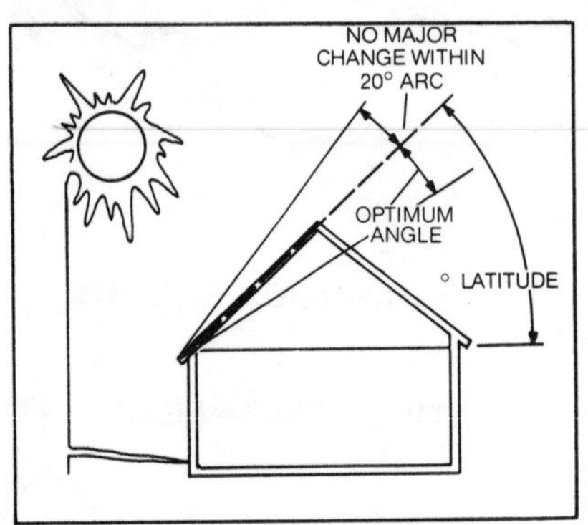


**Collector Tilt for Domestic Hot Water.** The optimum collector tilt for domestic water heating alone is usually equal to the site latitude. Variations of 10 degrees on either side of the optimum are acceptable.

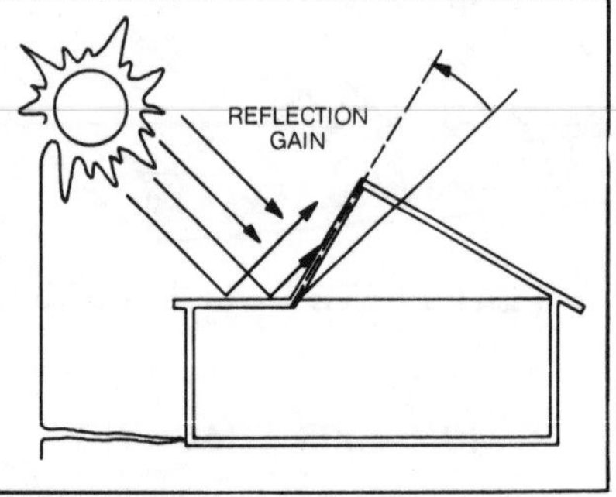


**Modification of Optimum Collector Tilt.** A greater gain in solar radiation collection sometimes may be achieved by tilting the collector away from the optimum in order to capture radiation reflected from adjacent ground or building surfaces. The corresponding reduction of radiation directly striking the collector, due to non-optimum tilt, should be recognized when considering this option.

**Snowfall Consideration.** The snowfall characteristics of an area may influence the appropriateness of these optimum collector tilts. Snow buildup on the collector, or drifting in front of the collector, should be avoided.

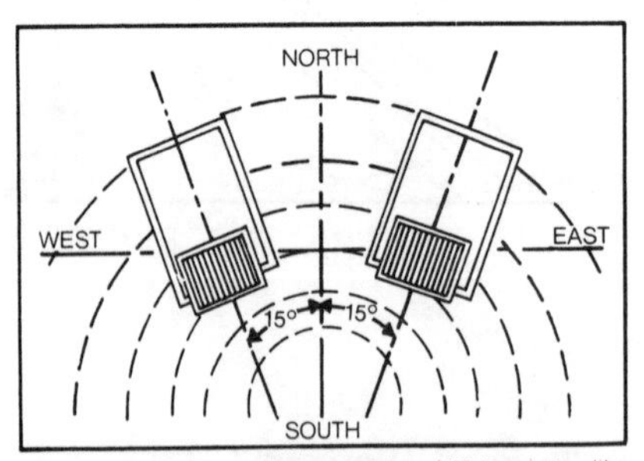


**Collector Orientation.** A collector orientation of 15 degrees to either side of true South is acceptable. However, local climate and collector type may influence the choice between East or West deviations.

**Figure 66: Solar Hot Water Systems: Collector Orientation and Tilt. (Source 22)**

# SOLAR HOT WATER SYSTEMS:

## Operational Description

The solar hot water system usually is designed to preheat water from the incoming water supply prior to passage through a conventional water heater. The domestic hot water preheat system can be combined with a solar heating system or designed as a separate system. Both situations are illustrated below.

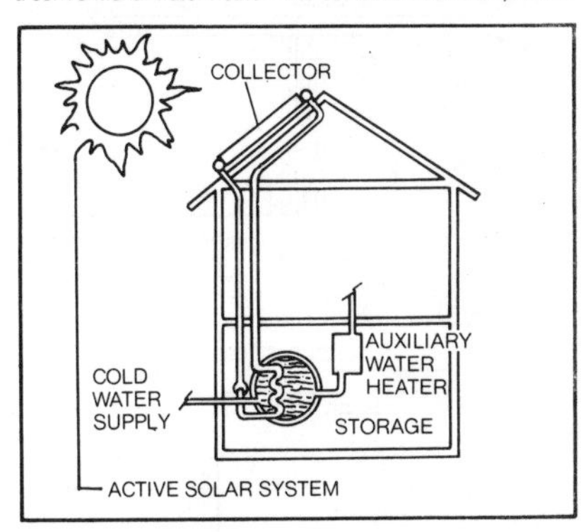


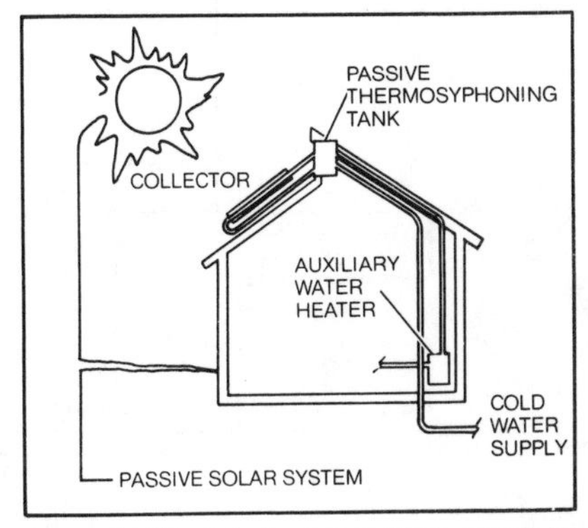


**Domestic Hot Water Preheating—Separate System.** Domestic hot water preheating may be the only solar system included in many designs. An active solar system is shown in the upper figure and a passive thermosyphoning arrangement in the lower.

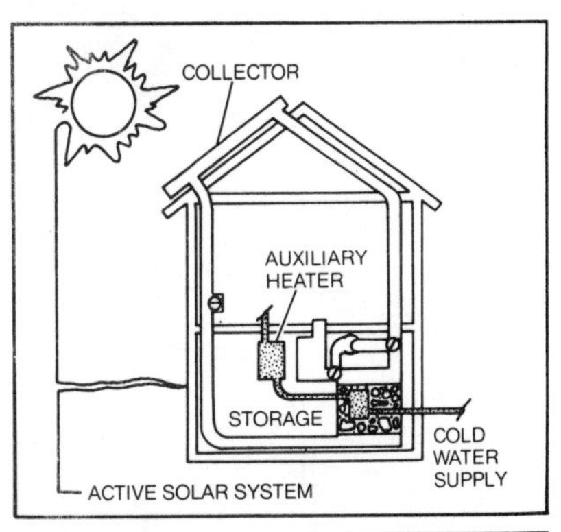


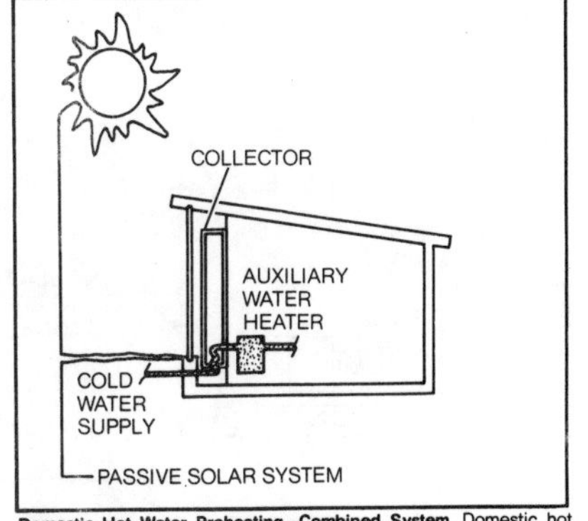


**Domestic Hot Water Preheating—Combined System.** Domestic hot water is preheated as it passes through heat storage enroute to the conventional water heater. An active solar system using air for heat transport is shown in the upper figure and a passive solar system in the lower.

**Figure 67: Active Solar Water and Space Heating System. (Source 26)**

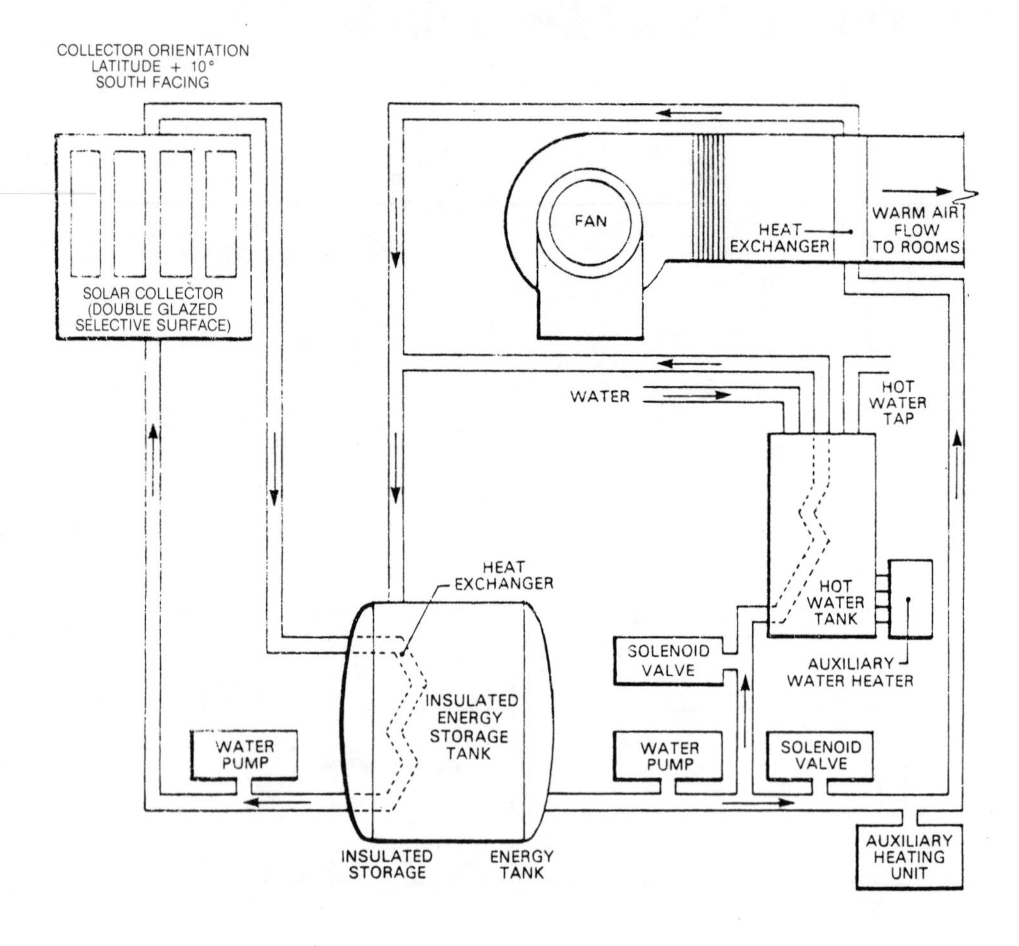

**Figure 68: Solar Hot Water Systems: Active Systems. (Source 22)**

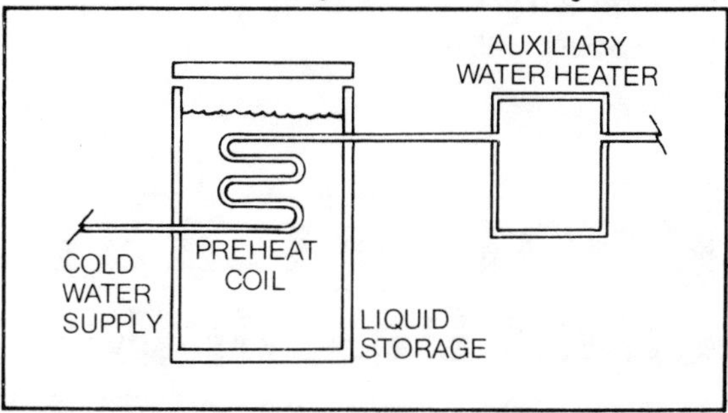


**Preheat Coil in Storage.** Water is passed through a suitably sized coil placed in storage enroute to the conventional water heater. Unless the preheat coil has a protective double wall construction, this method can only be used for solar systems employing non-toxic storage media.

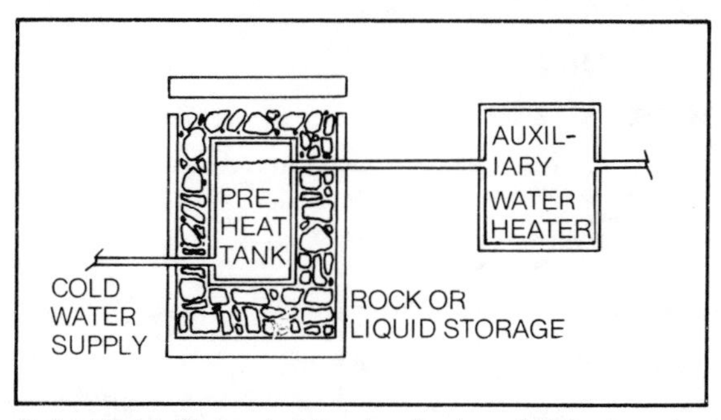


**Preheat Tank in Storage.** In this system, the domestic hot water preheat tank is located within the heat storage. The water supply passes through storage to the preheat tank where it is heated and stored, and later piped to a conventional water heater as needed. A protective double-wall construction again will be necessary unless a non-toxic storage medium is used.

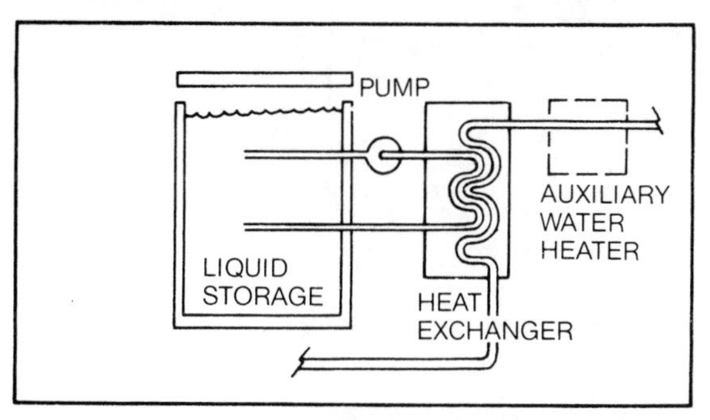


**Preheat Outside of Storage.** In this preheat method, the heat transfer liquid in storage is pumped through a separate heat exchanger to be used for domestic hot water preheating. This separate heat exchanger could be the conventional water heater itself. However, if the liquid from storage is toxic, the required separation of liquids is achieved by the use of a double wall exchanger, as diagrammed, in which the water supply simply passes through enroute to the conventional water heater.

**Figure 69: Solar Hot Water Systems: Component Description. (Source 22)**

# SOLAR HOT WATER SYSTEMS: Component Description

As noted in the functional description, a solar domestic hot water system is composed of numerous individual parts and pieces including: collectors; storage; a distribution network with pipes, pumps and valves; insulation; a system of manual or automatic controls; and possibly heat exchangers, expansion tanks and filters. These parts are assembled in a variety of combinations depending on function, component compatibility, climatic conditions, required performance, site characteristics and architectural requirements, to form a solar domestic hot water system. Some components that are unique to the collector system or that are used in an unconventional manner are briefly illustrated and discussed in the next few pages.

## Flat-Plate Collectors: An Exploded View

The flat-plate collector is a common solar collection device used for domestic water heating. Most collectors are designed to use liquid (usually treated water) as the heat transfer medium. However, air systems are available for domestic water heating. Most flat-plate collectors consist of the same general components, as illustrated below.

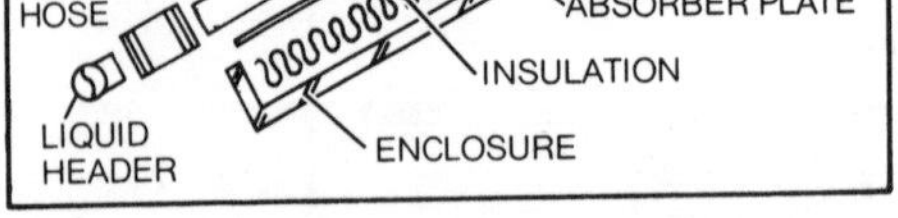

**Batten.** Battens serve to hold down the cover plate(s) and provide a weather tight seal between the enclosure and the cover.

**Cover Plate.** The cover plate usually consists of one or more layers of glass or plastic film or combinations thereof. The cover plate is separated from the absorber plate to reduce reradiation and to create an air space, which traps heat by reducing convective losses. This space between the cover and absorber can be evacuated to further reduce convective losses.

**Heat Transfer Fluid Passage.** Tubes are attached above, below or integral with an absorber plate for the purpose of transferring thermal energy from the absorber plate to a heat transfer medium. The largest variation in flat-plate collector design occurs with this component and its combination with the absorber plate. Tube on plate, integral tube and sheet, open channel flow, corrugated sheets, deformed sheets, extruded sheets and finned tubes are some of the techniques used.

**Absorber Plate.** Since the absorber plate must have a good thermal bond with the fluid passages, an absorber plate integral with the heat transfer media passages is common. The absorber plate is usually metallic, and normally treated with a surface coating which improves absorptivity. Black or dark paints or selective coatings are used for this purpose. The design of this passage and plate combination helps determine a solar system's effectiveness.

**Insulation.** Insulation is employed to reduce heat loss through the back of the collector. The insulation must be suitable for the high temperature that may occur under no-flow or dry-plate conditions, or even normal collection operation. Thermal decomposition and outgassing of the insulation must be prevented.

**Enclosure.** The enclosure is a container for all the above components. The assembly is usually weatherproof. Preventing dust, wind and water from coming in contact with the cover plate and insulation is essential to maintaining collector performance.

**Figure 70: Passive Solar Heating: Direct Gain. (Source 21)**

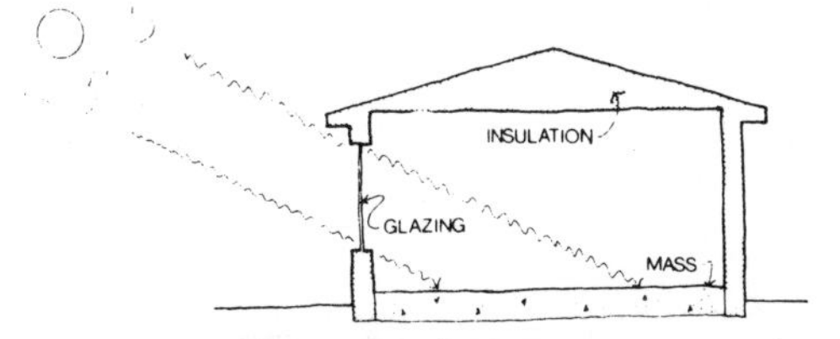


**Figure 71: Passive Solar Heating: Thermal Storage Wall. (Source 21)**

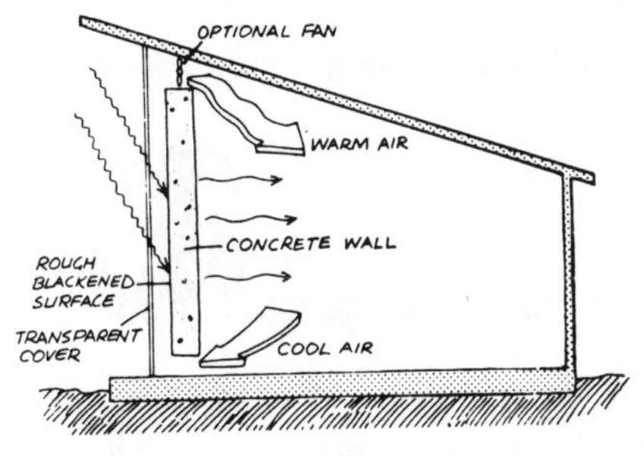


**Figure 72: Passive Solar Heating: Thermal Storage Roof. (Source 21)**

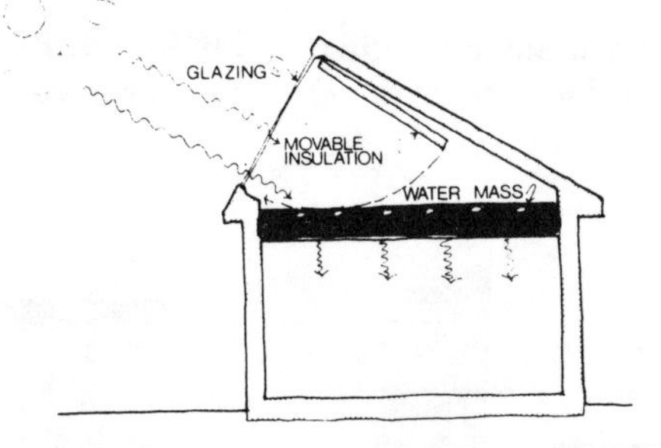


**Figure 73: Passive Solar Heating: Sunspace. (Source 21)**

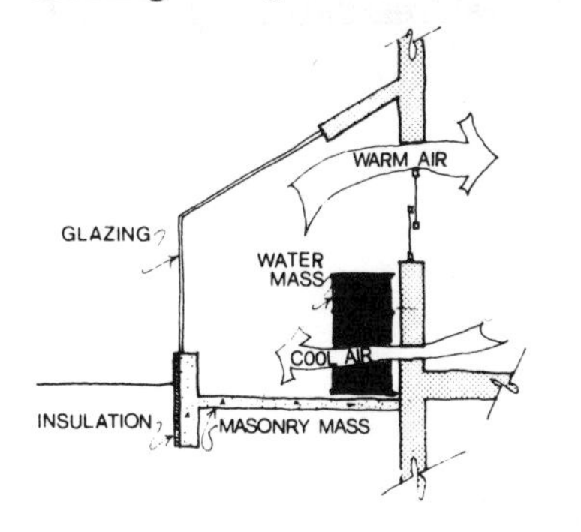


**Figure 74: Passive Solar Heating: Convective Loop. (Source 21)**

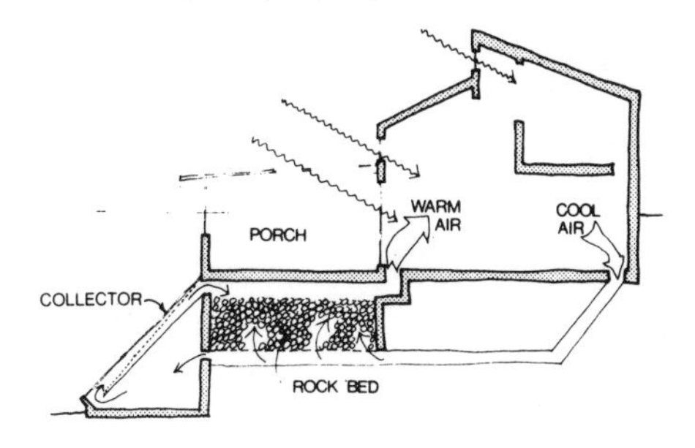

**Figure 75: Passive Solar Cooling: Night Air Cooling. (Source 21)**

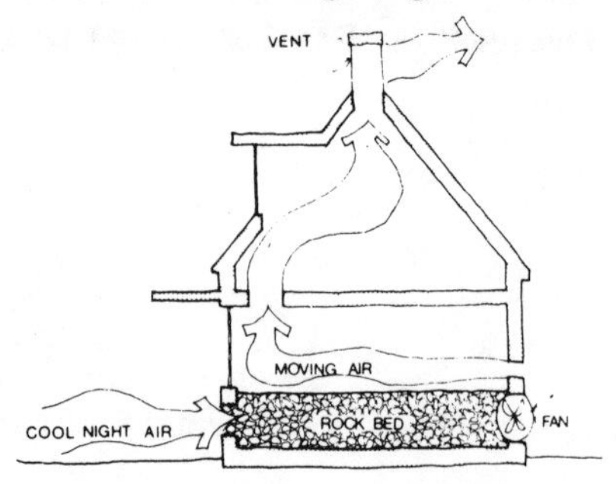


**Figure 76: Passive Solar Cooling: Night Sky Radiation. (Source 21)**

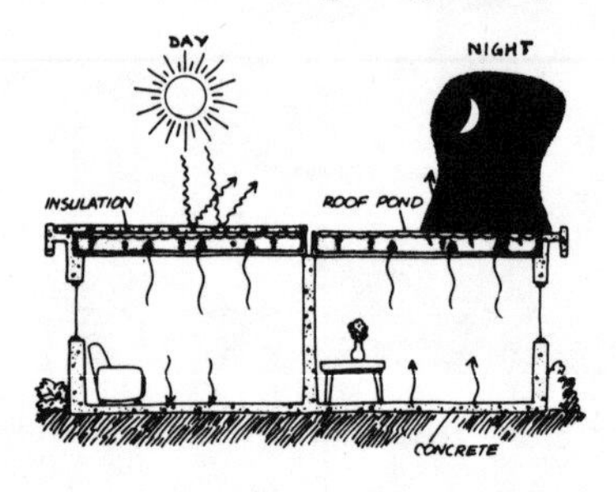


**Figure 77: Passive Solar Cooling: Ground Cooling. (Source 21)**

**Figure 78: Passive Solar Cooling: Evaporative Cooling. (Source 21)**

**Figure 79: Passive Solar Cooling: Natural Ventilation. (Source 21)**

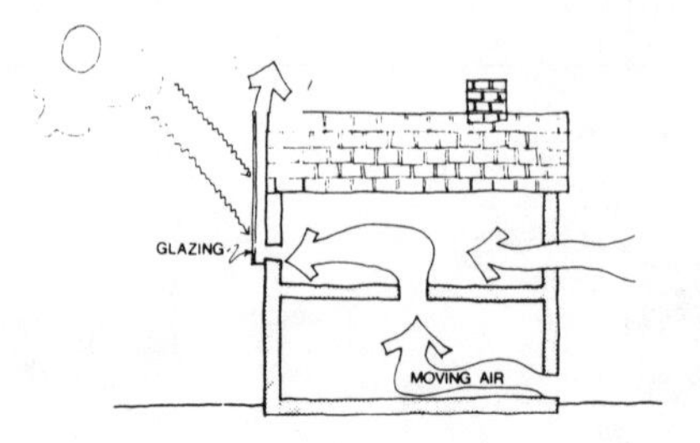

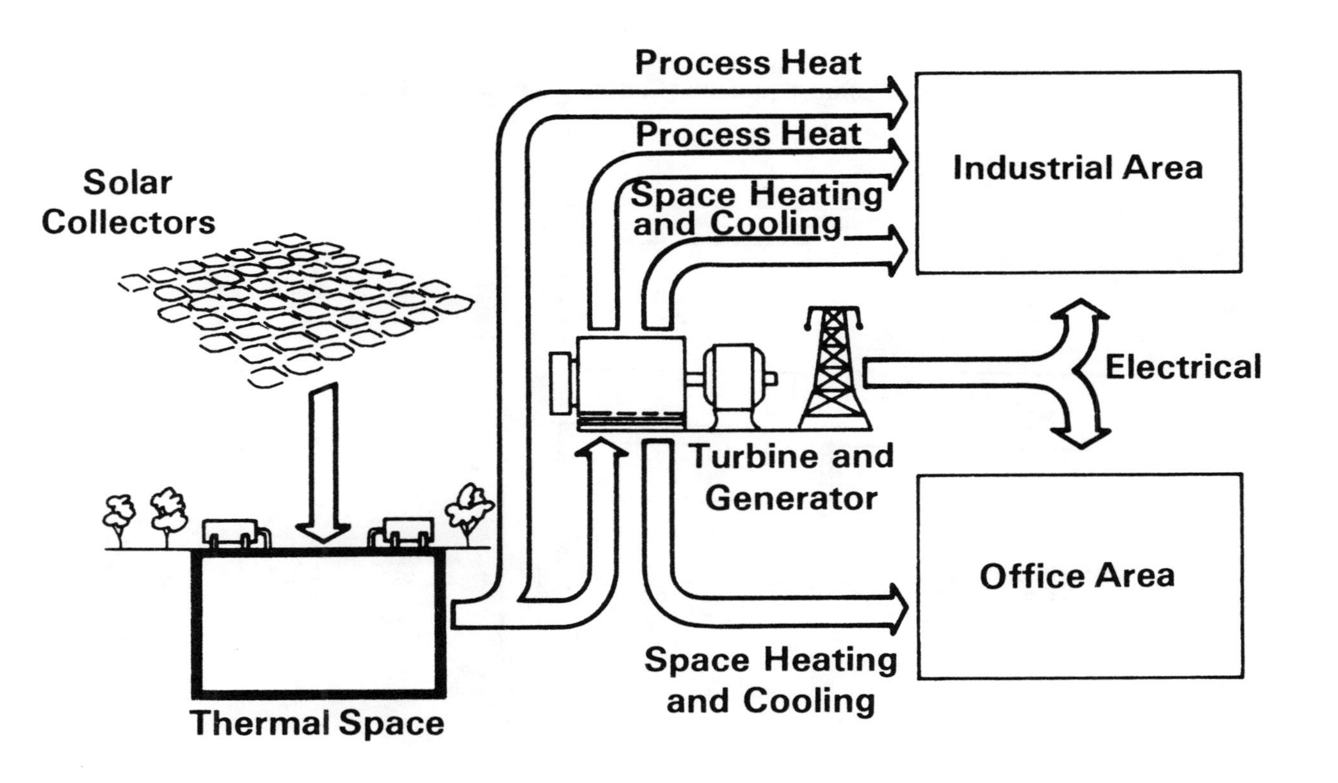


Figure 80: Solar Total Energy Application. (Source 20)

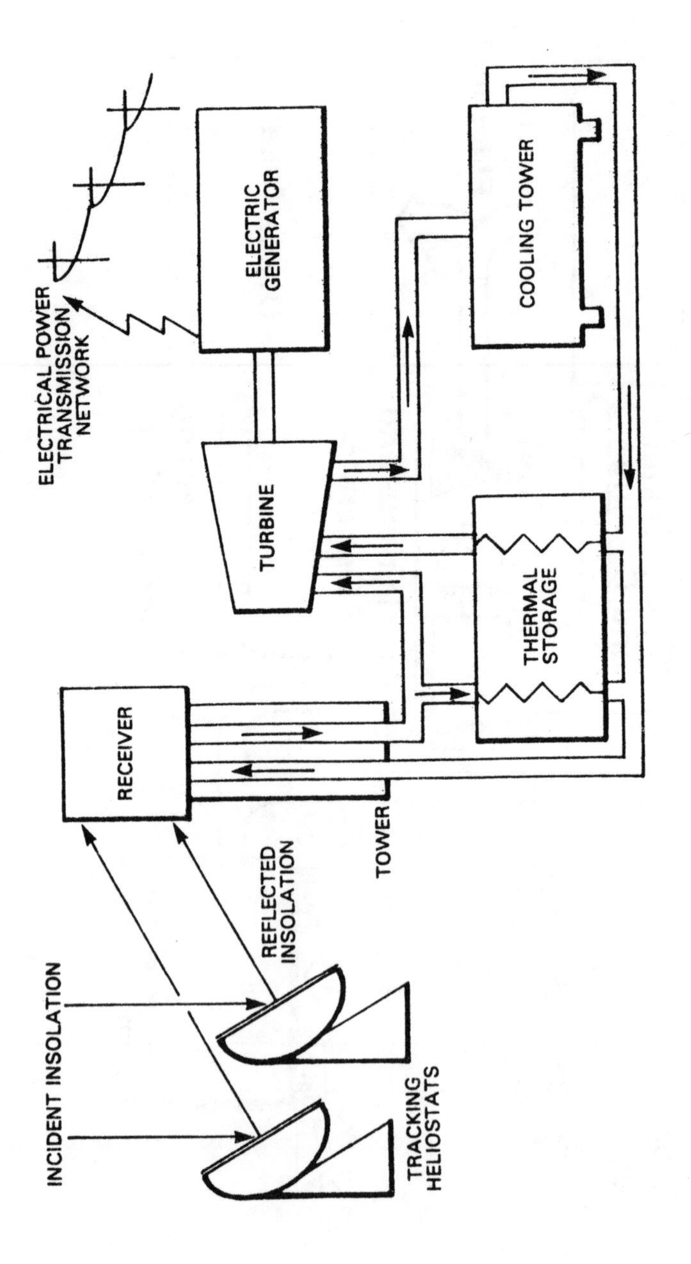


**Figure 81: Central Receiver Solar Thermal Power System. (Source 29)**

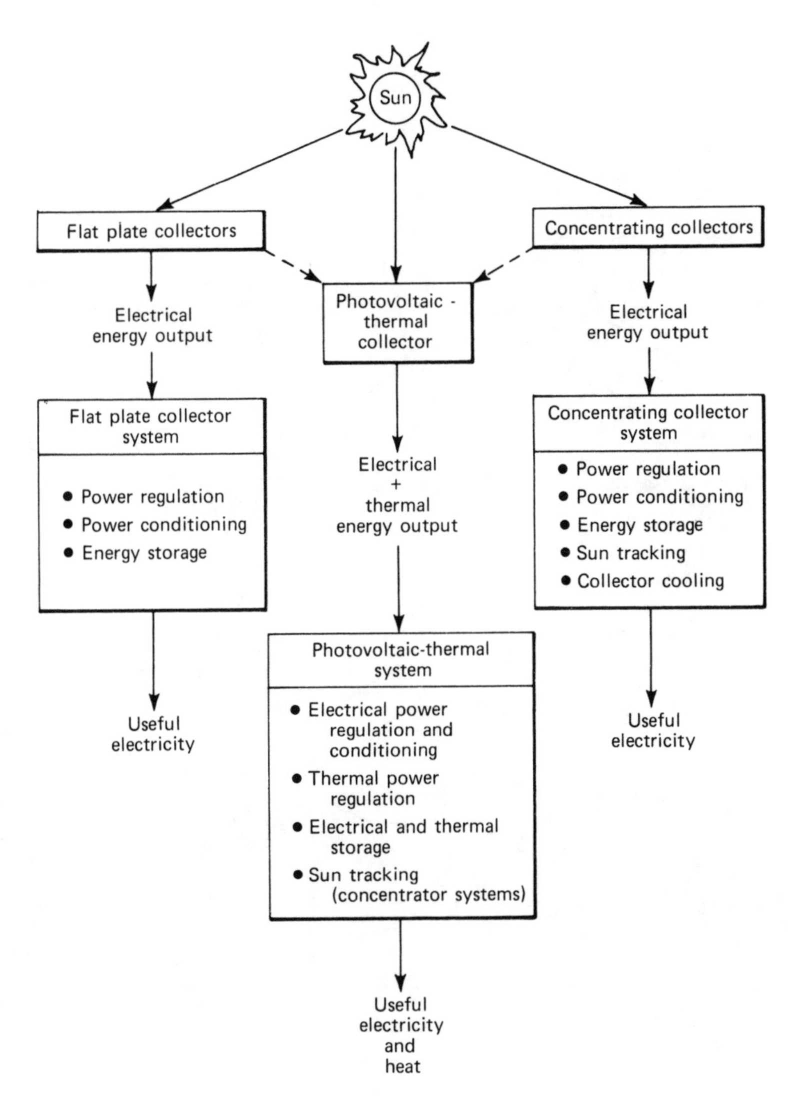


**Figure 82:**

**Terrestial Photovoltaic Solar Collector Systems. (Source 15)**

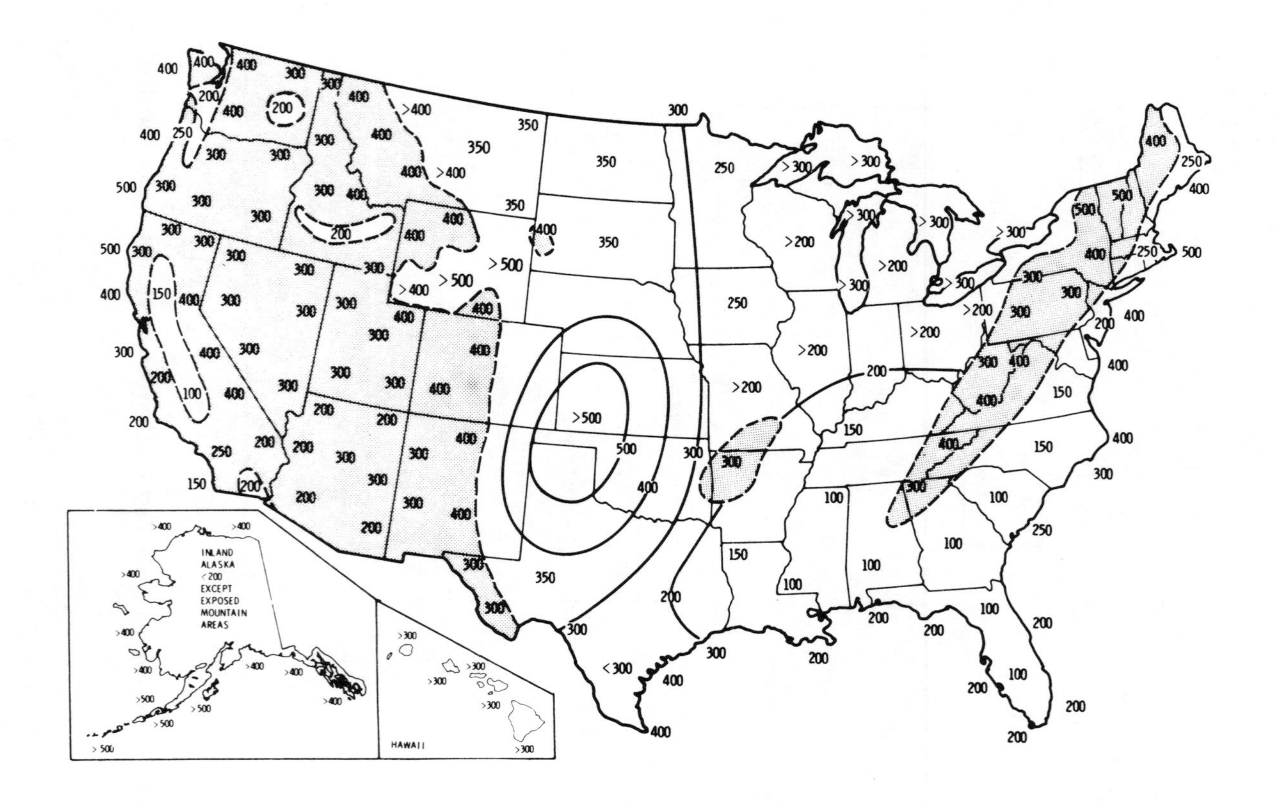


**Figure 83: Mean Annual Wind Power ($W/m^2$) Estimated at 50m Above Exposed Areas. (Source 4)**

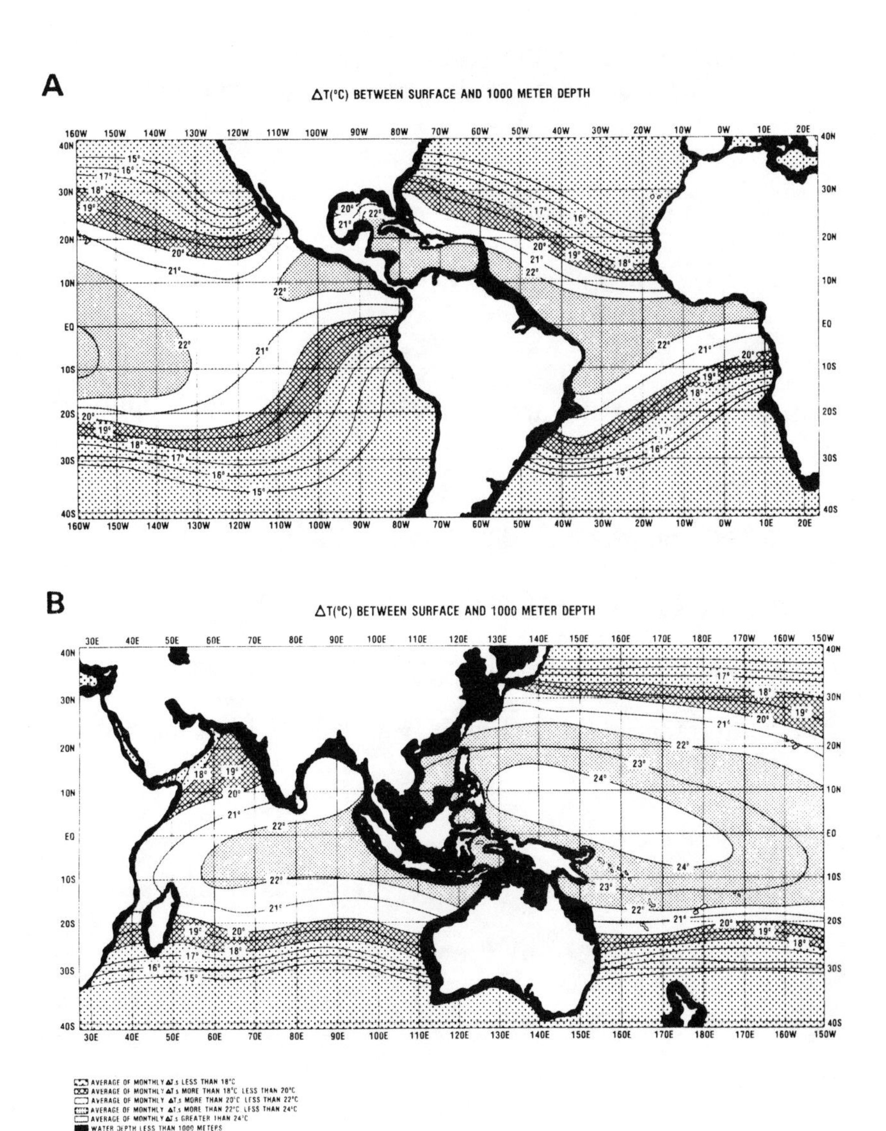


**Figure 84:**

**Contours Showing Annual Average Monthly Temperature Differences Between the Ocean Surface and Depth of 1,000 Meters. (Source 15)**

**Figure 85: OTEC Electric Energy Potential vs U.S. and Worldwide Demands (Source 17)**

Energy potentially available from OTEC

Ocean Area suitable for OTEC plantships = $60 \times 10^6$ $km^2$

($\Delta T \geq 22°C$)

Estimated operating area per 325 MW plantship = 2000 $km^2$

Total OTEC onboard power generation capability:

30,000 plantships x 325 MW $\simeq 10 \times 10^6$ MW

Onboard electrical energy $\simeq 90 \times 10^{12}$ kWh/annum

Electrical energy generation potential at land sites via

OTEC ammonia and fuel cells $\simeq 45 \times 10^{12}$ kWh/annum

Energy demand

Estimated worldwide electrical energy consumption

(1978) = $8 \times 10^{12}$ kWh/annum

US electrical energy consumption (1978) = $2.2 \times 10^{12}$

kWh/annum

**Figure 86:** **Geothermal Resource Potential in the United States for Direct Heat Applications. (Source 20)**

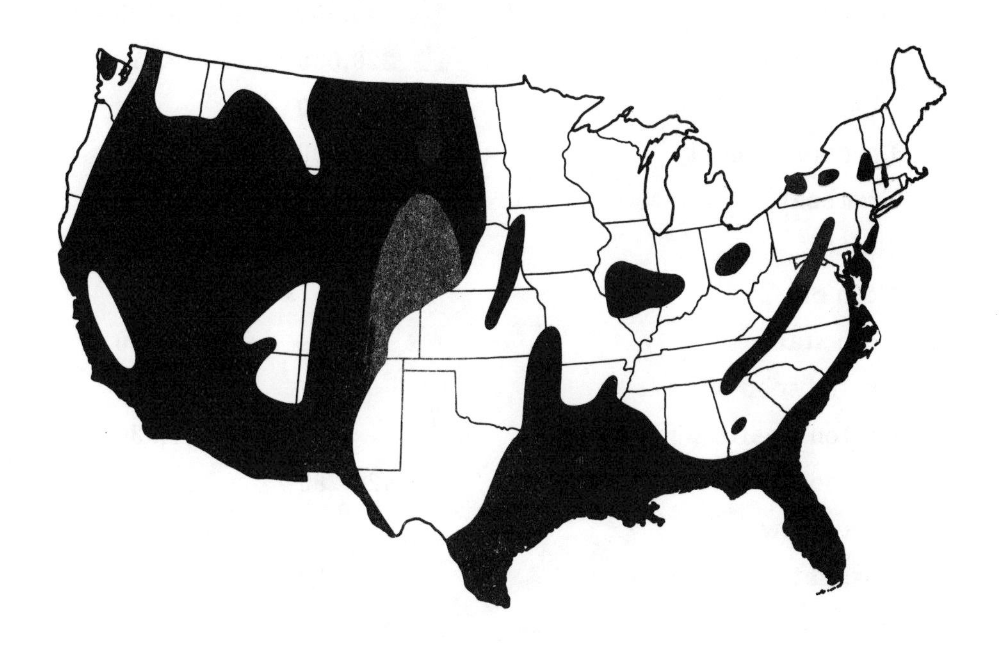

**Table 32:** **Potential Geothermal Energy Production. (Source 20)**

| | *1985* | *2000* | *2020* |
|---|---|---|---|
| | | (quads per year) | |
| Hydrothermal resources[1] | 0.34 | 3.0 | 10.0 |
| Geopressured resources[2] | 0.021 | 3.2 | 7.4 |
| Hot dry rock resources[1] | 0.002 | 0.007 | 1.1 |
| Total | 0.4 | 6.2 | 18.5 |

[1] Includes fossil fuel equivalent of electric power and direct heat applications.

[2] Includes methane and fossil fuel equivalent of electric power.

**Table 33: Approximate Heating Values of Various Fossil and Biomass Fuels (Source 10)**

| Product | $lb/ft^3$ | $Btu/ft^3$ | Btu/lb |
|---|---|---|---|
| Bagasse | 5.04 | 45,500 | 9,100 |
| Bark, wood | 24.0 | 220,000 | 9,010 |
| Biogas (58-70% methane) | | $\sim$620 | |
| Bituminous coal | 52 | 676,000 | 13,000 |
| Charcoal | 15.3 | 198,000 | 13,000 |
| Coffee | | | 11,420 |
| Corn oil | | | 17,133 |
| Corn stalks | | | $\sim$6,000 |
| Cotton seed oil | | | 16,825 |
| Cotton wastes (dry) | | | 7,930 |
| Ethanol | 49.2 | 628,280 | 12,770 |
| Flax straw | | | 8,590 |
| Gas, natural | | 1,000 | |
| Gasoline | 45 | 899,100 | 19,980 |
| Hogged fuel | 18 | 162,000 | 9,000 |
| Methane | .0417 | 1,025 | 24,580 |
| Methanol | 48.8 | 476,190 | 9,758 |
| Municipal wastes | 10 | 56,450 | 5,650 |
| Peanut husks | | | 8,487 |
| Peanut oil | | | 17,249 |
| Rice hulls | | | 6,280 |
| Soybean oil | | | 16,920 |
| Straw | | | 8,143 |
| Sunflower hulls | | | 8,600 |
| Wheat straw | | | 8,143 |
| Wood, dry | 28 | 238,000 | $\sim$8,500 |

**Table 34: Potential United States Energy Impacts from Biomass, Quads. (Source 10)**

(1 quad = $10^{15}$ Btu)

| Source | Year 2000 | Year 2020 | Conditions |
|---|---|---|---|
| Urban solid wastes | 0.25 | 0.25 | 75% of available urban wastes to be used by year 2000 |
| Agriculture residues (1.5 ton/acre annual average) | 0.68 | 1.88 | 30% of available residues available by year 2000; 50% by year 2020 |
| Forest residues (1.5 ton/acre) | 0.62 | 1.73 | 30% available residues to be used by year 2000; 50% by year 2020 |
| Terrestrial % aquatic biomass (yields to 30 ton/acre by year 2020) | 5.05 | 10.05 | 50 million acres available by 2000; 67 million acres available by 2020; 30% by year 2000; 50% by year 2020 |
| Marine biomass | 0.02 | 0.22 | One 100,000 acre farm by year 2000; ten 100,000 acre farms by 2020. |
| Total | 6.62 | 14.13 | |

Figure 87: Biomass Energy Paths. (Source 10)

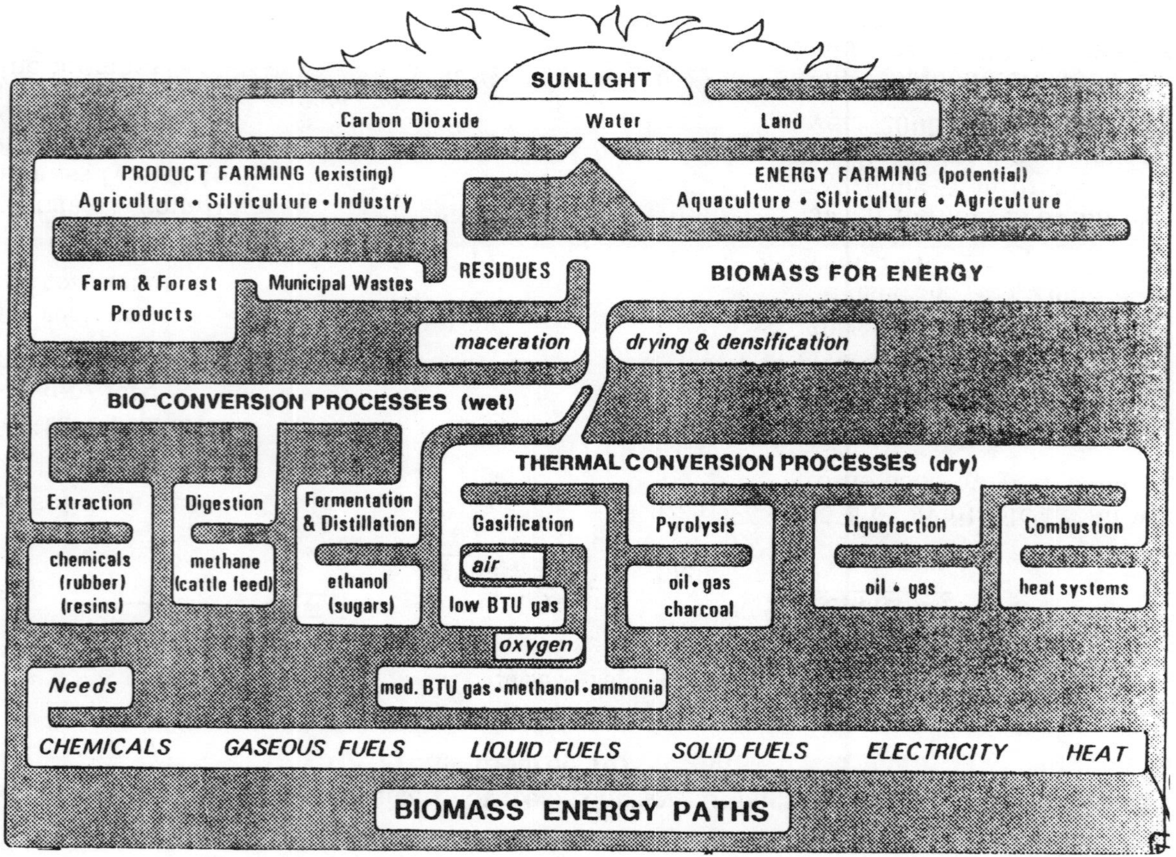

Figure 88: Time for Production of Biomass Energy and Fossil Fuel. (Source 10)

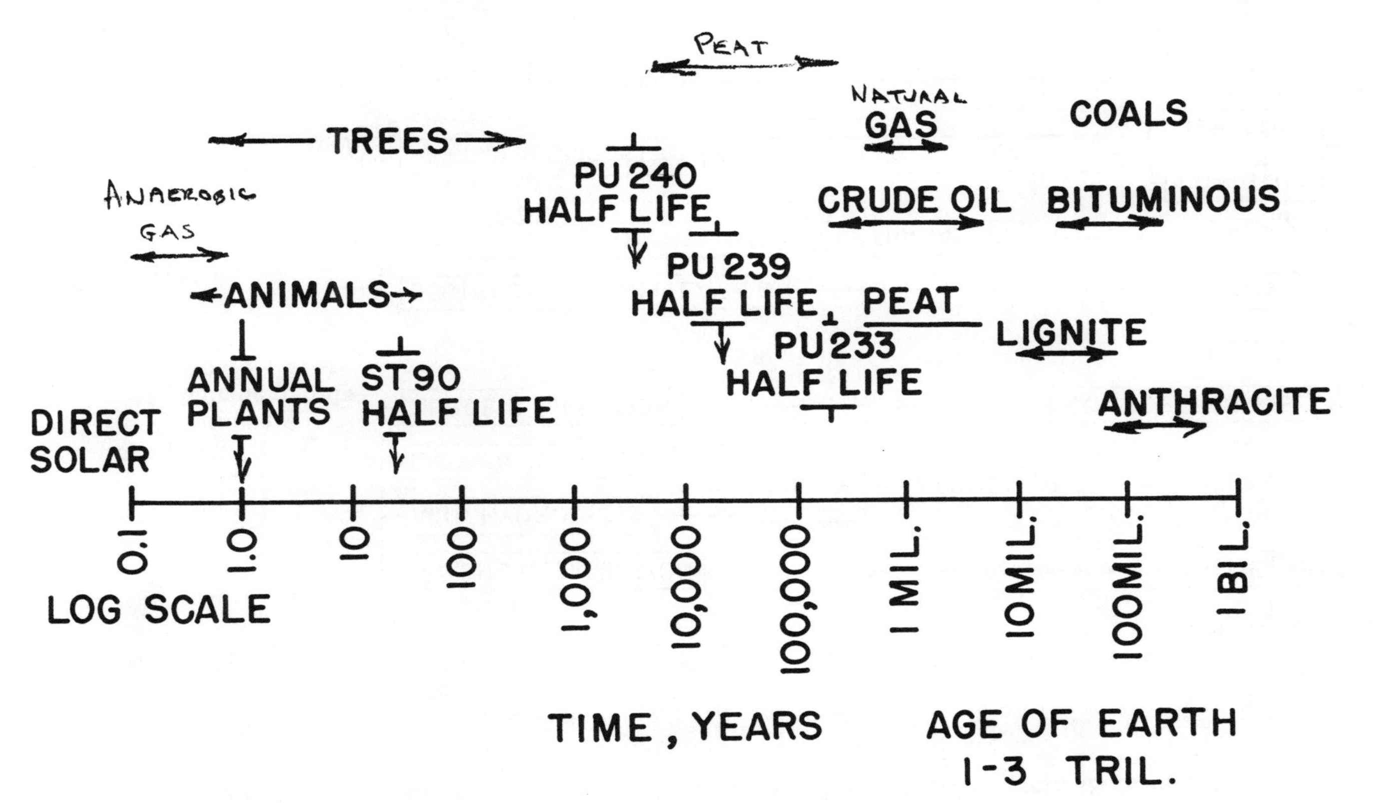

Figure 89: Major Processes for Conversion of Biomass for Fuel. (Source 10)

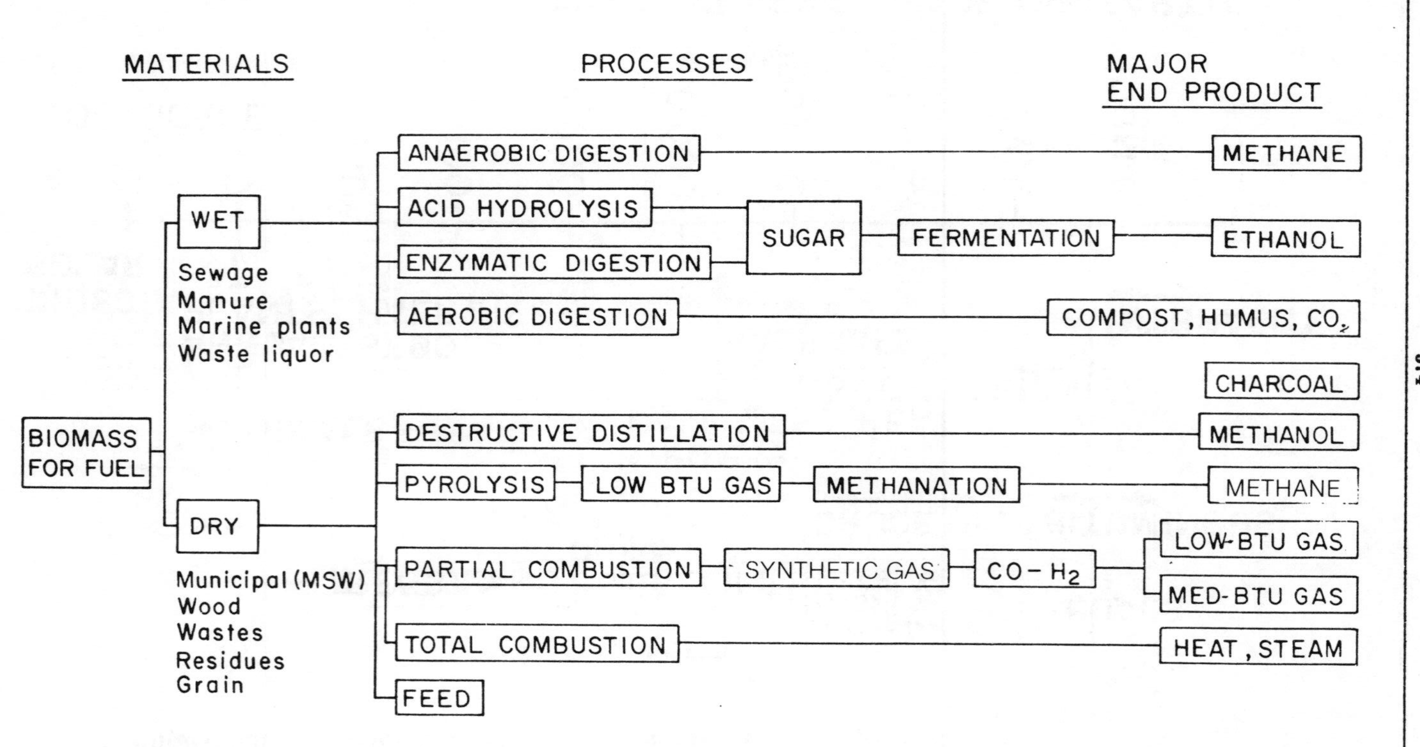

Figure 90: Biomass for Electricity and Gas Utility Production. (Source 10)

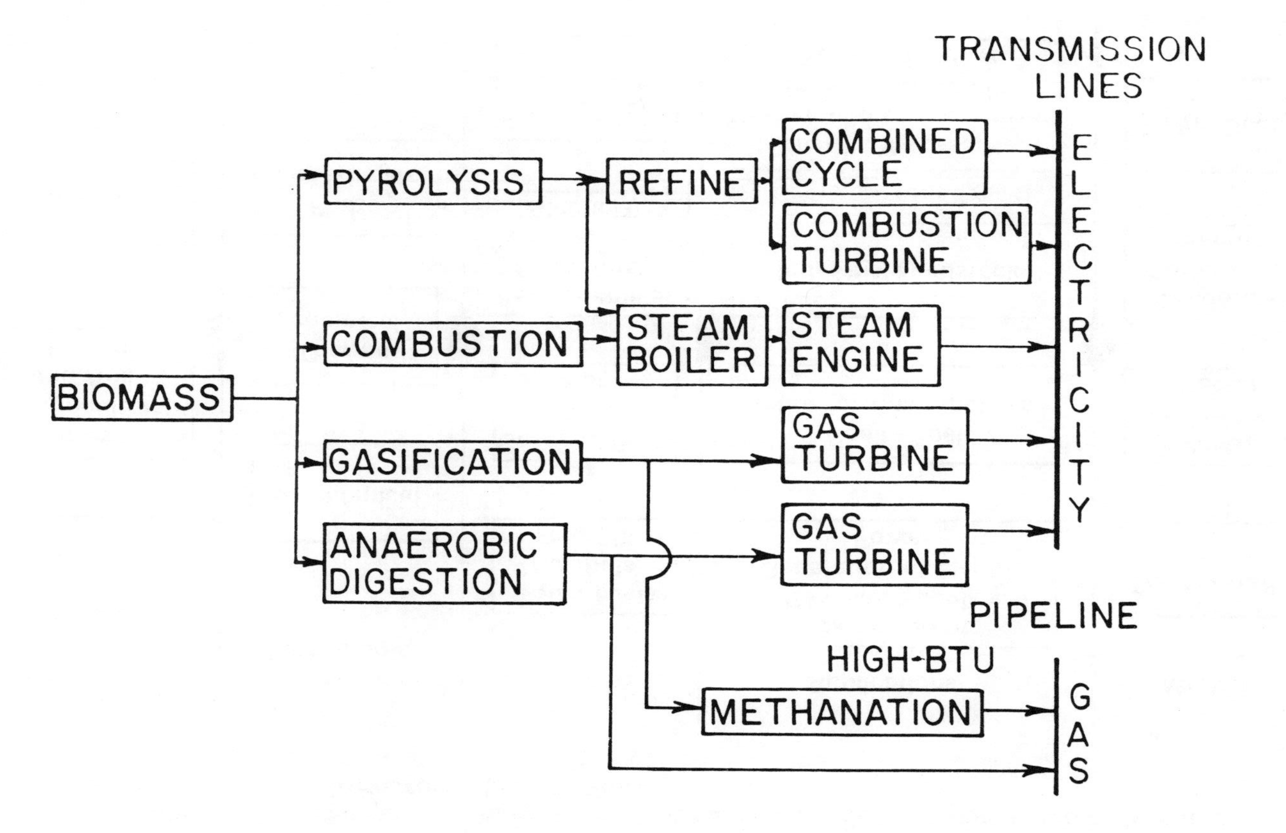

Figure 91: Scope of Markets and Applications for Products from Biomass Resources. (Source 20)

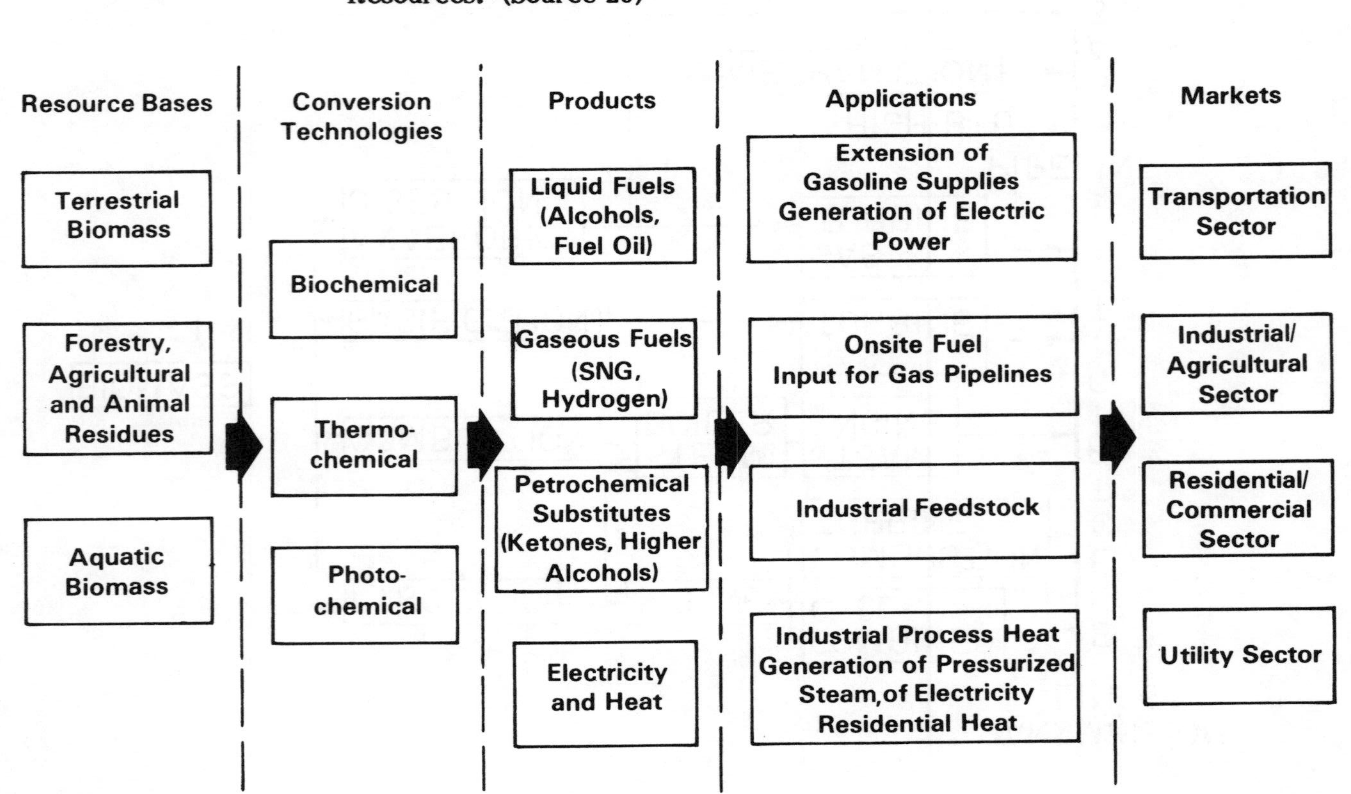

# Table 35: Properties of Methanol, Ethanol, Propanol, Butanol, and Gasoline. (Source 10)

| Property | Methanol | Ethanol | Propanol | Butanol | Gasoline |
|---|---|---|---|---|---|
| Formula | $CH_3OH$ | $C_2H_5OH$ | $C_3H_7OH$ | $C_4H_9OH$ | $C_8H_{17}$ |
| Specific gravity (SG) | 0.796 | 0.789 | 0.805 | 0.81 | 0.72-0.78 |
| Molecular weight | 32 | 46 | 60 | 74 | 113± |
| Carbon, % weight | 37.5 | 52.2 | 60 | 64.9 | 85-88 |
| Heating value (high), Btu/lb, 68° F | 8,570 | 11,500 | 14,436 | 15,556 | 18,900 |
| Latent heat of vaporization, Btu/lb, 68° F | 506 | 396 | 322 | 254 | 150 |
| Air-fuel ratio, stoichio-metric | 6.45 | 9.0 | | | 14.7± |
| Explosive limits, air volume% | 6-36.5 | 4.3-19 | 2.6-13.5 | 1.4-18 | 1.4-5.2 |
| Boiling point, °F (°C) | 149 (65) | 172 (78) | 206 (97) | 261 (127) | 365 (185)± |
| Flash point, °F (°C) | 64 (18) | 64 (18) | 95 (35) | 95 (35) | -50 (-45) |
| Autoignition temp., °F (°C) | 873 (467) | 752 (400) | 700 (371) | 650 (343) | 532 (278)± |
| Solubility in water | miscible | miscible | miscible | 7% solub. | no |

Specific heat is cal/g°C or Btu/lb°F. To convert to J/kgK, multiply by 4184.

To convert Btu/lb to J/kg, multiply by 2324

**Figure 92: Estimated Fuel Production Costs. (Source 10)**

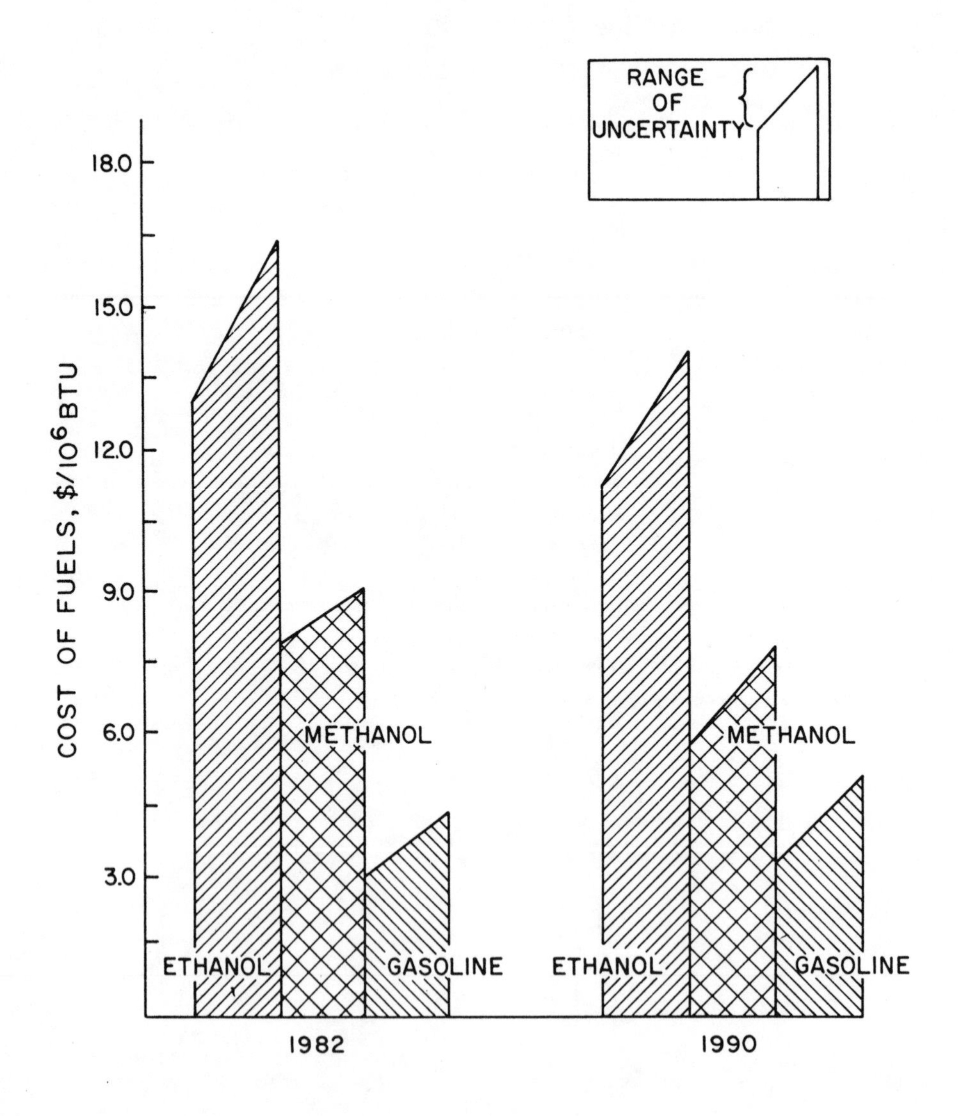

**Table 36: Biomass Equivalents. (Source 10)**

Carbohydrate plants

20,000 to 40,000 acres forest will support a 50 MW power plant using 1/3 for lumber, 1/3 for paper and 1/3 for fuel.

DOE suggests 3-5 quads of energy by biomass by year 2000 in U.S.

Annual global production of biomass is about 200 x $10^9$ dry tons.

200 million tons manure (MAF) produced by major farm animals in U.S. (Anderson, 1972).

By 1980, expect that 1.06 billion tons of organic wastes will be produced in the U.S. This has a net gas fuel potential of 10.6 Tcf.

Approximate conversion of solid wastes:
to oil - 1.25 bbl per ton of dry organic matter
to gas - 5.0 cu ft of methane for each pound of dry organic matter.

In U.S. 129 million tons of raw solid wastes are generated each year by urban sources which have an average moisture content of 21% and an average ash content of 25%. With a high heating value of 10 million Btu per ton, this amounts to 1.25 quads/yr. (Shafizadeh, Sarkanan, and Tillman, 1976).

Additionally, 75 million tons of solid wastes are produced annually from agricultural, forestry, and other activities.

For average solid carbohydrate biomass, could figure 8,000 Btu/lb MAF.

U.S. could supply 5% of total energy from biomass by year 2000.

One productive acre per person to supply energy needs by biomass in U.S.

Cost roughly $15 to $20 per ton to deliver crop residues to roadside, not including material costs. Transportation to center for use must be added (1979).

**Table 36: Biomass Equivalents. (cont'd)**

At $4.00 per bushel for grain, it costs $1.82 per gallon of 200 proof (100%) of ethanol.

1 ton of dry wood will produce about 80 gal of methanol or 60 gal of ethanol.

In U.S., there is a 1.73 pcd (pounds per capita per day) of wastes; for urban areas, 2.27 pcd, for all wastes; or,

1.25 ton/yr of solid waste produced per person in U.S. (Penner and Icerman, 1974).

300 MCF (at STP) of methane per day produced on feedlot of 10,000 head of cattle (Wise, 1977).

2000 lb glucose from 4000 lb newspapers.

In municipal solid waste, 28% of the organic matter is converted to methane in biochemical conversion.

1 person produces 1.2 $ft^3$/day of biogas.

1 cow's dung produces 80 $ft^3$/day of biogas.

8.8 lb of newspapers can be converted to 10 kWh.

10 ton yield of crops per acre can be converted to 20 bbl oil or 100,000 $ft^3$ of gas/acre.

1000 cord of wood per day for 100 MWe plant.

400 tons of city refuse per day will produce up to 15,000 kW.

600,000 metric tons per year of city refuse (4400 Btu/lb), producing steam, produces 64 MW electricity (Paris).

90 $mi^2$ required to produce sufficient wood for a 100 MW electric plant.

Hydrocarbon producing plants

*Euphorbiae*, produces 10 bbl/acre of oil per year (ES&T, 1980).

*Copaifera langdorfii*, produces 20-30 liter of oil every six months per tree.

These values are presented as rule-of-thumb estimates.

## Table 37: Biomass, Cellulose and Lignocellulose: Energy Equivalents and Conversion. (Source 10)

Equivalents:

1 lb dry wood = 8,500 Btu

1 lb cellulose = 5,070 Btu

1 lb cow manure = 7,000 Btu

1 lb straw = 5,000-6,500 Btu

1 lb sawdust (dry) = 8,100 Btu

1 lb peat (dry) = 7,040 Btu

1 cu ft = 11 lb, packaged

1 lb bagass = 7,200 Btu at 12% moisture

1 lb wood (11-13% moisture) = 7,200-8,000 Btu

1 cord of wood = 128 $ft^3$ bulk stacked

80 $ft^3$ solid wood, or 1.25 tons

1 ton MSW = $9.5 \times 10^6$ Btu

Conversions (Approximate)

1 cord (1600-3300 lb) of dry hardwood produces 55-65 lb of methanol, 140-185 lb of acetic acid, 1000-1050 lb of charcoal, plus gases, by carbonization.

100 kg of wood yields 19-20 liters of ethanol (5 gal).

1 ton of air-dried wood pulp, softwood, sulfite waste lignin yields 10-20 gal methanol (alcohol).

1 ton of wood (2240 lb) can yield 170.1 methanol (alcohol).

2 cords of hardwood will yield 1 ton of charcoal.

1 ton of wood waste will yield 1.5 bbl of low sulfur oil.

1 ton of cow dung will produce gas with 1 million Btu.

Converting wood to methanol is 34-37% efficient (thermal).

1 ton of oven dry wood will yield about 80-100 gal of methanol.

1 ton of manure produces 80 gal oil (USBM).

1 lb of dry peat yields 19.6 to 22.6 SCF gas (300Btu/SCF) (Lurgi).

1 ton dry wood will produce 16-40 gal of ethanol.

100 lb cellulosic material yields 32-43 lb oil.

1 lb cellulose will convert to 4.5 $ft^3$ (theoretical, 7$ft^3$).

1 lb organic matter in manure produces 4-5 cu ft of biogas (700 Btu/cu ft).

1 ton dry wood will yield 13,000 cu ft of natural gas.

1 ton solid waste will produce about 2.5 ton of steam.

100 lb of wood at 1-5% moisture will yield 40-50 lb oil.

1 cu ft hardwood = 32 lb

1 cu ft softwood = 27 lb

1 cu ft peat, packaged = 11 lb

To convert Btu/lb to J/kg, multiply by 2324.

## Table 38:

## Biomass, Sugar and Starch: Energy Equivalents and Conversion. (Source 10)

Equivalents

1 lb garbage, 8,800 Btu (dry), range 7,000-10,500 Btu
1 lb corn, shelled = 7,800-8,500 Btu
1 lb cotton seed = 7,100 Btu
1 lb animal fat = 16,000 Btu
1 lb wheat flour (11% water) = 6,300 Btu
1 lb glucose, dextrose, fructose = 6,700 Btu
1 lb sucrose = 7,092 Btu
1 lb molasses (24% water) - 4,325 Btu
1 lb oil of olive, castor bean, cottonseed, linseed, rapeseed = 16,930 Btu
1 lb rubber = 6,125 Btu; 1 lb pure rubber, $C_5H_8$ = 20,000 Btu
1 lb soybeans (7.5% water) = 6,325 Btu
1 lb sugar, granulated, = 7,170 Btu
1 lb potato = 1,030 Btu
1 lb fat = 16,000 Btu
1 lb protein = 7,200 Btu
1 lb starch = 7,528 Btu

Conversion, (Approximate) (by-products often have considerable value).

1 bu apples, peaches, and pears yield 0.3-0.4 gal ethanol.
2.3 to 2.7 gal of molasses will yield 1 gal (190 proof) alcohol.
12.8 lb of sugar will produce 1 gal of ethanol (alcohol).
1 bu of dry sorghum produces 2.2-3 gal (100% alcohol), 93% eff.
1 bu barley yields 1.9 gal ethanol.
1 bu wheat, corn, milo yields 2.5- 2.6 gal ethanol.
1 bu rice yields 1.8 gal ethanol.
2,000 lb dry waste yields 10,000 $ft^3$ SNG.
2,000 lb dry organic waste yields 2 bbl of oil (1.25 bbl net).
35% thermal efficiency in converting biomass to SNG.
1 bu corn has 14-16 lb cobs which yield 84,000 Btu.
1 ton vegetable matter fermented yields 20 gal of 95% alcohol.

**Table 38: Biomass, Sugar and Starch. (cont'd)**

Biomass to ethanol has a yield of 35% ± 5%.
Biomass to hydrocarbon liquids has a yield of 25% ± 5%
1 bu potatoes yields 0.7 gal; 1 bu sweet potatoes yields 0.9 gal ethanol.
1 cwt of Irish potatoes yields 1.4 gal ethanol.
1 cwt of sweet potatoes yields 1.8 gal of ethanol.
1 gal molasses yields 0.4 gal of ethanol.
1 lb starch yields 0.35 lb ethanol.
1 lb sucrose yields 0.54 lb ethanol at 100% yield.
1,000 lb cow produces 40 cu ft/day of gas (60% $CH_4$).
1 ton of grain yields 78 gal ethanol.
Oven dry wood to methanol has a yield of 38%.
1 ton sugar beets yields 21-22 gal ethanol.
1 ton potatoes yields 22-28 gal ethanol.
1 ton sugar cane yields 15-20 gal ethanol.
1 ton cassava yields 45-50 gal ethanol.
1 ton molasses yields 66-70 gal ethanol.
1 ton artichokes yields 28 gal ethanol.

To convert Btu/lb to J/kg, multiply by 2324.
1 kcal = 1 Cal = 3.968 Btu

Conversion

1 liter molasses (0.7-0.8 kg) will produce 175-200 g butanol.
1 $m^3$ corn (425 kg starch) = 77-100 kg butanol and 13-20 kg ethanol.
2.4 liter molasses produces 1 liter of ethanol (95%).
100 lb starch will yield 6 gal ethanol.

Biogas

16 $ft^3$ of biogas per brake horsepower-hour.
180-250 $ft^3$3 of biogas to do the work of 1 gallon of gasoline.

Shark

Oil shark, 50 lb, 4 ft long, yields 2/3 to 1 gal of oil from liver.
Basking sharks, 50-60 ft long, yield 120-160 gal of oil.
Mackeral sharks, 8-10 ft long, yield 7-11 gal of oil.

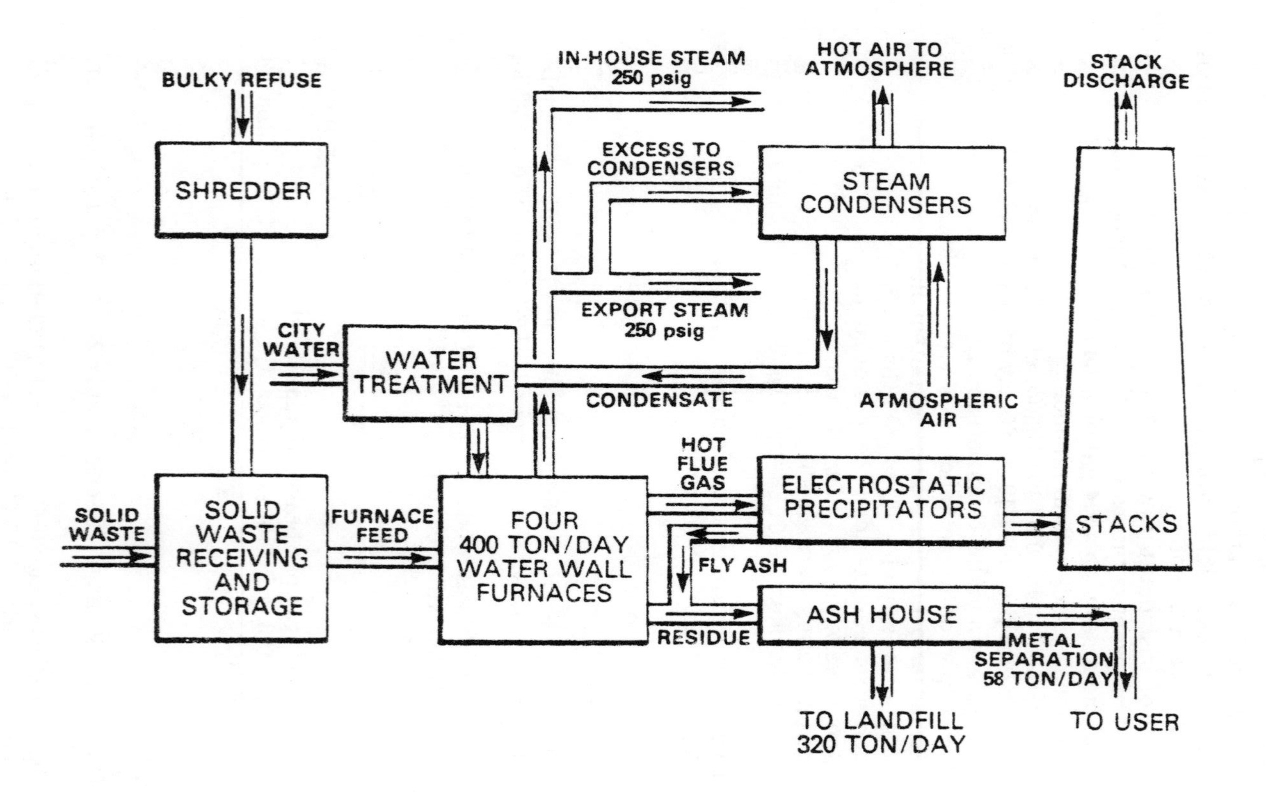


**Figure 93: Energy from Urban Waste, Chicago Incinerator. (Source 13)**

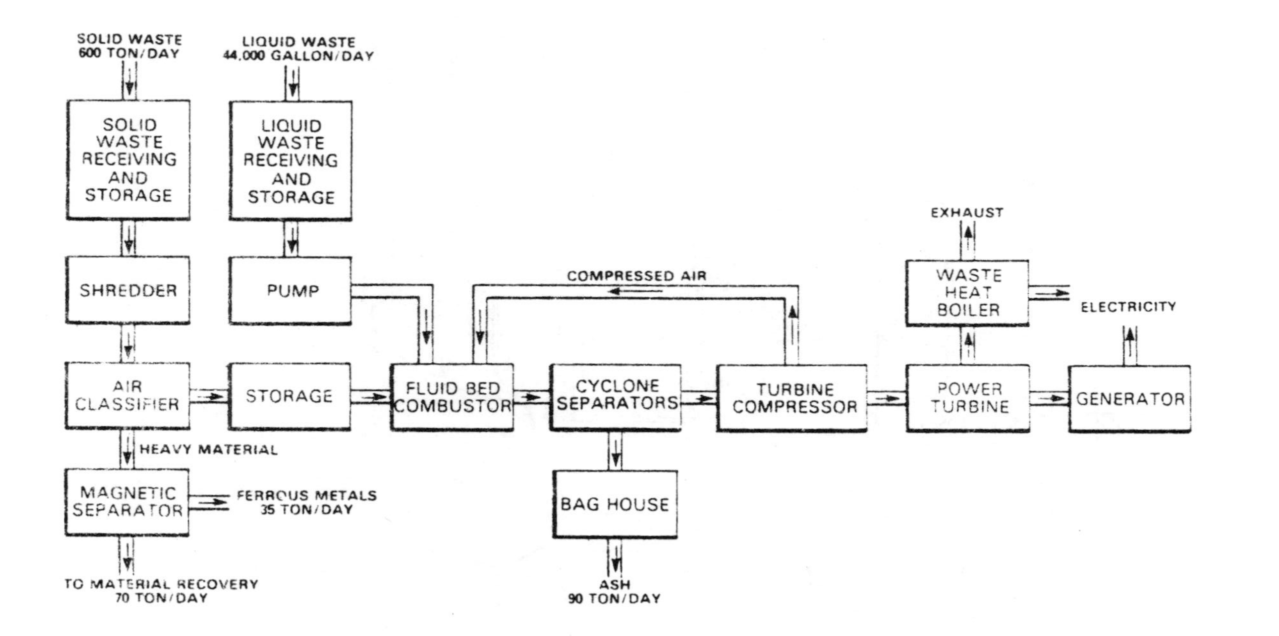


**Figure 94:** **Energy from Urban Waste, Combustion Power Company CPU-400 Process. (Source 13)**

**Figure 95: Cross Section of a Hydroelectric Power Plant. (Source 26)**

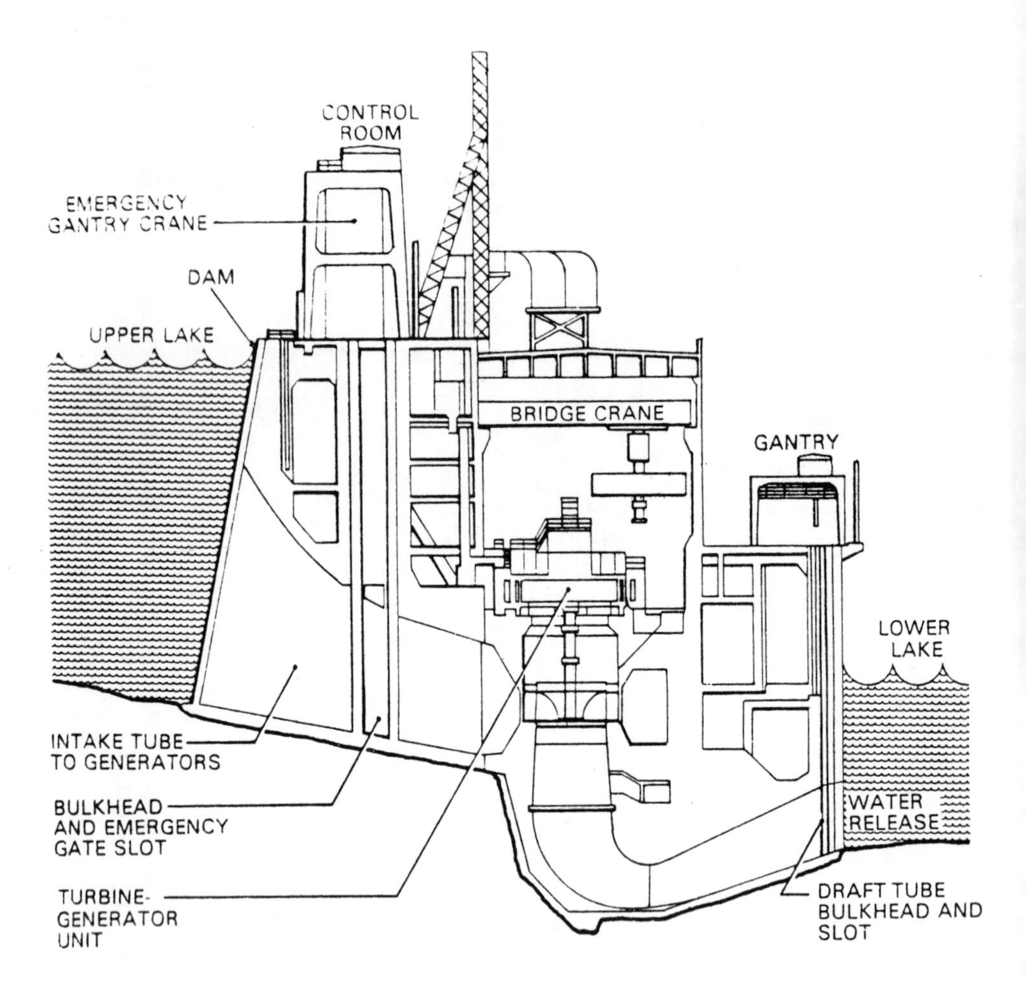

# MISCELLANEOUS

SECTION 5

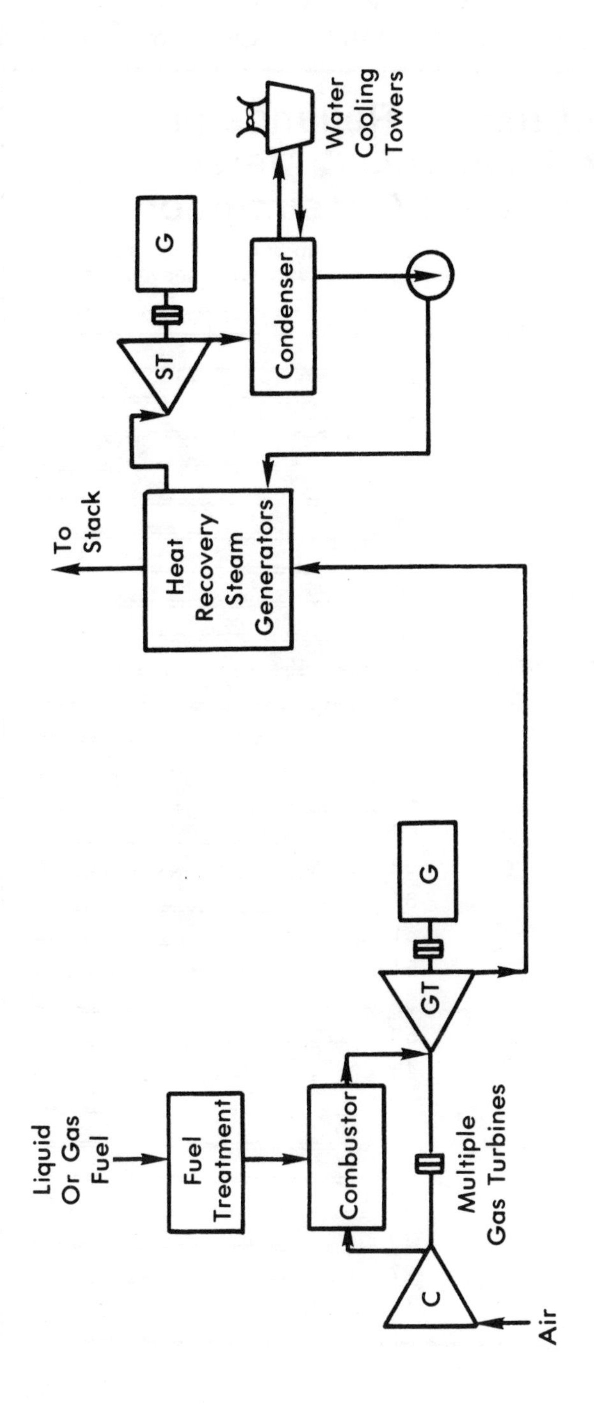


Figure 96: Gas Turbine/Steam Turbine Combined Cycle. (Source 14)

Figure 97: U.S. Net Import Reliance on Selected Minerals & Metals Consumption. (Source 23)

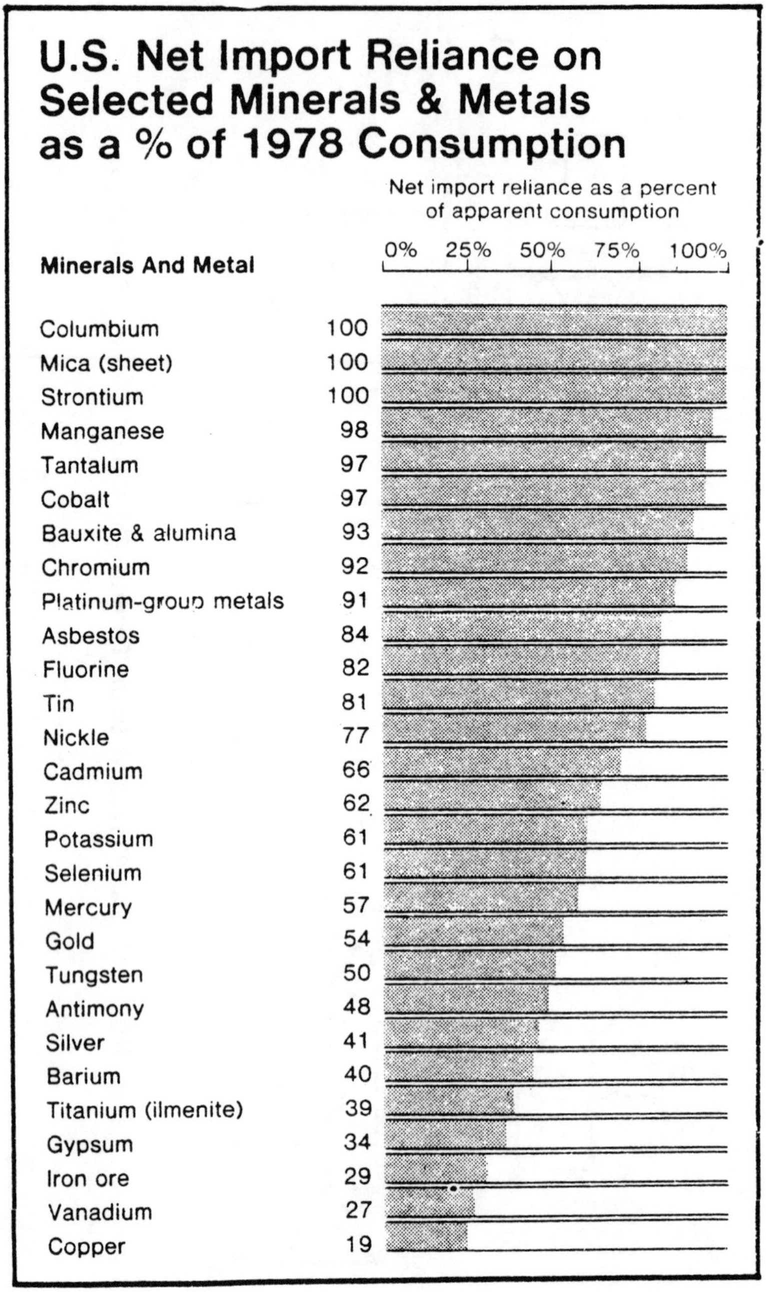

## U.S. Net Import Reliance on Selected Minerals & Metals as a % of 1978 Consumption

| Minerals And Metal | Net import reliance as a percent of apparent consumption |
|---|---|
| Columbium | 100 |
| Mica (sheet) | 100 |
| Strontium | 100 |
| Manganese | 98 |
| Tantalum | 97 |
| Cobalt | 97 |
| Bauxite & alumina | 93 |
| Chromium | 92 |
| Platinum-group metals | 91 |
| Asbestos | 84 |
| Fluorine | 82 |
| Tin | 81 |
| Nickle | 77 |
| Cadmium | 66 |
| Zinc | 62 |
| Potassium | 61 |
| Selenium | 61 |
| Mercury | 57 |
| Gold | 54 |
| Tungsten | 50 |
| Antimony | 48 |
| Silver | 41 |
| Barium | 40 |
| Titanium (ilmenite) | 39 |
| Gypsum | 34 |
| Iron ore | 29 |
| Vanadium | 27 |
| Copper | 19 |

## Table 39: Alphabetical List of Elements and Symbols. (Source 29)

| Element | Symbol | Atomic number | Atomic weight* |
|---|---|---|---|
| Actinium | Ac | 89 | 227 |
| Aluminum | Al | 13 | 26.98 |
| Americium | Am | 95 | 243 |
| Antimony | Sb | 51 | 121.75 |
| Argon | Ar | 18 | 39.942 |
| Arsenic | As | 33 | 74.91 |
| Astatine | At | 85 | 210 |
| Barium | Ba | 56 | 137.35 |
| Berkelium | Bk | 97 | 249 |
| Beryllium | Be | 4 | 9.013 |
| Bismuth | Bi | 83 | 208.99 |
| Boron | B | 5 | 10.82 |
| Bromine | Br | 35 | 79.913 |
| Cadmium | Cd | 48 | 112.40 |
| Calcium | Ca | 20 | 40.08 |
| Californium | Cf | 98 | 251 |
| Carbon | C | 6 | 12.010 |
| Cerium | Ce | 58 | 140.12 |
| Cesium | Cs | 55 | 132.90 |
| Chlorine | Cl | 17 | 35.455 |
| Chromium | Cr | 24 | 52.01 |
| Cobalt | Co | 27 | 58.94 |
| Copper | Cu | 29 | 63.54 |
| Curium | Cm | 96 | 247 |
| Dysporsium | Dy | 66 | 162.50 |
| Einsteinium | Es | 99 | 254 |
| Erbium | Er | 68 | 167.26 |
| Europium | Eu | 63 | 152.0 |
| Fermium | Fm | 100 | 253 |
| Fluorine | F | 9 | 19.00 |
| Francium | Fr | 87 | 223 |
| Gadolinium | Gd | 64 | 157.25 |
| Gallium | Ga | 31 | 69.72 |
| Germanium | Ge | 32 | 72.60 |
| Gold | Au | 79 | 197.0 |
| Hafnium | Hf | 72 | 178.49 |
| Helium | He | 2 | 4.003 |
| Holmium | Ho | 67 | 164.93 |
| Hydrogen | H | 1 | 1.0079 |
| Indium | In | 49 | 114.81 |
| Iodine | I | 53 | 126.90 |
| Iridium | Ir | 77 | 192.2 |
| Iron | Fe | 26 | 55.85 |
| Krypton | Kr | 36 | 83.80 |
| Lanthanum | La | 57 | 138.91 |
| Lawrencium | Lr | 103 | 257 |
| Lead | Pb | 82 | 207.20 |
| Lithium | Li | 3 | 6.940 |
| Lutetium | Lu | 71 | 174.98 |
| Magnesium | Mg | 12 | 24.32 |
| Manganese | Mn | 25 | 54.94 |
| Mendelevium | Md | 101 | 256 |
| Mercury | Hg | 80 | 200.60 |

## Table 39: Alphabetical List of Elements and Symbols. (cont'd)

| Element | Symbol | Atomic number | Atomic weight* |
|---|---|---|---|
| Molybdenum | Mo | 42 | 95.95 |
| Neodymium | Nd | 60 | 144.26 |
| Neon | Ne | 10 | 20.182 |
| Neptunium | Np | 93 | 237 |
| Nickel | Ni | 28 | 58.71 |
| Niobium (Columbium) | Nb | 41 | 92.91 |
| Nitrogen | N | 7 | 14.007 |
| Nobelium | No | 102 | 254 |
| Osmium | Os | 76 | 190.2 |
| Oxygen | O | 8 | 15.999 |
| Palladium | Pd | 46 | 106.4 |
| Phosphorus | P | 15 | 30.973 |
| Platinum | Pt | 78 | 195.08 |
| Plutonium | Pu | 94 | 242 |
| Polonium | Po | 84 | 210 |
| Potassium | K | 19 | 39.098 |
| Praseodymium | Pr | 59 | 140.91 |
| Promethium | Pm | 61 | 147 |
| Protactinium | Pa | 91 | 231 |
| Radium | Ra | 88 | 226 |
| Radon | Rn | 86 | 222 |
| Rhenium | Re | 75 | 186.21 |
| Rhodium | Rh | 45 | 102.90 |
| Rubidium | Rb | 37 | 85.48 |
| Ruthenium | Ru | 44 | 101.1 |
| Samarium | Sm | 62 | 150.34 |
| Scandium | Sc | 21 | 44.96 |
| Selenium | Se | 34 | 78.96 |
| Silicon | Si | 14 | 28.09 |
| Silver | Ag | 47 | 107.875 |
| Sodium | Na | 11 | 22.990 |
| Strontium | Sr | 38 | 87.63 |
| Sulfur | S | 16 | 32.064 |
| Tantalum | Ta | 73 | 180.94 |
| Technetium | Tc | 43 | 99 |
| Tellurium | Te | 52 | 127.60 |
| Terbium | Tb | 65 | 158.92 |
| Thallium | Tl | 81 | 204.38 |
| Thorium | Th | 90 | 232.04 |
| Thulium | Tm | 69 | 168.93 |
| Tin | Sn | 50 | 118.69 |
| Titanium | Ti | 22 | 47.90 |
| Tungsten (Wolfram) | W | 74 | 183.85 |
| Uranium | U | 92 | 238.06 |
| Vanadium | V | 23 | 50.95 |
| Xenon | Xe | 54 | 131.29 |
| Ytterbium | Yb | 70 | 173.03 |
| Yttrium | Y | 39 | 88.92 |
| Zinc | Zn | 30 | 65.38 |
| Zirconium | Zr | 40 | 91.22 |

* Atomic weight of the most abundant or best known isotope, or (in the case of radioactive isotopes) the isotope with the longest half-life, relative to atomic weight of Carbon-12 = 12.

## Table 40: Numerical Abbreviations. (Source 29)

Numerical abbreviations used in nuclear science are likely to be composed of two elements: first, an abbreviation of a numerical prefix expressing some multiple or fraction of unity, and second, an abbreviation of a unit which measures some basic property. Examples of both elements are:

PREFIXES

| Prefix | Meaning |
|---|---|
| pico | divide by 1 trillion ($10^{-12}$) |
| nano | divide by 1 billion ($10^{-9}$) |
| micro | divide by 1 million ($10^{-6}$) |
| milli | divide by 1 thousand ($10^{-3}$) |
| kilo | multiply by 1 thousand ($10^{3}$) |
| mega | multiply by 1 million ($10^{6}$) |
| giga | multiply by 1 billion ($10^{9}$) |

UNITS

| Unit | Abbreviation | Measured Property |
|---|---|---|
| angstrom | Å | length of radiation |
| barn | b | cross section |
| curie | c | radioactivity |
| electron volt | ev | energy |
| gram | g | mass |
| meter | m | length |
| rad | rad | radiation absorbed dose |
| roentgen | r | radiation dose |
| rem | rem | radiation dose |
| second | sec | time |
| ton | t | nuclear weapon energy in TNT equivalent |
| watt | w | power |

Knowing the two ingredients, it is easy to understand or to employ numerical abbreviations. Examples:

| Abbreviation | Full Term | Meaning |
|---|---|---|
| mb | millibarn | One thousandth of a barn |
| μc | microcurie | One millionth of a curie |
| kt | kiloton | One thousand tons of TNT equivalent |

# CONVERSION TABLES

SECTION 6

# Table 41: Constants. (Source 29)

CONSTANTS

The following values are supplied as useful reference values for students, for example, as recommended by the National Academy of Sciences-National Research Council, and adopted by the National Bureau of Standards:

| Constant | Symbol | Definition | Values |
|---|---|---|---|
| Speed of light in vacuum | C | - | $2.99795 \times 10^{10}$ centimeters/sec. |
| Avogadro number | N | Number of molecules in one gram-molecular weight of a substance. | $6.02252 \times 10^{23}$ |
| Faraday constant | F | Quantity of electricity to free chemical equivalent weight of a substance (in electrolysis). | $9.64870 \times 10^{4}$ coulombs |
| Planck constant | h | Energy of quantum of radiation in relation to frequency of source. | $6.62556 \times 10^{-27}$ erg-sec. |
| Elementary charge | e=F/N | Electric charge on one electron. | $4.80298 \times 10^{-10}$ e.s.u. (electrostatic units). |
| Electron rest mass | $m_e$ | - | $9.1091 \times 10^{-28}$ gram |
| Proton rest mass | $m_p$ | - | $1.67252 \times 10^{-24}$ gram |

## Table 42: Velocity. (Source 29)

| TO CONVERT / FROM \ TO | 1 FT/SEC | 2 FT/MIN | 3 MPH | 4 CM/SEC | 5 METER/MIN | 6 KNOTS |
|---|---|---|---|---|---|---|
| | MULTIPLY BY | | | | | |
| 1. FEET/SEC | 1.000 | 60.00 | 0.6818 | 30.48 | 18.29 | 0.5921 |
| 2. FEET/MIN | 0.01667 | 1.000 | 0.01136 | 0.5080 | 0.3048 | $9.872 \times 10^{-3}$ |
| 3. MILES/HR | 1.467 | 88.00 | 1.000 | 44.70 | 26.82 | 0.8684 |
| 4. CENTIMETER/SEC | 0.03281 | 1.969 | 0.02237 | 1.000 | 0.6000 | 0.01943 |
| 5. METER/MIN | 0.05468 | 3.281 | 0.03728 | 1.667 | 1.000 | 0.03238 |
| 6. KNOTS | 1.689 | 101.3 | 1.152 | 51.48 | 30.89 | 1.000 |

## Table 43: Thermal Conductance. (Source 29)

| TO CONVERT / FROM \ TO | 1 BTU/HR / (SQ FT) (°F) | 2 BTU/DAY / (SQ FT) (°F) | 3 WATTS / (SQ CM) (°C) | 4 CAL/SEC / (SQ CM) (°C) | 5 CAL/HR / (SQ CM) (°C) |
|---|---|---|---|---|---|
| | MULTIPLY BY | | | | |
| 1. (BTU/HR)/(SQ FT) (°F) | 1.000 | 24.00 | $5.678 \times 10^{-4}$ | $1.356 \times 10^{-4}$ | 0.4882 |
| 2. (BTU/DAY)/(SQ FT) (°F) | 0.04167 | 1.000 | $2.366 \times 10^{-5}$ | $5.651 \times 10^{-6}$ | 0.02034 |
| 3. WATTS/(SQ CM) (°C) | 1,761 | 42,270 | 1.000 | 0.2389 | 859.7 |
| 4. (CAL/SEC)/(SQ CM) (°C) | 7,373 | 177,000 | 4.186 | 1.000 | 3,600 |
| 5. (CAL/HR)/(SQ CM) (°C) | 2.048 | 49.16 | $1.163 \times 10^{-3}$ | $2.778 \times 10^{-4}$ | 1.000 |

## Table 44: Heat Flow. (Source 29)

| TO CONVERT / FROM \ TO | 1 BTU/HR / SQ FT | 2 BTU/DAY / SQ FT | 3 WATTS / SQ CM | 4 CAL SEC / SQ CM | 5 CAL/HR / SQ CM |
|---|---|---|---|---|---|
| | MULTIPLY BY | | | | |
| 1. (BTU/HR)/SQ FT | 1.000 | 24.00 | $3.154 \times 10^{-4}$ | $7.535 \times 10^{-5}$ | 0.2712 |
| 2. (BTU/DAY)/SQ FT | 0.04167 | 1.000 | $1.314 \times 10^{-5}$ | $3.139 \times 10^{-6}$ | 0.01130 |
| 3. WATTS/SQ CENTIMETER | 3,171 | 76,010 | 1.000 | 0.2389 | 860.0 |
| 4. (CALORIES/SEC)/SQ CM | 13,270 | 318,500 | 4.186 | 1.000 | 3,600 |
| 5. (CALORIES/HR)/SQ CM | 3.687 | 88.48 | $1.163 \times 10^{-3}$ | $2.778 \times 10^{-4}$ | 1.000 |

## Table 45: Viscosity. (Source 29)

VISCOSITY – ABSOLUTE VISCOSITY IS THE FORCE MOVING A UNIT AREA OF PLANE SURFACE AT UNIT VELOCITY RELATIVE TO ANOTHER PLANE SURFACE, THE TWO SURFACES BEING SEPARATED BY UNIT THICKNESS OF THE FLUID, LB FORCE/(SQ FT) [(FT/SEC)/FT] IF FORCE IS CONVERTED TO MASS BY NEWTON'S EQUATION (F = gc MA) THE NUMERATOR BECOMES RATE OF CHANGE OF MOMENTUM WITH TIME, (LB MASS) $(FT/SEC^2)$ gc/(SQ FT) [(FT/SEC)/FT]. THE TABLE GIVES NET UNITS AFTER CANCELLATIONS.

| TO CONVERT / FROM / TO | 1<br>CP | 2<br>POISES | 3<br>LB–SEC / SQ FT | 4<br>LB / FT–SEC | 5<br>LB / FT–HR |
|---|---|---|---|---|---|
| | MULTIPLY BY | | | | |
| 1. CENTIPOISES | 1.000 | 0.0100 | $2.090 \times 10^{-5}$ | $6.722 \times 10^{-4}$ | 2.42 |
| 2. POISES | 100.0 | 1.000 | $2.090 \times 10^{-3}$ | 0.06722 | 242 |
| 3. LB–SEC/SQ FT | 47,850 | 478.5 | 1.000 | 32.17 | 115,800 |
| 4. LB/FT–SEC | 1,488 | 14.88 | 0.03109 | 1.000 | 3,600 |
| 5. LB/FT–HR | 0.4133 | $4.133 \times 10^{-3}$ | $8.636 \times 10^{-6}$ | $2.778 \times 10^{-4}$ | 1.000 |

## Table 46: Kinematic Viscosity. (Source 29)

KINEMATIC VISCOSITY – ABSOLUTE VISCOSITY IN MASS UNITS DIVIDED BY MASS DENSITY, OR $FT^3$/(FT)(SEC). THE TABLE GIVES NET UNITS AFTER CANCELLATION.

| TO CONVERT / FROM / TO | 1<br>CENTISTOKES | 2<br>STOKES<br>$CM^2$/SEC | 3<br>$FT^2$/SEC | 4<br>$FT^2$/HR |
|---|---|---|---|---|
| | MULTIPLY BY | | | |
| 1. CENTISTOKES | 1.000 | 0.0100 | $1.076 \times 10^{-5}$ | $3.87 \times 10^{-2}$ |
| 2. STOKES, $CM^2$/SEC | 100.0 | 1.000 | $1.076 \times 10^{-3}$ | 3.87 |
| 3. $FT^2$/SEC | 92,900 | 929 | 1.000 | 3,600 |
| 4. $FT^2$/HR | 25.80 | 0.2580 | $2.778 \times 10^{-4}$ | 1.000 |

**Table 47: Equivalent Temperature Readings for Farenheit and Centigrade Scales. (Source 29)**

| Fahren- heit Degs. | Centi- grade Degs. | Fahren- heit Degs. | Centi- grade Degs. |
|---|---|---|---|
| -459.4 | -273. | -21. | -29.4 |
| -436. | -270. | -20.2 | -29. |
| -418. | -260. | -20. | -28.9 |
| -400. | -240. | -19. | -28.3 |
| -382. | -230. | -18.4 | -28. |
| -364. | -220. | -18. | -27.8 |
| -346. | -210. | -17. | -27.2 |
| -328. | -200. | -16.6 | -27. |
| -310. | -190. | -16. | -26.7 |
| -292. | -180. | -15. | -26.1 |
| -274. | -170. | -14.8 | -26. |
| -256. | -160. | -14. | -25.6 |
| -238. | -150. | -13. | -25. |
| -220. | -140. | -12. | -24.4 |
| -202. | -130. | -11.2 | -24. |
| -184. | -120. | -11. | -23.9 |
| -166. | -110. | -10. | -23.3 |
| -148. | -100. | - 9.4 | -23. |
| -139. | - 95. | - 9. | -22.8 |
| -130. | - 90. | - 8. | -22.2 |
| -121. | - 85. | - 7.6 | -22. |
| -112. | - 80. | - 7. | -21.7 |
| -103. | - 75. | - 6. | -21.1 |
| - 94. | - 70. | - 5.8 | -21. |
| - 85. | - 65. | - 5. | -20.6 |
| - 76. | - 60. | - 4. | -20. |
| - 67. | - 55. | - 3. | -19.4 |
| - 58. | - 50. | - 2.2 | -19. |
| - 49. | - 45. | - 2. | -18.9 |
| - 40. | - 40. | - 1. | -18.3 |
| - 39. | - 39.4 | - 0.4 | -18. |
| - 38.2 | - 39. | 0. | -17.8 |
| - 38. | - 38.9 | + 1. | -17.2 |
| - 37. | - 38.3 | 1.4 | -17. |
| - 36.4 | - 38. | 2. | -16.7 |
| - 36. | - 37.8 | 3. | -16.1 |
| - 35. | - 37.2 | 3.2 | -16. |
| - 34.6 | - 37. | 4. | -15.6 |
| - 34. | - 36.7 | 5. | -15. |
| - 33. | - 36.1 | 6. | -14.4 |
| - 32.8 | - 36. | 6.8 | -14. |
| - 32. | - 35.6 | 7. | -13.9 |
| - 31. | - 35. | 8. | -13.3 |
| - 30. | - 34.4 | 8.6 | -13. |
| - 29.2 | - 34. | 9. | -12.8 |
| - 29. | - 33.9 | 10. | -12.2 |
| - 28. | - 33.3 | 10.4 | -12. |
| - 27.4 | - 33. | 11. | -11.7 |
| - 27. | - 32.8 | 12. | -11.1 |
| - 26. | - 32.2 | 12.2 | -11. |
| - 25.6 | - 32. | 13. | -10.6 |
| - 25. | - 31.7 | 14. | -10. |
| - 24. | - 31.1 | 15. | - 9.4 |
| - 23.8 | - 31. | 15.8 | - 9. |
| - 23. | - 30.6 | 16. | - 8.9 |
| - 22. | - 30. | 17. | - 8.3 |

**Table 47: Equivalent Temperature Readings for Farenheit and Centigrade Scales. (cont'd)**

| Fahren-heit Degs. | Centi-grade Degs. | Fahren-heit Degs. | Centi-grade Degs. |
|---|---|---|---|
| + 17.6 | - 8. | 56. | 13.3 |
| 18. | - 7.8 | 57. | 13.9 |
| 19. | - 7.3 | 57.2 | 14. |
| 19.4 | - 7. | 58. | 14.4 |
| 20. | - 6.7 | 59. | 15. |
| 21. | - 6.1 | 60. | 15.6 |
| 21.2 | - 6. | 60.8 | 16. |
| 22. | - 5.6 | 61. | 16.1 |
| 23. | - 5. | 62. | 16.7 |
| 24. | - 4.4 | 62.6 | 17. |
| 24.8 | - 4. | 63. | 17.2 |
| 25. | - 3.9 | 64. | 17.8 |
| 26. | - 3.3 | 64.4 | 18. |
| 26.6 | - 3. | 65. | 18.3 |
| 27. | - 2.8 | 66. | 18.9 |
| 28. | - 2.2 | 66.2 | 19. |
| 28.4 | - 2. | 67. | 19.4 |
| 29. | - 1.7 | 68. | 20. |
| 30. | - 1.1 | 69. | 20.6 |
| 30.2 | - 1. | 69.8 | 21. |
| 31. | - 0.6 | 70. | 21.1 |
| 32. | 0. | 71. | 21.7 |
| 33. | + 0.6 | 71.6 | 22. |
| 33.8 | 1. | 72. | 22.2 |
| 34. | 1.1 | 73. | 22.8 |
| 35. | 1.7 | 73.4 | 23. |
| 35.6 | 2. | 74. | 23.3 |
| 36. | 2.2 | 75. | 23.9 |
| 37. | 2.8 | 75.2 | 24. |
| 37.4 | 3. | 76. | 24.4 |
| 38. | 3.3 | 77. | 25. |
| 39. | 3.9 | 78. | 25.6 |
| 32.2 | 4. | 78.8 | 26. |
| 40. | 4.4 | 79. | 26.1 |
| 41. | 5. | 80. | 26.7 |
| 42. | 5.6 | 80.6 | 27. |
| 42.8 | 6. | 81. | 27.2 |
| 43. | 6.1 | 82. | 27.8 |
| 44. | 6.7 | 82.4 | 28. |
| 44.6 | 7. | 83. | 28.3 |
| 45. | 7.2 | 84. | 28.9 |
| 46. | 7.8 | 84.2 | 29. |
| 46.4 | 8. | 85. | 29.4 |
| 47. | 8.3 | 86. | 30. |
| 48. | 8.9 | 87. | 30.6 |
| 48.2 | 9. | 87.8 | 31. |
| 49. | 9.4 | 88. | 31.1 |
| 50. | 10. | 89. | 31.7 |
| 51. | 10.6 | 89.6 | 32. |
| 51.8 | 11. | 90. | 32.2 |
| 52. | 11.1 | 91. | 32.8 |
| 53. | 11.7 | 91.4 | 33. |
| 53.6 | 12. | 92. | 33.3 |
| 54. | 12.2 | 93. | 33.9 |
| 55. | 12.8 | 93.2 | 34. |
| 55.4 | 13. | 94. | 34.4 |

**Table 47: Equivalent Temperature Readings for Farenheit and Centigrade Scales. (cont'd)**

| Fahren- heit Degs. | Centi- grade Degs. | Fahren- heit Degs. | Centi- grade Degs. |
|---|---|---|---|
| + 95. | + 35. | 134. | 56.7 |
| 96. | 35.6 | 134.6 | 57. |
| 96.8 | 36. | 135. | 57.2 |
| 97. | 36.1 | 136. | 57.8 |
| 98. | 36.7 | 136.4 | 58. |
| 98.6 | 37. | 137. | 58.3 |
| 99. | 37.2 | 138. | 58.9 |
| 100. | 37.8 | 138.2 | 59. |
| 100.4 | 38. | 139. | 59.4 |
| 101. | 38.3 | 140. | 60. |
| 102. | 38.9 | 141. | 60.6 |
| 102.2 | 39. | 141.8 | 61. |
| 103. | 39.4 | 142. | 61.1 |
| 104. | 40. | 143. | 61.7 |
| 105. | 40.6 | 143.6 | 62. |
| 105.8 | 41. | 144. | 62.2 |
| 106. | 41.1 | 145. | 62.8 |
| 107. | 41.7 | 145.4 | 63. |
| 107.6 | 42. | 146. | 63.3 |
| 108. | 42.2 | 147. | 63.9 |
| 109. | 42.8 | 147.2 | 64. |
| 109.4 | 43. | 148. | 64.4 |
| 110. | 43.3 | 149. | 65. |
| 111. | 43.9 | 150. | 65.6 |
| 111.2 | 44. | 150.8 | 66. |
| 112. | 44.4 | 151. | 66.1 |
| 113. | 45. | 152. | 66.7 |
| 114. | 45.6 | 152.6 | 67. |
| 114.8 | 46. | 153. | 67.2 |
| 115. | 46.1 | 154. | 67.8 |
| 116. | 46.7 | 154.4 | 68. |
| 116.6 | 47. | 155. | 68.3 |
| 117. | 47.2 | 156. | 68.9 |
| 118. | 47.8 | 156.2 | 69. |
| 118.4 | 48. | 157. | 69.4 |
| 119. | 48.3 | 158. | 70. |
| 120. | 48.9 | 159. | 70.6 |
| 120.2 | 49. | 159.8 | 71. |
| 121. | 49.4 | 160. | 71.1 |
| 122. | 50. | 161. | 71.7 |
| 123. | 50.6 | 161.6 | 72. |
| 123.8 | 51. | 162. | 72.2 |
| 124. | 51.1 | 163. | 72.8 |
| 125. | 51.7 | 163.4 | 73. |
| 125.6 | 52. | 164. | 73.3 |
| 126. | 52.2 | 165. | 73.9 |
| 127. | 52.8 | 165.2 | 74. |
| 127.4 | 53. | 166. | 74.4 |
| 128. | 53.3 | 167. | 75. |
| 129. | 53.9 | 168. | 75.6 |
| 129.2 | 54. | 168.8 | 76. |
| 130. | 54.4 | 169. | 76.1 |
| 131. | 55. | 170. | 76.7 |
| 132. | 55.6 | 170.6 | 77. |
| 132.8 | 56. | 171. | 77.2 |
| 133. | 56.1 | 172. | 77.8 |

**Table 47: Equivalent Temperature Readings for Farenheit and Centigrade Scales. (cont'd)**

| Fahren-heit Degs. | Centi-grade Degs. | Fahren-heit Degs. | Centi-grade Degs. |
|---|---|---|---|
| + 172.4 | + 78. | 211. | 99.4 |
| 173. | 78.3 | 212. | 100. |
| 174. | 78.9 | 213. | 100.6 |
| 174.2 | 79. | 213.8 | 101. |
| 175. | 79.4 | 214. | 101.1 |
| 176. | 80. | 215. | 101.7 |
| 177. | 80.6 | 215.6 | 102. |
| 177.8 | 81. | 216. | 102.2 |
| 178. | 81.1 | 217. | 102.8 |
| 179. | 81.7 | 217.4 | 103. |
| 179.6 | 82. | 218. | 103.3 |
| 180. | 82.2 | 219. | 103.9 |
| 181. | 82.8 | 219.2 | 104. |
| 181.4 | 83. | 220. | 104.4 |
| 182. | 83.3 | 221. | 105. |
| 183. | 83.9 | 222. | 105.6 |
| 183.2 | 84. | 222.8 | 106. |
| 184. | 84.4 | 223. | 106.1 |
| 185. | 85. | 224. | 106.7 |
| 186. | 85.6 | 224.6 | 107. |
| 186.8 | 86. | 225. | 107.2 |
| 187. | 86.1 | 226. | 107.8 |
| 188. | 86.7 | 226.4 | 108. |
| 188.6 | 87. | 227. | 108.3 |
| 189. | 87.2 | 228. | 108.9 |
| 190. | 87.8 | 228.2 | 109. |
| 190.4 | 88. | 229. | 109.4 |
| 191. | 88.3 | 230. | 110. |
| 192. | 88.9 | 231. | 110.6 |
| 192.2 | 89. | 231.8 | 111. |
| 193. | 89.4 | 232. | 111.1 |
| 194. | 90. | 233. | 111.7 |
| 195. | 90.6 | 233.6 | 112. |
| 195.8 | 91. | 234. | 112.3 |
| 196. | 91.1 | 235. | 112.8 |
| 197. | 91.7 | 235.4 | 113. |
| 197.6 | 92. | 236. | 113.3 |
| 198. | 92.2 | 237. | 113.9 |
| 199. | 92.8 | 237.2 | 114. |
| 199.4 | 93. | 238. | 114.4 |
| 200. | 93.3 | 239. | 115. |
| 201. | 93.9 | 240. | 115.6 |
| 201.2 | 94. | 240.8 | 116. |
| 202. | 94.4 | 241. | 116.1 |
| 203. | 95. | 242. | 116.7 |
| 204. | 95.6 | 242.6 | 117. |
| 204.8 | 96. | 243. | 117.2 |
| 205. | 96.1 | 244. | 117.8 |
| 206. | 96.7 | 244.4 | 118. |
| 206.6 | 97. | 245. | 118.3 |
| 207. | 97.2 | 246. | 118.9 |
| 208. | 97.8 | 246.2 | 119. |
| 208.4 | 98. | 247. | 119.4 |
| 209. | 98.3 | 248. | 120. |
| 210. | 98.9 | 249. | 120.6 |
| 210.2 | 99. | 249.8 | 121. |

## Table 48: Energy Conversion Factors. (Source 29)

| Energy Content of Fuels | Crude Oil Equiv., Barrels | British Thermal Units (Btu) | Kilowatt-Hours (kwh) |
|---|---|---|---|
| Anthracite coal, short ton | 4.38 | 25,400,000 | 7440.0 |
| Bituminous coal, short ton | 4.24 | 24,580,000 | 7240.0 |
| Average coal, short ton | | 24,020,000 | 7040.0 |
| Automotive gasoline, gallon | 0.0216 | 125,000 | 36.6 |
| Aviation gasoline, gallon | 0.0216 | 125,000 | 36.6 |
| Jet fuel kerosene type, gallon | 0.0234 | 135,000 | 39.5 |
| Jet fuel naphtha type, gallon | 0.0219 | 127,000 | 37.2 |
| Kerosene, gallon | 0.0234 | 135,000 | 39.5 |
| Diesel oil, gallon | 0.0239 | 138,700 | 40.7 |
| Distillate fuel oil (#2), gallon | 0.0239 | 138,700 | 40.7 |
| Distillate fuel oil (#2), barrel | 1.004 | 5,825,000 | 1,707.0 |
| Residual fuel oil, gallon | 0.0258 | 149,700 | 43.9 |
| Residual fuel oil, barrel | 1.084 | 6,287,000 | 1,843.0 |
| Natural gas, standard cubic foot (SCF) | 0.000178 | 1,031 | 0.302 |
| Liquified petroleum gas, SCF (including propane and butane) | | 2,522 | |
| Electricity, Btu of fuel consumed at power plant per kwh delivered to consumer (assume 10,536 Btu/kwh station heat rate for all stations, 9% line loss as reported for 1971 by Edison Electric Institute) | 0.0020 | 11,600 | 3.40 |

## Table 48: Energy Conversion Factors. (cont'd)

| Energy Content of Fuels | | Crude Oil Equiv., Barrels | | British Thermal Units (Btu) | | Kilowatt-Hours (kwh) |
|---|---|---|---|---|---|---|
| Steam, Btu of fuel consumed at boiler plant per pound of steam delivered to consumer (assume 1000 Btu/lb of steam generated, 82% boiler efficiency and 12% line loss) | | 0.000196 | | 1,390 | | 0.407 |
| 1 kwh | = | $3.600 \times 10^{6}$ joules (J) | = | 859.9 kilocalorie (kcal) | = | 3412 Btu |
| 1 horsepower-hour (hp-hr) | = | 0.746 kwh | = | 2545 Btu | | |
| 1 J | = | $2.778 \times 10^{-7}$ kwh | = | .2388 cal | = | $9.478 \times 10^{-4}$ Btu |
| 1 Btu | = | $1.055 \times 10^{3}$ J | = | $2.931 \times 10^{-4}$ kwh | = | .2520 kcal |

**Table 49: Approximate Conversion Factors for Crude Oil. (Source 29)**

| From | Into: Metric tons | Long tons | Short tons | Barrels | Kiloliters (cubic meters) | 1,000 gallons (Imperial) | 1,000 gallons (U.S.) |
|---|---|---|---|---|---|---|---|
| | Multiply by | | | | | | |
| Metric tons | 1 | 0.984 | 1.102 | 7.33 | 1.16 | 0.256 | 0.308 |
| Long tons | 1.016 | 1 | 1.120 | 7.45 | 1.18 | .261 | .313 |
| Short tons | .907 | .893 | 1 | 6.05 | 1.05 | .233 | .279 |
| Barrels | .136 | .134 | .150 | 1 | .159 | .035 | .042 |
| Kiloliters (cubic meters) | .863 | .849 | .951 | 6.29 | 1 | .220 | .264 |
| 1,000 gallons (Imperial) | 3.91 | 3.83 | 4.29 | 28.6 | 4.55 | 1 | 1.201 |
| 1,000 gallons (U.S.) | 3.25 | 3.19 | 3.58 | 23.8 | 3.79 | .833 | 1 |

* Based on world average (excluding natural gas liquids).

## Table 50: The Btu. (Source 29)

A BTU is the amount of heat required to raise the temperature of one pound of water one degree Fahrenheit. The BTU is a very small unit of measurement and when one adds up large quantities of energy, one must count in large multiples of the BTU.

The BTU equivalents of common fuels are as follows:

| *FUEL AND COMMON MEASURES* | *BTU's* |
|---|---|
| Crude oil--Barrel (Bbl.) | 5,800,000 |
| Natural Gas--Cubic Foot (CF) | 1,032 |
| Coal--Ton | 24,000,000 to 28,000,000 |
| Electricity--Kilowatt Hour (KWH) | 3,412 |

Two trillion BTU's per year are approximately equal to 1,000 barrels per day of crude oil.

## Table 51: Approximate Calorific Equivalents of Oil. (Source 29)

One million tons of oil equals approximately--
- Heat Units:
  - 41 Million million BTU's
  - 415 million therms
  - 10,500 Teracalories
- Solid Fuels:
  - 1.5 million tons of coal
  - 4.9 million tons of lignite
  - 3.3 million tons of peat
- Natural Gas (1 cu. ft equals 1,000 BTU's) 1 cu. meter equals 4,200 Kcals):
  - 2.5 thousand million cu. meters
  - 88.3 thousand million cu. ft.
  - 242 million cu. ft./day for a year
- Electricity (1 Kwh equals 3,412 BTU's) (1 Kwh equals 860 Kcals):
  - 12 thousand million Kwh

## Table 52: Heat Units and Other Fuels Expressed in Terms of Million Tons of Oil. (Source 29)

| | Million tons of oil |
|---|---|
| 10 million million BTUs approximates to | 0.24 |
| 100 million therms approximates to | .24 |
| 10,000 Teracalories approximates to | .95 |
| 1 million tons of coal approximates to | .67 |
| 1 million tons of lignite approximates to | .20 |
| 1 million tons of peat approximates to | .30 |
| 1 thousand million cu. meters approximates to | .86 |
| 10 thousand million cu. ft. approximates to | .24 |
| 100 million cu. ft./day for a year approximates to | .88 |
| 1 thousand million cu. meters approximates to | .40 |
| 10 thousand million cu. ft. approximates to | .11 |
| 100 million cu. ft./day for a year approximates to | .41 |
| 10 thousand million Kwh approximates to | .82 |

(one million tons of oil produces about 4,000 millions units (Kwh) of electricity in a modern power station)

## Table 53: Relative Amounts of Energy Which Theoretically Can be Released from Storage in Matter. (Source 29)

(in British Thermal Units, btu, per pound of matter)

| | |
|---|---|
| From a pound of wound clocksprings | 0.17 |
| By letting a pound of water drop over falls 1000 feet high | 1.3 |
| By letting air escape from a cylinder at 200 lbs. per square inch pressure | 2.7 |
| From a flywheel spinning just below its bursting speed | 4.9 |
| By letting a pound of twisted rubber bands untwist | 25. |
| From a fully-charged storage battery | 92. |
| By letting water freeze | 144. |
| By letting steam condense | 965. |
| By exploding gunpowder | 1.25 thousand |
| By exploding dynamite | 2.3 thousand |
| By burning firewood | 5.8 to 7.7 thousand** |
| By burning alcohol | 12. thousand** |
| By burning fuel gas | 6. to 10. thousand** |
| By digesting butter | 14. thousand** |
| By burning crude petroleum | 16. thousand** |
| By burning butter | 16.5 thousand** |
| By burning rubberbands | 19.5 thousand** |
| By burning gasoline | 20. thousand** |
| By burning hydrogen gas | 51.5 thousand** |
| From the fission of the $U_{235}$ isotope in one pound of natural uranium | 251 million |
| From the complete fission of a pound of uranium | 34,100 million |
| From the fusion to helium of a pound of hydrogen | 260,000 million |
| From the total annihilation of matter of any kind | 38,700,000 million |

*Although the values in this table have been calculated carefully and checked several times, many depend on assumptions which may vary with circumstances. They should therefore be taken only as illustrative of orders of magnitude.

**Does not include weight of oxygen needed to release energy, since this can usually be taken from the earth's atmosphere.

## Table 54: Conversion Factors for Coals of Various Calorific Contents.

| Type of Coal | Typical Calorific Content | Quantity Equivalent to 1 tce |
|---|---|---|
| Bituminous | 12,000 Btu/lb | 1.05 tons |
| Subbituminous | 9,000 Btu/lb | 1.4 tons |
| Lignite | 7,000 Btu/lb | 1.8 tons |

## Table 55: Illustrative Scaling Comparisons for Various Quantities of Coal.

| Quantity of Coal | Indicator of Amount of Energy Provided |
|---|---|
| 2 mtce | Annual primary fuel requirement for a 1,000 MWe electric power plant if it operates at a 65 percent capacity factor and generates 5.7 billion kWh per year electricity |
| 5-7 mtce | Annual coal feedstock requirement for a 50,000 barrels per day synthetic liquids plant or a 250 million cubic feet per day synthetic gas facility |
| 34 mtce | Amount of energy provided by 1 exajoule ($10^{18}$ joules) |
| 76 mtce | Amount of energy supplied annually by 1 million barrels per day of oil |
| 100-140 mtce | Annual coal feedstock requirement for production of 1 mbdoe synthetic liquids |

## Table 56: Cost of Coal, Natural Gas, and Heat Equivalent to Various Oil Prices.

| | Oil Price | | |
|---|---|---|---|
| Fuel Type | $20/barrel | $30/barrel | $40/barrel |
| Coal ($ per tce) | $95 | $142 | $190 |
| Natural Gas ($ per thousand cf) | $3.30 | $5.00 | $6.60 |
| Heat ($ per million Btu) | $3.40 | $5.00 | $6.80 |

## Table 57: Selected Conversion Factors.

| To convert from | To | Multiply by |
|---|---|---|
| Acres | Hectares | 0.40468564 |
| | Square feet | 43560 |
| | Square kilometers | 0.0040468564 |
| | Square meters | 4046.8564 |
| | Square miles | 0.0015625 |
| Atmospheres | Bars | 1.01325 |
| | Inches of Hg (32° F) | 29.9213 |
| | Kilograms/square centimeter | 1.03323 |
| | Millimeters of Hg (0° C) | 760 |
| | Newtons/square meter | 101320 |
| | Pounds/square inch | 14.6960 |
| Barrels (petroleum,US) | Cubic feet | 5.614583 |
| | Cubic meters | 0.15898 |
| | Gallons (U.S.) | 42 |
| | Liters | 158.98284 |
| Bars | Atmospheres | 0.986923 |
| | Inches of Hg (32° F) | 29.5300 |
| | Kilograms/square centimeter | 1.019716 |
| | Newtons/square meter | 100,000 |
| | Pounds/square inch | 14.5038 |
| Btu | Btu (IST)[1] | 0.999346 |
| | Calories, gram (IST) | 251.831 |
| | Foot-pounds | 777.649 |
| | Horsepower-hours | 0.000392752 |
| | Joules | 1054.35 |
| | Joules (international) | 1054.18 |
| | Kilowatt-hours | 0.000292875 |
| | Kilowatt-hours (international) | 0.000292827 |
| | Tons of refrigeration (U.S. standard) | $3.46995 \times 10^{-3}$ |
| Btu (IST) | Btu | 1.00065 |
| Calories, gram[2] | Btu | 0.0039683207 |
| | Btu (IST) | 0.00396573 |
| | Calories, gram (IST) | 0.999346 |
| | Foot-pounds | 3.08596 |
| | Horsepower-hours | $1.55857 \times 10^{-6}$ |
| | Joules | 4.184 |
| | Joules (international) | 4.18331 |
| | Kilowatt-hours | $1.162222 \times 10^{-6}$ |
| Centimeters | Feet | 0.032808399 |
| | Inches | 0.39370079 |
| | Meters | 0.01 |
| | Microns | 10000 |
| | Millimeters | 10 |
| | Millimicrons | $1 \times 10^{7}$ |

## Table 57: Selected Conversion Factors. (cont'd)

| To convert from | To | Multiply by |
|---|---|---|
| Cubic centimeters | Cubic feet | $3.5314667 \times 10^{-5}$ |
| | Cubic inches | 0.061023744 |
| | Cubic meters | $1 \times 10^{-6}$ |
| | Gallons (liquid, U.S.) | 0.00026417205 |
| | Liters | 0.000999972 |
| Cubic feet | Barrels (petroleum, U.S.) | 0.17811 |
| | Cubic centimeters | 28316.847 |
| | Cubic meters | 0.028316847 |
| | Gallons (liquid, U.S.) | 7.4805195 |
| | Liters | 28.31605 |
| Cubic inches | Cubic centimeters | 16.387064 |
| | Cubic feet | 0.00057870370 |
| | Gallons (liquid, U.S.) | 0.0043290043 |
| | Liters | 0.01638661 |
| Cubic meters | Barrels (liquid, U.S.) | 8.3864145 |
| | Barrels (petroleum, U.S.) | 6.2900 |
| | Cubic centimeters | $1 \times 10^{6}$ |
| | Cubic feet | 35.314667 |
| | Cubic inches | 61023.74 |
| | Gallons (liquid, U.S.) | 264.17205 |
| | Liters | 999.972 |
| | Tons (long, petroleum) | 0.849 |
| | Tons (metric, petroleum) | 0.863 |
| | Tons (short, petroleum) | 0.957 |
| Feet | Centimeters | 30.48 |
| | Inches | 12 |
| | Meters | 0.3048 |
| | Microns | 304800 |
| | Miles (statute) | 0.000189393 |
| Foot-pounds | Btu (IST) | 0.00128509 |
| | Calories, gram (IST) | 0.323836 |
| | Horsepower-hours | $5.05050 \times 10^{-7}$ |
| | Joules | 1.35582 |
| | Kilowatt-hours | $3.76616 \times 10^{-7}$ |
| Gallons (liquid, U.S.) | Barrels (liquid, U.S.) | 0.031746032 |
| | Barrels (petroleum, U.S.) | 0.023809524 |
| | Cubic centimeters | 3785.4118 |
| | Cubic feet | 0.133680555 |
| | Cubic inches | 231 |
| | Cubic meters | 0.0037854118 |
| | Liters | 3.785306 |
| | Tons (long, petroleum) | 0.00319 |
| | Tons (metric, petroleum) | 0.00325 |
| | Tons (short, petroleum) | 0.00358 |
| Hectares | Acres | 2.4710538 |
| | Square centimeters | $1 \times 10^{8}$ |
| | Square feet | 107639.10 |

## Table 57: Selected Conversion Factors. (cont'd)

| To convert from | To | Multiply by |
|---|---|---|
| Hectares (cont.) | Square kilometers | 0.01 |
| | Square meters | 10000 |
| | Square miles | 0.0038610216 |
| Horsepower-hours | Btu | 2546.14 |
| | Btu (IST) | 2544.47 |
| | Calories, gram (IST) | 641196 |
| | Foot-pounds | $1.98 \times 10^6$ |
| | Joules | $2.68452 \times 10^6$ |
| | Kilowatt-hours | 0.745700 |
| Inches | Centimeters | 2.54 |
| | Feet | 0.083333 |
| | Meters | 0.0254 |
| Inches of Hg (32° F) | Atmospheres | 0.0334211 |
| | Bars | 0.0338639 |
| | Kilograms/square meter | 345.316 |
| | Millimeters of Hg (60°) | 25.4 |
| | Newtons/square meter | 3386.5 |
| | Pounds/square inch | 0.49115 |
| Joules (absolute) | Btu (IST) | 0.000947831 |
| | Calories, gram (IST) | 0.238849 |
| | Ergs | $1 \times 10^7$ |
| | Foot-pounds | 0.737562 |
| | Horsepower-hours | $3.72506 \times 10^{-7}$ |
| | Joules (international) | 0.999835 |
| | Kilowatt-hours | $2.7777 \times 10^{-7}$ |
| Kilograms | Pounds (avoirdupois) | 2.2046226 |
| | Tons (long) | 0.00098420653 |
| | Tons (metric) | 0.001 |
| | Tons (short) | 0.0011023113 |
| Kilograms/square meter | Atmospheres | $9.67841 \times 10^{-5}$ |
| | Bars | $9.80665 \times 10^{-5}$ |
| | Inches of Hg (32° F) | 0.00289590 |
| | Millimeters of Hg (0° C) | 0.0735559 |
| | Pounds/square inch | 0.0014223343 |
| Kilometers | Centimeters | 100000 |
| | Feet | 3280.8399 |
| | Meters | 1000 |
| | Miles (statute) | 0.62137119 |
| Kilowatt-hours | Foot-pounds | $2.65522 \times 10^5$ |
| | Horsepower-hours | 1.34102 |
| | Joules | $3.6 \times 10^6$ |
| Kilowatt-hours (international) | Btu | 3412.7 |
| | Calories, gram (IST) | 860000 |
| | Foot-pounds | $2.65566 \times 10^6$ |
| | Horsepower-hours | 1.34124 |

## Table 57: Selected Conversion Factors. (cont'd)

| To convert from | To | Multiply by |
|---|---|---|
| Kilowatt-hours (cont.) (international) | Joules | $3.60059 \times 10^6$ |
| | Joules (international) | $3.6 \times 10^6$ |
| Liters | Barrels (petroleum, U.S.) | 0.00629 |
| | Cubic centimeters | 1000.028 |
| | Cubic feet | 0.03531566 |
| | Cubic inches | 61.02545 |
| | Cubic meters | 0.001000028 |
| | Gallons (liquid, U.S.) | 0.2641794 |
| Meters | Centimeters | 100 |
| | Feet | 3.2808399 |
| | Inches | 39.370079 |
| | Kilometers | 0.001 |
| | Miles (statute) | 0.00062137119 |
| | Millimeters | 1000 |
| | Millimicrons | $1 \times 10^9$ |
| Mhos | $Ohms^{-1}$ | 1 |
| | Siemen's units | 1 |
| Microns | Centimeters | 0.0001 |
| | Inches | $3.9370079 \times 10^{-5}$ |
| | Millimeters | 0.001 |
| | Millimicrons | 1000 |
| Miles (statute) | Feet | 5280 |
| | Kilometers | 1.609344 |
| | Meters | 1609.344 |
| Miles/hour | Centimeters/second | 44.704 |
| | Feet/hour | 5280 |
| | Feet/minute | 88 |
| | Kilometers/hour | 1.609344 |
| | Meters/second | 0.44704 |
| Millimeters of Hg (0° C) | Atmospheres | 0.0013157895 |
| | Bars | 0.00133322 |
| | Inches of Hg (32° F) | 0.039370 |
| | Kilograms/square meter | 13.5951 |
| | Newtons/square meter | 133.32 |
| | Pounds/square inch | 0.0193368 |
| Millimicrons | Centimeters | $1 \times 10^{-7}$ |
| | Inches | $3.9370079 \times 10^{-8}$ |
| | Microns | 0.001 |
| | Millimeters | $1 \times 10^{-6}$ |
| Newtons/square meter | Atmospheres | 0.0000098692 |
| | Bars | 0.00001 |
| | Inches of Hg (32° F) | 0.0002953 |
| | Millimeters of Hg (0° C) | 0.007506 |
| | Pounds/square inch | 0.00014504 |

## Table 57: Selected Conversion Factors. (cont'd)

| To convert from | To | Multiply by |
|---|---|---|
| Pounds (avoirdupois) | Grams | 453.59237 |
| | Kilograms | 0.45359237 |
| | Ounces (avoirdupois) | 16 |
| | Tons (long) | 0.00044642857 |
| | Tons (metric) | 0.00045359237 |
| | Tons (short) | 0.0005 |
| Pounds/square inch | Atmospheres | 0.0680460 |
| | Bars | 0.0689476 |
| | Inches of Hg (32° F) | 2.03602 |
| | Kilograms/square centimeter | 0.070306958 |
| | Millimeters of Hg (0° C) | 51.7149 |
| | Newtons/square meter | 6894.9 |
| Square centimeters | Square feet | 0.0010763910 |
| | Square inches | 0.15500031 |
| | Square meters | 0.0001 |
| | Square millimeters | 100 |
| Square feet | Acres | $2.295684 \times 10^{-5}$ |
| | Square centimeters | 929.0304 |
| | Square inches | 144 |
| | Square meters | 0.09290304 |
| | Square miles | $3.5870064 \times 10^{-8}$ |
| Square inches | Square centimeters | 6.4516 |
| | Square feet | 0.0069444 |
| | Square meters | 0.00064516 |
| | Square millimeters | 645.16 |
| Square kilometers | Acres | 247.10538 |
| | Hectares | 100 |
| | Square feet | $1.0763910 \times 10^{7}$ |
| | Square meters | $1 \times 10^{6}$ |
| | Square miles | 0.38610216 |
| Square meters | Hectares | 0.0001 |
| | Square centimeters | 10000 |
| | Square feet | 10.763910 |
| | Square inches | 1550.0031 |
| Square miles | Acres | 640 |
| | Hectares | 258.99881 |
| | Square feet | $2.7878288 \times 10^{7}$ |
| | Square kilometers | 2.5899881 |
| | Square meters | 2589988.1 |
| Tons (long) | Barrels (petroleum, U.S.) | 7.45 |
| | Kilograms | 1016.0469 |
| | Pounds (avoirdupois) | 2240 |
| | Tons (metric) | 1.0160469 |
| | Tons (short) | 1.12 |

## Table 57: Selected Conversion Factors. (cont'd)

| To convert from | To | Multiply by |
|---|---|---|
| Tons (metric) | Barrels (petroleum, U.S.) | 7.33 |
| | Kilograms | 1000 |
| | Pounds (avoirdupois) | 2204.6226 |
| | Tons (long) | 0.98420653 |
| | Tons (short) | 1.1023113 |
| Tons (short) | Barrels (petroleum, U.S.) | 6.65 |
| | Kilograms | 907.18474 |
| | Pounds (avoirdupois) | 2000 |
| | Tons (long) | 0.89285714 |
| | Tons (metric) | 0.90718474 |
| Tons of refrigeration (standard, U.S.) | Btu (IST) | 288000 |

[1]IST - International Steam Table.

[2]This is the calorie as defined by the U.S. National Bureau of Standards and is equal to 4.18400 joules.

**Table 58: Quad Equivalents. (Source 10)**

1 Quad

= $10^{15}$ Btu = $10^{18}$ Joule = $10^{15}$ kJ = $10^{6}$ TJ = $10^{-3}$ Q units
= Total amount of energy used in U.S. every 5 days
= Total amount of energy used in world every day
= Energy in 500,000 bbl/day of oil for one year
= Energy in 40 x $10^{6}$ ton of bituminous coal, or
5,000 trainloads of 100 cars (100 ton/car) per train
= Energy in all the wheat produced in the U.S. in one year; or
1.75 x $10^{9}$ bu cereal grain; or
one-half of total corn produced in one year in the U.S.
= Energy in 1 trillion cubic feet of natural gas
= Energy in 182 million barrels of oil
= Energy in 42 million tons of bituminous coal
= Energy in 293 billion kWh of electricity
= Energy in 4 x $10^{6}$ lb uranium in LWR
= Energy in 50 x $10^{3}$ lb uranium in nuclear breeder (60%)
= Energy in 62.5 x $10^{6}$ tons wood (dry)
= Energy in 105 x $10^{6}$ tons MSW
= Eleven 1,000 MWe power plants operating continuously for
1 year at 100% capacity

## Table 59: Conversion Efficiency of Various Fuels.

| Initial Material | Final Product | Efficiency, % |
|---|---|---|
| Coal | Powdered coal | 95 |
| | Solvent refined (SRC), process | 87 |
| | Gas by gasifier | 84 |
| | Producer gas | 75-80 |
| | Synthetic oil | 70 (42% energy efficiency) |
| | Distillate | 63 |
| | Gasoline/Diesel fuel | 55 |
| | Cogas plant efficiency | 43-45 |
| | Methanol | 43 |
| | Fischer-Tropsch liquids | 37 |
| | Methane | 50-55% (max. 75%) |
| | Hydrogasification, process | 65-70% (max. 90%) |
| | Electricity to users | 28-30 |
| Oil shale | Hydrogenated shale oil | 75 |
| | Distillate | 67 |
| | Gas, process | 57-72 |
| | Gasoline/Diesel fuel | 63 |
| Biomass | Ethanol (grain) | 62%; Btu/Btu 35% |
| | Ethanol (wood) | 65 |
| | Methanol (wood), process | 30-37% thermal |
| | Hydrocarbon liquids | 25 |
| | Pyrolytic oil | 35 (yield, 25%) |
| | Methane (fermentation) | 33 |
| | Low-/Medium- Btu gas | 70 |
| | MSW gasification, process | 66 |
| | Gasification of wood, process | 70 |
| | Animal manure to gas, anaerobic | 33 |
| | Hydrogasification of animal manure | 40 |
| Petroleum | Clean crude oil | 98 |
| | Distillate | 95 |
| | Gasoline/Diesel fuel | 87/90 |
| | Electric heating | 17 |
| | Electricity | 30-38 |
| | Methanol | 65 |